Classical Mechanics

Classical Mechanics

Edwin Roman

Willford Press
www.willfordpress.com

Published by Willford Press,
118-35 Queens Blvd., Suite 400,
Forest Hills, NY 11375, USA

ISBN: 978-1-68285-741-0

Cataloging-in-Publication Data

Classical mechanics / Edwin Roman.
p. cm.
Includes bibliographical references and index.
ISBN 978-1-68285-741-0
1. Mechanics. 2. Physics. I. Roman, Edwin.
QC125.2 .C53 2019
531--dc23

For information on all Willford Press publications
visit our website at www.willfordpress.com

Willford Press

Printed in China.

Contents

Preface ..... **VII**

Chapter 1 **Classical Mechanics: Prominent Branches and Fundamental Concepts** **1**

i. Classical Mechanics 1
ii. Celestial Mechanics 10
iii. Continuum Mechanics 20
iv. Kinematics 34
v. Statics 50
vi. Newton's Laws of Motion 54
vii. D'Alembert's Principle 59
viii. Momentum 63
ix. Angular Momentum 79

Chapter 2 **Rotational and Circular Motion** **81**

i. Rotational Motion 81
ii. Rotating Reference Frame 108
iii. Centrifugal Force 115
iv. Centripetal Force 122
v. Circular Motion 124

Chapter 3 **Translational and Oscillatory Motion** **140**

i. Translational Motion 140
ii. Oscillatory Motion 151

Chapter 4 **Formulations of Classical Mechanics** **159**

i. Constraint 159
ii. Koopman-von Neumann Classical Mechanics 162
iii. Analytical Mechanics 169
iv. Lagrangian Mechanics 180
v. Hamiltonian Mechanics 203
vi. Canonical Transformation 210
vii. Appell's Equation of Motion 214
viii. Udwadia-Kalaba Equation 221
ix. Euler-Lagrange Equation 226

Chapter 5 **Understanding Special Relativity** **231**

i. Galilean Invariance 237

ii. Frame of Reference 242

iii. Relativity of Simultaneity 255

iv. Relativistic Mass 260

Permissions

Index

Preface

Classical mechanics is the science that is concerned with the motion of macroscopic objects, such as projectiles and celestial objects, such as planets, stars, galaxies and spacecraft. It is the approach that is generally adopted for the study of massive objects that move at speeds smaller than the speed of light. In classical mechanics, objects are modeled as point particles, which are characterized by parameters like mass, position and the forces that act on it. Statics, dynamics and kinematics are the primary branches of classical mechanics. Based on mathematical formalism, classical mechanics can be classified as Newtonian, Lagrangian and Hamiltonian mechanics. This book is a valuable compilation of topics, ranging from the basic to the most complex theories and principles in the field of classical mechanics. It presents this complex subject in the most comprehensible and easy to understand language. For someone with an interest and eye for detail, this book covers the most significant topics in classical mechanics.

A foreword of all chapters of the book is provided below:

Chapter 1- The study of the motion of macroscopic objects is under the scope of the field of classical mechanics. The Newton's laws of motion are fundamental to the development of this field. This chapter introduces all the fundamental concepts of classical mechanics and also discusses the prominent branches such as statics, kinematics, celestial mechanics and continuum mechanics; **Chapter 2-** The motion of an object with respect to a central point in a circular path is called circular motion. The circular motion of an object around a center of rotation or axis of rotation is called rotational motion. This chapter discusses the fundamentals of rotational and circular motion such as rotating reference frame, centrifugal force and centripetal force; **Chapter 3-** A translation is a displacement in the position of an object. It can be a linear or an angular displacement. Translational motion can be rectilinear or curvilinear. The repetitive variation of some measure about a point of equilibrium is termed as an oscillation. Simple harmonic motion is a type of oscillatory motion. This chapter has been carefully written to provide an easy understanding of translational and oscillatory motion; **Chapter 4-** In order to develop a complete understanding of classical mechanics, it is necessary to understand the formulations and concepts central to it, such as constraints, Lagrangian and Hamiltonian mechanics, Euler-Lagrange equation, canonical transformation, Koopman-von Neumann classical mechanics, analytical mechanics, etc. All these topics have been presented in detail in this chapter; **Chapter 5-** Special relativity is a theory in physics, which defines the relationship of space and time. This chapter discusses in detail the principal topics, such as relativity of simultaneity, relativistic mass, Galilean invariance and frame of reference, in order to provide an extensive understanding of this important theory.

At the end, I would like to thank all the people associated with this book devoting their precious time and providing their valuable contributions to this book. I would also like to express my gratitude to my fellow colleagues who encouraged me throughout the process.

Edwin Roman

1

Classical Mechanics: Prominent Branches and Fundamental Concepts

The study of the motion of macroscopic objects is under the scope of the field of classical mechanics. The Newton's laws of motion are fundamental to the development of this field. This chapter introduces all the fundamental concepts of classical mechanics and also discusses the prominent branches such as statics, kinematics, celestial mechanics and continuum mechanics.

Classical mechanics is the study of the *motion* of bodies (including the special case in which bodies remain at rest) in accordance with the general principles first enunciated by Sir Isaac Newton in his *Philosophiae Naturalis Principia Mathematica* (1687), commonly known as the *Principia*. Classical mechanics was the first branch of Physics to be discovered, and is the foundation upon which all other branches of Physics are built. Moreover, classical mechanics has many important applications in other areas of science, such as Astronomy (*e.g.*, celestial mechanics), Chemistry (*e.g.*, the dynamics of molecular collisions), Geology (*e.g.*, the propagation of seismic waves, generated by earthquakes, through the Earth's crust), and Engineering (*e.g.*, the equilibrium and stability of structures). Classical mechanics is also of great significance outside the realm of science. After all, the sequence of events leading to the discovery of classical mechanics--starting with the ground-breaking work of Copernicus, continuing with the researches of Galileo, Kepler, and Descartes, and culminating in the monumental achievements of Newton--involved the complete overthrow of the Aristotelian picture of the Universe, which had previously prevailed for more than a millennium, and its replacement by a recognizably modern picture in which humankind no longer played a privileged role.

In our investigation of classical mechanics we shall study many different types of motion, including:

- *Translational motion*--motion by which a body shifts from one point in space to another (*e.g.*, the motion of a bullet fired from a gun).
- *Rotational motion*--motion by which an extended body changes orientation, with respect to other bodies in space, without changing position (*e.g.*, the motion of a spinning top).
- *Oscillatory motion*--motion which continually repeats in time with a fixed period (*e.g.*, the motion of a pendulum in a grandfather clock).
- *Circular motion*--motion by which a body executes a circular orbit about another fixed body [*e.g.*, the (approximate) motion of the Earth about the Sun].

Of course, these different types of motion can be combined: for instance, the motion of a properly bowled bowling ball consists of a combination of translational and rotational motion, whereas wave propagation is a combination of translational and oscillatory motion. Furthermore, the above mentioned types of motion are not entirely distinct: *e.g.*, circular motion contains elements of both rotational and oscillatory motion.

Description of the Theory

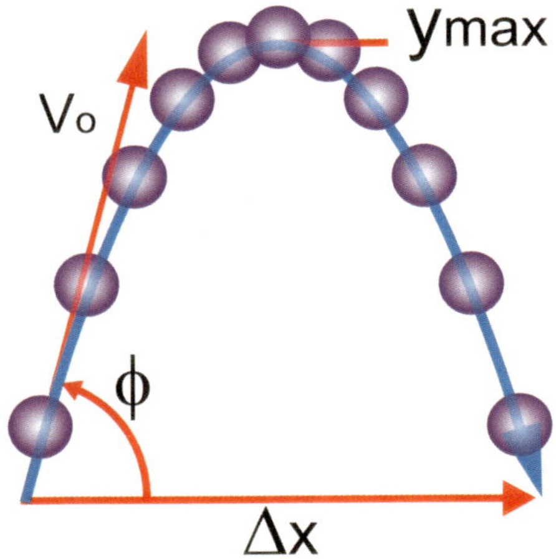

The analysis of projectile motion is a part of classical mechanics.

The following introduces the basic concepts of classical mechanics. For simplicity, it often models real-world objects as point particles (objects with negligible size). The motion of a point particle is characterized by a small number of parameters: its position, mass, and the forces applied to it. Each of these parameters is discussed in turn.

In reality, the kind of objects that classical mechanics can describe always have a non-zero size. (The physics of *very* small particles, such as the electron, is more accurately described by quantum mechanics.) Objects with non-zero size have more complicated behavior than hypothetical point particles, because of the additional degrees of freedom, e.g., a baseball can spin while it is moving. However, the results for point particles can be used to study such objects by treating them as composite objects, made of a large number of collectively acting point particles. The center of mass of a composite object behaves like a point particle.

Classical mechanics uses common-sense notions of how matter and forces exist and interact. It assumes that matter and energy have definite, knowable attributes such as location in space and speed. Non-relativistic mechanics also assumes that forces act instantaneously.

Position and its Derivatives

The SI derived "mechanical" (that is, not electromagnetic or thermal) units with kg, m and s

position	m
angular position/angle	unitless (radian)
velocity	$m \cdot s^{-1}$

angular velocity	s^{-1}
acceleration	$m \cdot s^{-2}$
angular acceleration	s^{-2}
jerk	$m \cdot s^{-3}$
"angular jerk"	s^{-3}
specific energy	$m^2 \cdot s^{-2}$
absorbed dose rate	$m^2 \cdot s^{-3}$
moment of inertia	$kg \cdot m^2$
momentum	$kg \cdot m \cdot s^{-1}$
angular momentum	$kg \cdot m^2 \cdot s^{-1}$
force	$kg \cdot m \cdot s^{-2}$
torque	$kg \cdot m^2 \cdot s^{-2}$
energy	$kg \cdot m^2 \cdot s^{-2}$
power	$kg \cdot m^2 \cdot s^{-3}$
pressure and energy density	$kg \cdot m^{-1} \cdot s^{-2}$
surface tension	$kg \cdot s^{-2}$
spring constant	$kg \cdot s^{-2}$
irradiance and energy flux	$kg \cdot s^{-3}$
kinematic viscosity	$m^2 \cdot s^{-1}$
dynamic viscosity	$kg \cdot m^{-1} \cdot s^{-1}$
density (mass density)	$kg \cdot m^{-3}$
density (weight density)	$kg \cdot m^{-2} \cdot s^{-2}$
number density	m^{-3}
action	$kg \cdot m^2 \cdot s^{-1}$

The *position* of a point particle is defined in relation to a coordinate system centered on an arbitrary fixed reference point in space called the origin *O*. A simple coordinate system might describe the position of a particle *P* with a vector notated by an arrow labeled r that points from the origin *O* to point *P*. In general, the point particle does not need to be stationary relative to *O*. In cases where *P* is moving relative to *O*, r is defined as a function of *t*, time. In pre-Einstein relativity (known as Galilean relativity), time is considered an absolute, i.e., the time interval that is observed to elapse between any given pair of events is the same for all observers. In addition to relying on absolute time, classical mechanics assumes Euclidean geometry for the structure of space.

Velocity and Speed

The *velocity*, or the rate of change of position with time, is defined as the derivative of the position with respect to time:

$$\mathbf{v} = \frac{d\mathbf{r}}{dt}.$$

In classical mechanics, velocities are directly additive and subtractive. For example, if one car travels east at 60 km/h and passes another car traveling in the same direction at 50 km/h, the slower car perceives the faster car as traveling east at 60 – 50 = 10 km/h. However, from the perspective of the faster car, the slower car is moving 10 km/h to the west, often denoted as -10 km/h where the sign implies opposite direction. Velocities are directly additive as vector quantities; they must be dealt with using vector analysis.

Mathematically, if the velocity of the first object in the previous discussion is denoted by the vector u = *u*d and the velocity of the second object by the vector v = *v*e, where *u* is the speed of the first object, *v* is the speed of the second object, and d and e are unit vectors in the directions of motion of each object respectively, then the velocity of the first object as seen by the second object is

$$\mathbf{u}' = \mathbf{u} - \mathbf{v}.$$

Similarly, the first object sees the velocity of the second object as

$$\mathbf{v}' = \mathbf{v} - \mathbf{u}.$$

When both objects are moving in the same direction, this equation can be simplified to

$$\mathbf{u}' = (u - v)\mathbf{d}.$$

Or, by ignoring direction, the difference can be given in terms of speed only:

$$u' = u - v.$$

Acceleration

The *acceleration*, or rate of change of velocity, is the derivative of the velocity with respect to time (the second derivative of the position with respect to time):

$$\mathbf{a} = \frac{\mathrm{d}\mathbf{v}}{\mathrm{d}t} = \frac{\mathrm{d}^2\mathbf{r}}{\mathrm{d}t^2}.$$

Acceleration represents the velocity's change over time. Velocity can change in either magnitude or direction, or both. Occasionally, a decrease in the magnitude of velocity "*v*" is referred to as *deceleration*, but generally any change in the velocity over time, including deceleration, is simply referred to as acceleration.

Frames of Reference

While the position, velocity and acceleration of a particle can be described with respect to any observer in any state of motion, classical mechanics assumes the existence of a special family of reference frames in which the mechanical laws of nature take a comparatively simple form. These special reference frames are called inertial frames.

An inertial frame is a frame of reference within which an object interacting with no forces (an idealized situation) appears either at rest or moving uniformly in a straight line. This is the fundamental definition of an inertial frame. These are characterized by the requirement that all forces entering the observer's physical laws originate from identifiable sources caused by fields, such as electro-static field (caused by static electrical charges), electro-magnetic field (caused by moving charges), gravitational field (caused by mass), and so forth.

A key concept of inertial frames is the method for identifying them. For practical purposes, reference frames that do not accelerate with respect to distant stars (an extremely distant point) are regarded as good approximations to inertial frames. Non-inertial reference frames accelerate in relation to an existing inertial frame. They form the basis for Einstein's relativity. Due to the relative motion, particles in the non-inertial frame appear to move in ways not explained by forces from existing fields in the reference frame. Hence, it appears that there are other forces that enter the equations of motion solely as a result of the relative acceleration. These forces are referred to as fictitious forces, inertia forces, or pseudo-forces.

Consider two reference frames *S* and *S'*. For observers in each of the reference frames an event has space-time coordinates of (x,y,z,t) in frame *S* and (x',y',z',t') in frame *S'*. Assuming time is measured the same in all reference frames, and if we require $x = x'$ when $t = 0$, then the relation between the space-time coordinates of the same event observed from the reference frames *S'* and *S*, which are moving at a relative velocity of u in the x direction is:

$$x' = x - ut$$

$$y' = y$$

$$z' = z$$

$$t' = t.$$

This set of formulas defines a group transformation known as the Galilean transformation (informally, the *Galilean transform*). This group is a limiting case of the Poincaré group used in special relativity. The limiting case applies when the velocity u is very small compared to c, the speed of light.

The transformations have the following consequences:

- $v' = v - u$ (the velocity v' of a particle from the perspective of S' is slower by u than its velocity v from the perspective of S);
- $a' = a$ (the acceleration of a particle is the same in any inertial reference frame);
- $F' = F$ (the force on a particle is the same in any inertial reference frame);
- the speed of light is not a constant in classical mechanics, nor does the special position given to the speed of light in relativistic mechanics have a counterpart in classical mechanics.

For some problems, it is convenient to use rotating coordinates (reference frames). Thereby one can either keep a mapping to a convenient inertial frame, or introduce additionally a fictitious centrifugal force and Coriolis force.

Forces; Newton's Second Law

Newton was the first to mathematically express the relationship between force and momentum. Some physicists interpret Newton's second law of motion as a definition of force and mass, while others consider it a fundamental postulate, a law of nature. Either interpretation has the same mathematical consequences, historically known as "Newton's Second Law":

$$\mathbf{F} = \frac{d\mathbf{p}}{dt} = \frac{d(m\mathbf{v})}{dt}.$$

The quantity mv is called the (canonical) momentum. The net force on a particle is thus equal to the rate of change of the momentum of the particle with time. Since the definition of acceleration is a = dv/dt, the second law can be written in the simplified and more familiar form:

$$\mathbf{F} = m\mathbf{a}.$$

So long as the force acting on a particle is known, Newton's second law is sufficient to describe the motion of a particle. Once independent relations for each force acting on a particle are available, they can be substituted into Newton's second law to obtain an ordinary differential equation, which is called the *equation of motion*.

As an example, assume that friction is the only force acting on the particle, and that it may be modeled as a function of the velocity of the particle, for example:

$$\mathbf{F}_R = -\lambda\mathbf{v},$$

where λ is a positive constant, the negative sign states that the force is opposite the sense of the velocity. Then the equation of motion is

$$-\lambda\mathbf{v} = m\mathbf{a} = m\frac{d\mathbf{v}}{dt}.$$

This can be integrated to obtain

$$v = \mathbf{v}_0 e^{\frac{-\lambda t}{m}}$$

where v_0 is the initial velocity. This means that the velocity of this particle decays exponentially to zero as time progresses. In this case, an equivalent viewpoint is that the kinetic energy of the particle is absorbed by friction (which converts it to heat energy in accordance with the conservation of energy), and the particle is slowing down. This expression can be further integrated to obtain the position r of the particle as a function of time.

Important forces include the gravitational force and the Lorentz force for electromagnetism. In addition, Newton's third law can sometimes be used to deduce the forces acting on a particle: if it is known that particle A exerts a force F on another particle B, it follows that B must exert an equal and opposite *reaction force*, −F, on A. The strong form of Newton's third law requires that F and −F act along the line connecting A and B, while the weak form does not. Illustrations of the weak form of Newton's third law are often found for magnetic forces.

Work and Energy

If a constant force F is applied to a particle that makes a displacement Δr, the *work done* by the force is defined as the scalar product of the force and displacement vectors:

$$W = \mathbf{F} \cdot \Delta\mathbf{r}.$$

More generally, if the force varies as a function of position as the particle moves from r_1 to r_2 along a path C, the work done on the particle is given by the line integral

$$= \int \mathbf{F}(\mathbf{r}) \cdot d\mathbf{r}.$$

If the work done in moving the particle from r_1 to r_2 is the same no matter what path is taken, the force is said to be conservative. Gravity is a conservative force, as is the force due to an idealized spring, as given by Hooke's law. The force due to friction is non-conservative.

The kinetic energy E_k of a particle of mass m travelling at speed v is given by

$$E_k = \tfrac{1}{2}mv^2.$$

For extended objects composed of many particles, the kinetic energy of the composite body is the sum of the kinetic energies of the particles.

The work–energy theorem states that for a particle of constant mass m, the total work W done on the particle as it moves from position r_1 to r_2 is equal to the change in kinetic energy E_k of the particle:

$$W = \Delta E_k = E_{k,2} - E_{k,1} = \tfrac{1}{2}m\left(v_2^2 - v_1^2\right).$$

Conservative forces can be expressed as the gradient of a scalar function, known as the potential energy and denoted E_p:

$$\mathbf{F} = -\nabla E_p.$$

If all the forces acting on a particle are conservative, and E_p is the total potential energy (which is defined as a work of involved forces to rearrange mutual positions of bodies), obtained by summing the potential energies corresponding to each force

$$\mathbf{F} \cdot \Delta\mathbf{r} = -\nabla E_p \cdot \Delta\mathbf{r} = -\Delta E_p.$$

The decrease in the potential energy is equal to the increase in the kinetic energy

$$-\Delta E_p = \Delta E_k \Rightarrow \Delta(E_k + E_p) = 0.$$

This result is known as *conservation of energy* and states that the total energy,

$$\sum E = E_k + E_p,$$

is constant in time. It is often useful, because many commonly encountered forces are conservative.

Beyond Newton's Laws

Classical mechanics also describes the more complex motions of extended non-pointlike objects. Euler's laws provide extensions to Newton's laws in this area. The concepts of angular momentum rely on the same calculus used to describe one-dimensional motion. The rocket equation extends the notion of rate of change of an object's momentum to include the effects of an object "losing mass".

There are two important alternative formulations of classical mechanics: Lagrangian mechanics and Hamiltonian mechanics. These, and other modern formulations, usually bypass the concept of "force", instead referring to other physical quantities, such as energy, speed and momentum, for describing mechanical systems in generalized coordinates.

The expressions given above for momentum and kinetic energy are only valid when there is no significant electromagnetic contribution. In electromagnetism, Newton's second law for current-carrying wires breaks down unless one includes the electromagnetic field contribution to the momentum of the system as expressed by the Poynting vector divided by c^2, where c is the speed of light in free space.

Limits of Validity

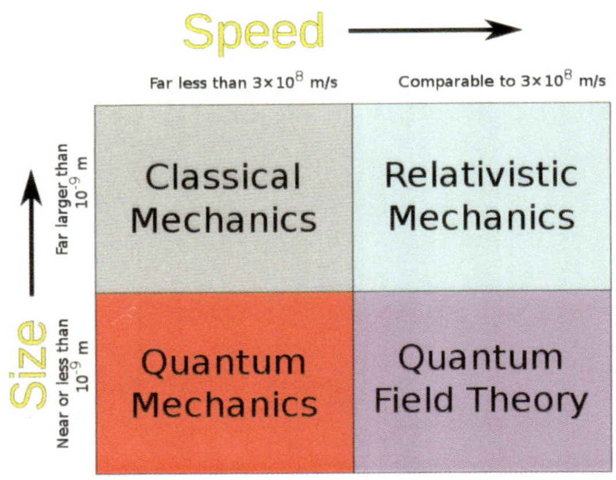

Domain of validity for Classical Mechanics

Many branches of classical mechanics are simplifications or approximations of more accurate forms; two of the most accurate being general relativity and relativistic statistical mechanics. Geometric optics is an approximation to the quantum theory of light, and does not have a superior "classical" form.

When both quantum mechanics and classical mechanics cannot apply, such as at the quantum level with many degrees of freedom, quantum field theory (QFT) is of use. QFT deals with small distances and large speeds with many degrees of freedom as well as the possibility of any change in the number of particles throughout the interaction. When treating large degrees of freedom at the macroscopic level, statistical mechanics becomes useful. Statistical mechanics describes the behavior of large (but countable) numbers of particles and their interactions as a whole at the macroscopic level. Statistical mechanics is mainly used in thermodynamics for systems that lie outside the bounds of the assumptions of classical thermodynamics. In the case of high velocity objects approaching the speed of light, classical mechanics is enhanced by special relativity. In case that objects become extremely heavy (i.e. their Schwarzschild radius is not negligibly small for a given application), deviations from Newtonian mechanics become apparent and can be quantified by using the Parameterized post-Newtonian formalism. In that case, General relativity (GR) becomes applicable. However, until now there is no theory of Quantum gravity unifying GR and QFT in the sense that it could be used when objects become extremely small and heavy.

The Newtonian Approximation to Special Relativity

In special relativity, the momentum of a particle is given by

$$\mathbf{p} = \frac{m\mathbf{v}}{\sqrt{1 - v^2 / c^2}},$$

where m is the particle's rest mass, v its velocity, v is the modulus of v, and c is the speed of light.

If v is very small compared to c, v^2/c^2 is approximately zero, and so

$$\mathbf{p} \approx m\mathbf{v}.$$

Thus the Newtonian equation p = mv is an approximation of the relativistic equation for bodies moving with low speeds compared to the speed of light.

For example, the relativistic cyclotron frequency of a cyclotron, gyrotron, or high voltage magnetron is given by

$$f = f_c \frac{m_0}{m_0 + T / c^2},$$

where f_c is the classical frequency of an electron (or other charged particle) with kinetic energy T and (rest) mass m_0 circling in a magnetic field. The (rest) mass of an electron is 511 keV. So the frequency correction is 1% for a magnetic vacuum tube with a 5.11 kV direct current accelerating voltage.

The Classical Approximation to Quantum Mechanics

The ray approximation of classical mechanics breaks down when the de Broglie wavelength is not much smaller than other dimensions of the system. For non-relativistic particles, this wavelength is

$$\lambda = \frac{h}{p}$$

where h is Planck's constant and p is the momentum.

Again, this happens with electrons before it happens with heavier particles. For example, the electrons used by Clinton Davisson and Lester Germer in 1927, accelerated by 54 V, had a wavelength of 0.167 nm, which was long enough to exhibit a single diffraction side lobe when reflecting from the face of a nickel crystal with atomic spacing of 0.215 nm. With a larger vacuum chamber, it would seem relatively easy to increase the angular resolution from around a radian to a milliradian and see quantum diffraction from the periodic patterns of integrated circuit computer memory.

More practical examples of the failure of classical mechanics on an engineering scale are conduction by quantum tunneling in tunnel diodes and very narrow transistor gates in integrated circuits.

Classical mechanics is the same extreme high frequency approximation as geometric optics. It is more often accurate because it describes particles and bodies with rest mass. These have more momentum and therefore shorter De Broglie wavelengths than massless particles, such as light, with the same kinetic energies.

Branches

Classical mechanics was traditionally divided into three main branches:

- Statics, the study of equilibrium and its relation to forces.
- Dynamics, the study of motion and its relation to forces.
- Kinematics, dealing with the implications of observed motions without regard for circumstances causing them.

Another division is based on the choice of mathematical formalism:

- Newtonian mechanics.
- Lagrangian mechanics.
- Hamiltonian mechanics.

Alternatively, a division can be made by region of application:

- Celestial mechanics, relating to stars, planets and other celestial bodies.
- Continuum mechanics, for materials modelled as a continuum, e.g., solids and fluids (i.e., liquids and gases).
- Relativistic mechanics (i.e. including the special and general theories of relativity), for bodies whose speed is close to the speed of light.
- Statistical mechanics, which provides a framework for relating the microscopic properties of individual atoms and molecules to the macroscopic or bulk thermodynamic properties of materials.

Celestial Mechanics

Celestial mechanics, in the broadest sense, the application of classical mechanics to the motion of celestial bodies acted on by any of several types of forces. By far the most important force experienced by these bodies, and much of the time the only important force, is that of their mutual gravitational attraction. But other forces can be important as well, such as atmospheric drag on artificial satellites, the pressure of radiation on dust particles, and even electromagnetic forces on dust particles if they are electrically charged and moving in a magnetic field.

The term celestial mechanics is sometimes assumed to refer only to the analysis developed for the motion of point mass particles moving under their mutual gravitational attractions, with emphasis on the general orbital motions of solar system bodies. The term astrodynamics is often used to refer to the celestial mechanics of artificial satellite motion. Dynamic astronomy is a much broader term, which, in addition to celestial mechanics and astrodynamics, is usually interpreted to include all aspects of celestial body motion (e.g., rotation, tidal evolution, mass and mass distribution determinations for stars and galaxies, fluid motions in nebulas, and so forth).

The Pre-newtonian Celestial Kinematics

The story of the mathematical representation of celestial motions starts in the antiquity and, notwithstanding the prevalent wrong ideas placing the Earth at the center of the universe, the prediction of the planetary motions were very accurate allowing, for instance, to forecast eclipses and to keep calendars synchronized with the motion of the Earth around the Sun. The epicycles, introduced by Apollonius of Perga around 200 BC, allowed the observed motions to be represented by series of circular functions. They were used to predict celestial motions for almost two millennia. Their long life was certainly related to the stagnation that prevailed in the western world during the dark ages between the end of the Helenic civilization and the Renaissance. In the 16th century, the Copernican revolution put the Sun in center of the Universe. However, the breakthrough in our knowledge of celestial motions was rather related to Tycho Brahe and Johannes Kepler. Tycho, in his Uraniborg observatory, accurately measured the position of the planets in the sky for more than 20 years. Kepler inherited the data gathered by Tycho and used them to discover the three laws that bear his name.

- First Law or Elliptical Orbits Law (1609): The planets move on ellipses with one focus in the Sun.
- Second Law or Law of Equal Areas (1609): The planets move with constant areal velocity (equal areas are swept in equal times); in modern words: with constant angular momentum.
- Third Law or Harmonic Law (1619): The ratio of the cube of the semi-major axes of the ellipses to the square of the periods of the planetary motions is constant and the same for all planets.

The work of Kepler is a monument to the human genius. First of all, Tycho's data on Mars could not be fitted to a heliocentric uniform motion. With respect to a uniform motion, sometimes Mars was in advance, sometimes in retard. Kepler decided to tackle the problem from scratch! Remember that mathematics had remained stagnant since antiquity and the tools inherited from the Greeks, geometry and arithmetic, were the only available. Kepler considered as working hypotheses that the Earth was uniformly moving on a circle and that the motion of Mars was periodic and coplanar with the motion of the Earth. Then he used Tycho's observations to determine the orbit of Mars. Tycho's observations were apparent positions of the planets on the celestial sphere. The resulting datum is a direction (only recently had celestial distances been measured, and only in a few cases). Kepler constructed triangles. After having determined the period of the motion of Mars around the Sun, he looked for observations in dates separated by just one period. Then he constructed triangles, each having as vertices one position of Mars in space (assumed to be the same – after one period Mars returns to the same position) and the position of the Earth in the two dates. The angles of the triangle were obtained from the measurements done by Tycho, and these triangles allowed the position of Mars relative to the Earth to be determined. He thus found that the orbit of Mars was not a circle but rather an ellipse with one focus in the Sun. After that, he inverted the process. He assumed that the real motion of Mars followed an ellipse with constant areal velocity, and started looking for observations separated by one year (in one year the Earth is back to the same place). The triangles now have two vertices on the orbit of Mars (assumed as known) and one vertex on the position of the Earth at those dates. The triangles thus obtained allowed one to determine the position of the Earth with respect to the orbit of Mars. In this way Kepler discovered that the Earth, also, was moving on an ellipse with constant areal velocity. The two first laws were thus discovered. The third law remained elusive for about one more decade, but was finally unraveled.

Newton's Celestial Mechanics

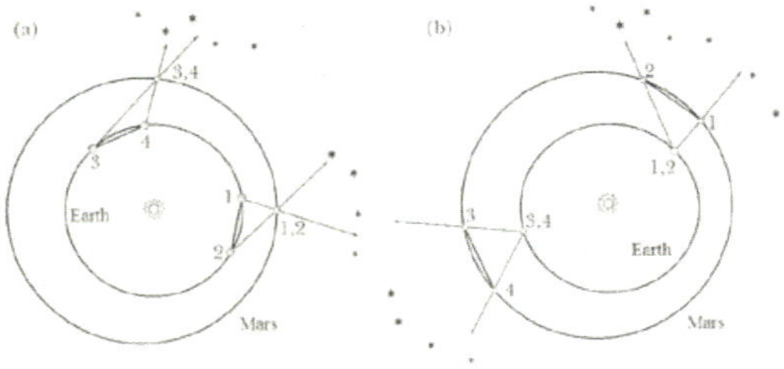

Figure: Kepler's technique to determine the orbits of the Earth and Mars. (a) The position of Mars is determined from two observations of Mars done at times separated by one integer number of Mars revolutions (i.e. in the two points Mars is in the same position in its orbit). (b) The position of the Earth is determined from two observations of Mars done at times separated by one year (i.e. in the two points, the Earth is in the same position in its orbit). The figure shows 2 triangles constructed in this way. The observed directions of Mars were taken from Tycho Brahe's observations.

Newton's theory of universal gravitation resulted from experimental and observational facts. The observational facts were those encompassed in the three Kepler laws. The experimental facts were those reported by Galileo in his book Discorsi intorno à due nuove scienze ("Discourses Relating to Two New Sciences", which should not be confounded with his most celebrated "Dialogue Concerning the Two Chief World Systems"). The basis of Newton theory arose from the perception that the force keeping the Moon in orbit around the Earth is the same that, on Earth, commands the fall of the bodies.

- Law of Universal Gravitation (1687) – Bodies attract themselves mutually with a force proportional to their masses and inversely proportional to the square of the distance between them.

In other words, if two bodies have masses m_1 and m_2 and are separated by a distance r, they attract one another with the force $\left|\vec{f}\right| = \frac{Gm_1m_2}{r^2}$ where G is a constant $\left(G = 6.678 \times 10^{-8} cm^3 g^{-1} s^{-2}\right)$. This constant is universal and does not depend of the nature of the bodies or on where in the Universe they are found, here or elsewhere.

This law inaugurated the Celestial Mechanics (even if the name came to be used only after Laplace's work). Newton initially studied the problem of the motion followed by two bodies in mutual attraction (for instance, the Sun and one planet). He showed that under ideal conditions (when no other forces disturb the motions of the two bodies), the relative motion obeys laws which, in some sense, include the first two laws of Kepler.

The first result, easy to obtain, was that the angular momentum of the planet is conserved. The angular momentum is the vector

$$\vec{A} = m\vec{r} \times \vec{v}$$

where $\vec{r}$ is the heliocentric position vector of the planet, $\vec{v}$ the velocity of the planet, and m its mass. If the vector $\vec{A}$ remains constant, this means that the plane formed by $\vec{r}$ and $\vec{v}$ is always the same (the motion of the planet is planar) and the areal velocity $\frac{1}{2}\vec{r} \times \vec{v}$ is constant as given by Kepler's second law. The interesting point concerning this result is that it does not depend on the explicit form of the attraction forces. They arise for all attraction laws in which the two bodies attract themselves with forces aligned with the line passing by them (the so-called central forces). Another result found by Newton is that the mechanical energy is conserved. The mechanical energy of the planet is the sum of its heliocentric kinetic and potential energies:

$$E = \frac{1}{2}mv^2 - \frac{G(M+m)m}{r}$$

where M is the mass of the Sun. These two conservation laws may be combined into a first-order differential equation in the distance r having as independent variable the position angle of the planet in the plane of its heliocentric motion. This equation is easily solved and gives

$$r = \frac{p}{1 + e\cos\theta}$$

This equation is the equation of a conic section in the polar coordinates r, θ and the constants p and e are its parameter and eccentricity, which are related to the planet energy and angular momentum through

$$e = \sqrt{1 + \frac{2EA^2}{G^2(M+m)^2 m^3}}, \text{ and}$$

$$p = \frac{A^2}{G(M+m)m^2}$$

This conic section is an ellipse when $0 < e < 1$, a hyperbola when $e > 1$, a parabola when $e = 1$ and one circle. Therefore, Newton's result generalizes the first of Kepler's laws showing that, indeed, the motion of one body attracted by the Sun may be an ellipse, as the orbit of the planets, but may also be a hyperbola as the motion of some comets. The type of the conic section will be determined by its energy. If the energy is negative, the above equations give $e < 1$ and the motion is an ellipse. If the energy is positive, the above equations give $e > 1$ and the motion is a hyperbola. The circumference and the parabola are the limiting cases in which the energy is exactly equal to $-\frac{G^2(M+m)^2 m^3}{2A^2}$ or zero, respectively.

In the case of an ellipse, the semi-major axis may be obtained from the parameter p through

$$a = p\sqrt{1-e^2}.$$

In addition, in the case of elliptic motion, the combination of the various equations allows us to find the relationship between the semi-major axis of the ellipse and the orbital period P:

$$\frac{a^3}{P^2} = \frac{G(M+m)}{4\pi^2},$$

which is not the result given by Kepler's third law. In Kepler's third law, it is said that the ratio of the cube of the semi-major axes to the square of the periods is the same for all planets. Newton's theory shows that this quantity is in fact proportional to the sum of the masses $M+m$. However, planetary masses are small compared with the mass of the Sun (the mass of the largest planet is 1/1047 of the mass of the Sun) and Kepler's third law is a very good approximation of the actual result. It is enough to adopt the approximation $M+m \sim M$, to transform the above equation in Kepler's third law.

Newton did not limit himself to the problem of the motion of two attracting bodies. He considered in his work also the problem of the motion of the Moon around the Earth under the joint attractive forces of the Earth and the Sun. One of the greatest achievements of Newton's theory was due to one of his disciples, Edmond Halley. Halley understood that Newton's theory should be universal and applied it to study the motion of comets at a time in which the nature of these objects was not yet known. He analyzed the list of bright comets observed since the antiquity and verified that many of these observations were separated, one from another, by about 3/4 of century. He concluded that the comets of 1531, 1607 and 1682 were, in reality, one and the same comet and predicted its return in 1758. The comet that is today known as Halley's actually appeared at the predicted time. This event brought great renown to Newton's theory. However, the mathematics invented by Newton to compose his work (the theory of fluxions, or today, the differential calculus) was not understood by many other scientists and its adoption was only gradual. Even so, his gravitation theory was successfully used in next century for the construction of theories of the motion of planets, satellites, and comets. His theory was able to explain the polar flattening of the Earth, the variations of the gravity acceleration g on the surface of the Earth, the tides, etc. The success was so extensive that many people started to believe that it would be able to explain everything. The deterministic view of natural phenomena grew. After this view, it would be enough to know exactly the present situation to determine the future evolution. With some humor, the imaginary being which would be determining in an unequivocal way the motion of all bodies is sometime times called Laplace's demon! But, as we know today, one of the tricks of gravitation is that the determinism of its equations is not enough to make their solutions predictable for ever.

Newton's gravitation theory allows the construction of sets of differential equations whose solutions are the motions followed by the bodies. These equations are not easy to solve, and the great achievement of 18th century, to which we may associate the names of Leonhard Euler and Joseph-Louis Lagrange, among many others, was the construction of theories to obtain the solution of Newton's equations, the Theory of Perturbations. The most notorious achievement of the Theory of Perturbations was recorded in 1846. Its story began in 1781, when a new planet, Uranus, was discovered by William Herschel. It was the first planet noticed with a telescope and thus, the first planet discovered since remote antiquity. It was observed in the sky year after year for many decades and eventually the observations were enough to allow the construction of an accurate theory of its motion, which should be fully explained with Newton's equations. But this was not so. The motion of Uranus did not follow the results given by the theory. Bessel was the first to consider the problem posed by these discrepancies, but the arc drawn in the sky by the motion of Uranus since its discovery was too short and he could not find the explanation. His ideas were reconsidered some decades later, around 1840, by Adams in England and Leverrier in France. Adams concluded in 1845 that the observed discrepancies in the motion of Uranus were due to a yet unknown planet, and gave details on its orbit. However, his results did not get the attention it deserved from English astronomers. In France, Leverrier obtained the same results and published them in 1845 and 1846. In addition, when he believed that his results were accurate enough to allow a search, he sent a letter to Galle, in Berlin's observatory, and asked him to look for the planet. His request was promptly accepted and in the next evening Galle discovered Neptune less than one degree afar from the position indicated by Leverrier.

Uranus was not the only problematic planet. The motion of Mercury also was showing discrepancies. The very accurate calculations done by Leverrier showed that the perturbations of the other planets on the motion of Mercury were such that its ellipse was not kept fixed, but was precessing. The major axis of the ellipse should have a slow rotation: 530 arc seconds per century (277 arc seconds due to the attraction of Venus, 153 due to Jupiter, 90 due to the Earth and 11 due to Mars and the other planets). But the observed motion of Mercury was showing not 530, but about 570 arc seconds per century. Just as for Neptune, Leverrier looked for the possibility of an extra, intra-mercurial, planet which could be responsible for the excess of approximately 40 arc seconds per century. This planet was several times "discovered" and even got a name: Vulcan. However, these "discoveries" were never confirmed and the discrepancy could not be explained by Newton's Celestial Mechanics and had to wait for the next progresses on our knowledge of gravitation.

Celestial Mechanics after Einstein

In 1915 Einstein published his first results on a new theory of gravitation which became known as General Relativity Theory (GRT). The application to Celestial Mechanics done by him showed that the two-body motion laws introduced by Newton (and Kepler) should be corrected. The planets were not moving on fixed ellipses but on ellipses whose axes were slowly rotating. In the case of Mercury, the rotation indicated by Einstein's theory was 43 arc seconds per century. This was almost exactly the value of the discrepancy between the theories founded on Newton's gravitation and the observations.

Celestial Mechanics was one of the first branches of science to explore the consequences of the GRT. As soon as 1916, the first articles on the construction of the equations of planetary motion were published by Droste and by De Sitter. In Newton's theory, the law expressing the attraction force between two bodies is fully independent of the equations of the motion of these bodies. In GRT, the field and the motion appear together in only one set of equations: Einstein's field equations. These equations allow the calculation of a matrix giving the curvature of the space-time at every point, and GRT predicts that the motion of bodies will follow space-time geodesics.

This is a very intricate mathematical problem that could be completely solved only in particular cases as the field created by a fixed spherical body (Schwarzschild, 1916) and, later on, a fixed spheroid (Levi-Civita, 1917) and a rotating sphere (Kerr, 1943). The GRT equations of the motion of point masses in gravitational interaction was published in 1938, almost simultaneously in 2 papers, one by Eddington and Clark and the other by Einstein, Infeld and Hoffmann.

In the simplest of these cases, the Schwarzschild solution, the field has spherical symmetry and the distance between two points in this field is simply written using spherical coordinates. But this brings embedded one of the most difficult problems of GRT. To give an example, we may compute the planetary orbits and find Einstein's precessing ellipses. We may also look for the relationship between the size of the ellipse (its semi-major axis) and the period of the motion and find a given function, a relativistic version of Kepler's harmonic law. But the hypothesis necessary to reach these results is the spherical symmetry of the field. If it is deformed (inflated for instance) in such a way that the spherical symmetry is kept, the results may not change. The way in which the distance from one point in space to the center of the field is defined can not affect the solutions. To solve the problem, it is necessary to construct, in parallel to the theory of the motion, the theory of the processes used to measure the distances – e.g. using light rays propagating between the points – and use the distances measured with the same "rule". Only after that step it is possible to compare the complete results of the theory to those observed.

Until the second half of the 20th century, astronomers did not need to calculate any relativistic effects beyond the precession of the planetary ellipses given by GRT (42.98 arc seconds per century for Mercury, 8.62 arcseconds per century for Venus, 3.84 arc seconds per century for the Earth, 1.35 arc seconds per century for Mars). However, the progresses of Astrometry, including laser, radar, and VLBI (very-long baseline interferometry) forced astronomers to tackle the problem of the full consideration of the equations of GRT for the motion of the celestial bodies and also the measurement of times and distance.

Chaos

Nowadays, it is no longer possible to talk of Celestial Mechanics without mentioning chaos. Many problems in Celestial Mechanics are characterized by an evolution due only to gravitational forces with conservation of total energy and angular momentum for times of the order of millions or billions of years. Long evolutions under these conditions propitiate the rise of chaotic phenomena. The solutions of the equations of CM show great sensitivity to initial conditions: very close initial conditions may lead to totally different evolutions.

Many cases of chaotic evolution are found among the asteroids when they move in orbits with periods commensurable with the orbital period of Jupiter (11.8 years). For instance, the asteroids of the 3/1 resonance (i.e. with periods equal to 1/3 of the period of Jupiter) show three main regimes of motion, as shown in figure. Chaos means, in this case, that these three regimes of motion are not strictly separated and that there are solutions transiting from one regime to the next in time scales shorter than 1 million years. The transition from the regime (a) to the regime (b), the first discovered in this problem, allowed the explanation of the almost non existence of asteroids in this resonance. While the asteroid is in the low-eccentricity regime (a), the motion is nearly regular on a precessing ellipse whose eccentricity may show small periodic variations, but remains below 0.2 – 0.3. In the medium-eccentricity regime (b), on the contrary, the orbital eccentricity may reach values as high as 0.6 – 0.7. When the eccentricity becomes higher than 0.3 – 0.4, the minimal distance of the asteroid to the Sun becomes smaller than the average radius of the orbit of Mars. This means that, once per period, the asteroid will cross the orbit of Mars. If by chance, in one of these occasions, the asteroid crosses the orbit of Mars when the planet is close to the crossing point, the attraction of Mars will disturb the motion of the asteroid and may greatly change its period so that it will leave the resonance (after the close approach to Mars it will be on an orbit whose period is no longer 1/3 of the period of Jupiter). Further progress was made with the discovery of the existence of the high-eccentricity regime (c) in which the eccentricity of the asteroid may reach 0.9. Even if the transitions to this regime are not expected to have the same frequency as the transitions to (b), they are full of consequences for the asteroid's fate. Mars has a very small mass and an asteroid may remain in the regime (b) for millions of years without having an approach to Mars close enough to disturb its motion. But in regime (c), the asteroid will not only cross the orbit of Mars but also the orbits of the Earth and Venus, which are 10 times more massive than Mars, and can have its motion disturbed by these planets even in a less close approach. These results were obtained with simplified models considering Jupiter moving in accordance with Kepler's laws. However, more complex models, where the real motion of Jupiter was considered, showed that the reality is still more drastic: the regime (c) is not bounded, and the asteroid may enter into stretched orbits, crossing the orbits of the inner planets and allowing the asteroid to collide with the Sun. In this case we have more than a change in the orbit of the asteroid but also its physical destruction.

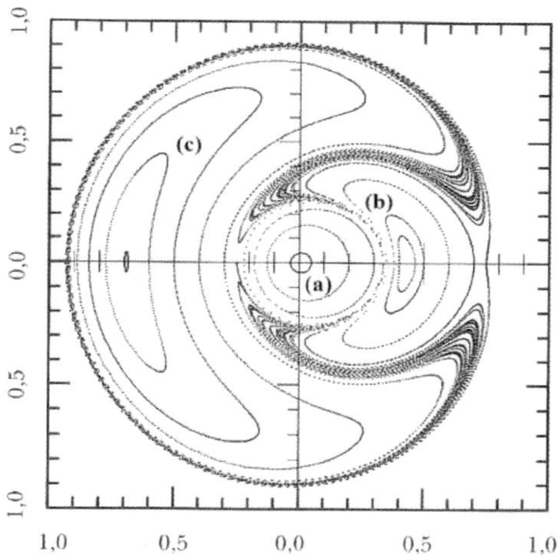

Figure: Regimes of motion of an asteroid whose mean orbital period is 1/3 of Jupiter's period.

In this figure are represented, in polar coordinates, the eccentricity (radius vector) and the longitude of the perihelion (polar angle) of the asteroid's orbit. The curves are evolutionary paths. The longitudes are reckoned from the direction of the perihelion of Jupiter's orbit. Regimes: (a) The orbit of the asteroid is almost circular (small eccentricities) and the perihelion rotates; (b) The orbit has great variations in eccentricity (between 0.2 and 0.4) and the perihelion remains oscillating around 0 (i.e about the direction of Jupiter's perihelion); (c) The eccentricity of the asteroid shows large variations and may become very large (in the outermost curves, the eccentricity goes beyond 0.9). The asteroid perihelion remains oscillating around 180 degrees (i.e the direction opposite to Jupiter's perihelion). The three regions are not perfectly insulated and the asteroid may chaotically transit from one regime of motion to another.

The possibility of chaos, i.e of transitions between different regimes of motion, affects directly the evolution of the Solar System. The apparent harmony that we observe, resulting from 5 billion years of evolution, will not last forever. Chaotic instabilities act very slowly as in the example above described. The rotation axis of Mars and Venus undergo large chaotic variations. Fortunately for us, the Moon forces the rotation of the Earth to be more regular thus keeping the delicate climatic equilibrium of our planet. Mercury, which has been parading with the other planets for 5 billion years, can have a close approach to Venus in the next 10 billion years with unpredictable consequences for its future motion. Violent events such as the collisions of bodies creating big craters, the formation of the Moon around the Earth, and other events that are difficult to imagine, may have occurred because minor bodies under the continuous disturbing influence of the larger ones evolved chaotically from very regular primordial motions.

In conclusion, Celestial Mechanics is a four centuries old science. During this long time it developed in many directions, impossible to consider in a short introduction. We may mention that several chapters of mathematics appeared aiming at the study of the solutions of the motion of *N* bodies under mutual gravitational attraction. The three body problem became a classic (q.v.). The KAM theory founded on the famous Kolmogorov's theorem, aimed at solving problems raised by the Theory of Perturbations. In the domain of applications, we have to mention Astrodynamics, one chapter of Celestial Mechanics that experienced a great development in the 20th century when the theory of the motion of artificial satellites was established, as well as the theories of the maneuvers necessary to transfer a spaceship from one orbit to another. One last point to keep in mind is that present-day Celestial Mechanics can not be restricted to gravitational forces. There is a panoply of non-gravitational forces acting on natural and artificial celestial bodies that perturb their motion in a significant way: gas drag, thermal emissions, interactions between radiation and matter, comet jets, tidal friction, etc.

Examples of Problems

Celestial motion without additional forces such as thrust of a rocket, is governed by gravitational acceleration of masses due to other masses. A simplification is the *n*-body problem, where the problem assumes some number *n* of spherically symmetric masses. In that case, the integration of the accelerations can be well approximated by relatively simple summations.

Examples:

- 4-body problem: spaceflight to Mars (for parts of the flight the influence of one or two bodies is very small, so that there we have a 2- or 3-body problem;
- 3-body problem:
    - Quasi-satellite
    - Spaceflight to, and stay at a Lagrangian point

In the case that $n=2$ (two-body problem), the situation is much simpler than for larger *n*. Various explicit formulas apply, where in the more general case typically only numerical solutions are possible. It is a useful simplification that is often approximately valid.

Examples:

- A binary star, e.g., Alpha Centauri (approx. the same mass)
- A binary asteroid, e.g., 90 Antiope (approx. the same mass)

A further simplification is based on the "standard assumptions in astrodynamics", which include that one body, the orbiting body, is much smaller than the other, the central body. This is also often approximately valid.

Examples:

- Solar system orbiting the center of the Milky Way
- A planet orbiting the Sun
- A moon orbiting a planet
- A spacecraft orbiting Earth, a moon, or a planet (in the latter cases the approximation only applies after arrival at that orbit)

Perturbation Theory

Perturbation theory comprises mathematical methods that are used to find an approximate solution to a problem which cannot be solved exactly. (It is closely related to methods used in numerical analysis, which are ancient.) The earliest use of perturbation theory was to deal with the otherwise unsolveable mathematical problems of celestial mechanics: Newton's solution for the orbit of the Moon, which moves noticeably differently from a simple Keplerian ellipse because of the competing gravitation of the Earth and the Sun.

Perturbation methods start with a simplified form of the original problem, which is carefully chosen to be exactly solvable. In celestial mechanics, this is usually a Keplerian ellipse, which is correct when there are only two gravitating bodies (say, the Earth and the Moon), or a circular orbit, which is only correct in special cases of two-body motion, but is often close enough for practical use. The solved, but simplified problem is then *"perturbed"* to make its starting conditions closer to the real problem, such as including the gravitational attraction of a third body (the Sun). The slight changes that results in, which themselves may have been simplified yet again, are used as corrections. Because of simplifications introduced along every step of the way, the corrections are never perfect, but even one cycle of corrections often provides a remarkably better approximate solution to the real problem.

There is no requirement to stop at only one cycle of corrections. A partially corrected solution can be re-used as the new starting point for yet another cycle of perturbations and corrections. The common difficulty with the method is that usually the corrections progressively make the new solutions very much more complicated, so each cycle is much more difficult to manage than the previous cycle of corrections. Newton is reported to have said, regarding the problem of the Moon's orbit *"It causeth my head to ache."*

This general procedure – starting with a simplified problem and gradually adding corrections that make the starting point of the corrected problem closer to the real situation – is a widely used mathematical tool in advanced sciences and engineering. It is the natural extension of the "guess, check, and fix" method used anciently with numbers.

Continuum Mechanics

Continuum mechanics is the mathematical description of deformation and related stresses. The fundamental assumption inscribed in the name is that materials are assumed to be homogeneous, isotropic, continuous and independent of any particular coordinate system. This assumption is based on the intuitive perception that, although nanoscale studies established the discontinuous (atomic, granular, molecular) nature of matter, the macroscopic appearance of solid objects is continuous, i.e. their properties seem to vary gradually, without discontinuity.

There are two classical techniques describing the motion of fluids and material points. In Lagrangian formulation the coordinates are glued to and move with a material point through space and time. Therefore, the coordinates of both the material point and the attached variable, such as temperature, do not change along their trajectory; they are given for a reference time and are time invariant. This formulation has the disadvantage that the reference points move with the flowing fluid. It is therefore difficult to know the state of the fluid at a given point in space and time. In Eulerian formulation one assumes a fixed coordinate system through which material points move as time passes. Therefore, the position of material points and any related quantity changes as a function of time. This change is then described by a partial derivative. A traditional image is a boat going down a river. The Lagrangian description describes the flow as seen from the boat; the Eulerian description describes the flow as seen from a fixed place on the river bank. A less bucolic illustration is to compare a flowing fluid to a car race, one car constituting a particle. Obviously the movement of every car should be defined at any one time to describe the race, i.e. the flow. The initial configuration is the starting line. The Lagrangian approach places a camera on each car to film the race. It is valid if one needs to follow the trajectory of the car that carries the camera; this motion is a function of time since the starting position, but the car looks immobile in the camera; the car velocity is nil, the outside world is in motion. Continuity is expressed by the spatial and temporal continuity between the starting position of the car and its position at the considered instant. The Eulerian approach fixes cameras along the track and records which car passes which camera at which time. Any car has velocity, the outside world is immobile. The velocity of all cars at each instant defines the flow. Continuity is expressed in the differentiability in space and time, at any moment, of physical quantities. Both Lagrangian and Eulerian approaches yield mathematically the same result, but the Eulerian formulation is often found more practical. Describing the motion of a fluid through volume elements at fixed locations (any of the "cameras") in function of time is termed Eulerian to honour the Swiss mathematician Leonhard Euler (1707-1783) who introduced this type of specification to study fluid motion and the mechanics (dynamics and kinematics) of deforming materials.

Considering the object of study as a closed system, the application of continuum mechanics requires respecting three fundamental physical principles:

- Conservation of mass,

- Conservation of momentum, both linear and angular,

typically comes from experiments. For car traffic it is well established that cars typically travel at a speed depending upon density, $u = V(\rho)$ for some experimentally determined function V that is a decreasing function of density. For example, experiments in the Lincoln Tunnel, New York, found that a good fit (except at low density) is obtained by $u = V(\rho) = 27.5\ln(142/\rho)$ (km/hr for density in cars/km).

Thus the basic continuum model for car traffic is the PDE

$$\frac{\partial \rho}{\partial t} + \frac{\partial}{\partial x}[\rho V(\rho)] = 0$$

for the car density $\rho(x,t)$ on the highway.

Major Areas

<table>
<tr><td rowspan="4">Continuum mechanics
The study of the physics of continuous materials</td><td rowspan="2">Solid mechanics
The study of the physics of continuous materials with a defined rest shape.</td><td colspan="2">Elasticity
Describes materials that return to their rest shape after applied stresses are removed.</td></tr>
<tr><td>Plasticity
Describes materials that permanently deform after a sufficient applied stress.</td><td rowspan="2">Rheology
The study of materials with both solid and fluid characteristics.</td></tr>
<tr><td rowspan="2">Fluid mechanics
The study of the physics of continuous materials which deform when subjected to a force.</td><td>Non-Newtonian fluids do not undergo strain rates proportional to the applied shear stress.</td></tr>
<tr><td colspan="2">Newtonian fluids undergo strain rates proportional to the applied shear stress.</td></tr>
</table>

An additional area of continuum mechanics comprises elastomeric foams, which exhibit a curious hyperbolic stress-strain relationship. The elastomer is a true continuum, but a homogeneous distribution of voids gives it unusual properties.

Formulation of Models

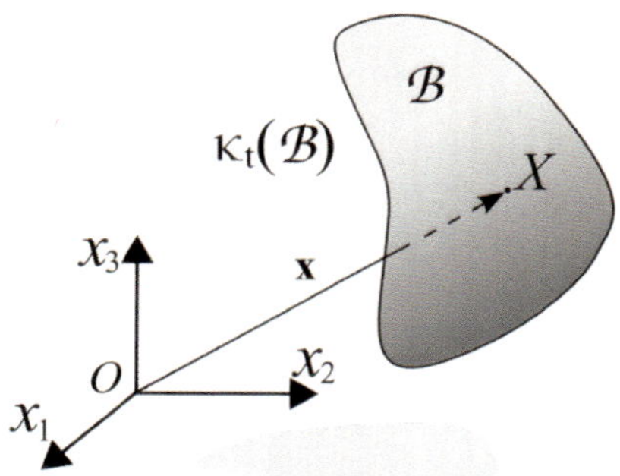

Figure: Configuration of a continuum body

Continuum mechanics models begin by assigning a region in three-dimensional Euclidean space to the material body $\mathcal{B}$ being modeled. The points within this region are called particles or material points. Different *configurations* or states of the body correspond to different regions in Euclidean space. The region corresponding to the body's configuration at time t is labeled $\kappa_t(\mathcal{B})$.

A particular particle within the body in a particular configuration is characterized by a position vector

$$\mathbf{x} = \sum_{i=1}^{3} x_i \mathbf{e}_i,$$

where $\mathbf{e}_i$ are the coordinate vectors in some frame of reference chosen for the problem. This vector can be expressed as a function of the particle position $\mathbf{X}$ in some *reference configuration*, for example the configuration at the initial time, so that

$$\mathbf{x} = \kappa_t(\mathbf{X}).$$

This function needs to have various properties so that the model makes physical sense. $\kappa_t(\cdot)$ needs to be:

- continuous in time, so that the body changes in a way which is realistic,
- globally invertible at all times, so that the body cannot intersect itself,
- orientation-preserving, as transformations which produce mirror reflections are not possible in nature.

For the mathematical formulation of the model, $\kappa_t(\cdot)$ is also assumed to be twice continuously differentiable, so that differential equations describing the motion may be formulated.

Forces in a Continuum

Continuum mechanics deals with deformable bodies, as opposed to rigid bodies. A solid is a deformable body that possesses shear strength, *sc.* a solid can support shear forces (forces parallel to the material surface on which they act). Fluids, on the other hand, do not sustain shear forces. For the study of the mechanical behavior of solids and fluids these are assumed to be continuous bodies, which means that the matter fills the entire region of space it occupies, despite the fact that matter is made of atoms, has voids, and is discrete. Therefore, when continuum mechanics refers to a point or particle in a continuous body it does not describe a point in the interatomic space or an atomic particle, rather an idealized part of the body occupying that point.

Following the classical dynamics of Newton and Euler, the motion of a material body is produced by the action of externally applied forces which are assumed to be of two kinds: surface forces $\mathbf{F}_C$ and body forces $\mathbf{F}_B$. Thus, the total force $\mathcal{F}$ applied to a body or to a portion of the body can be expressed as:

$$\mathcal{F} = \mathbf{F}_B + \mathbf{F}_C$$

Surface forces or *contact forces*, expressed as force per unit area, can act either on the bounding surface of the body, as a result of mechanical contact with other bodies, or on imaginary internal surfaces that bound portions of the body, as a result of the mechanical interaction between the parts of the body to either side of the surface (Euler-Cauchy's stress principle). When a body is acted upon by external contact forces, internal contact forces are then transmitted from point to point inside the body to balance their action, according to Newton's third law of motion of conservation of linear momentum and angular momentum (for continuous bodies these laws are called the Euler's equations of motion). The internal contact forces are related to the body's deformation through constitutive equations. The internal contact forces may be mathematically described by how they relate to the motion of the body, independent of the body's material makeup.

The distribution of internal contact forces throughout the volume of the body is assumed to be continuous. Therefore, there exists a *contact force density* or *Cauchy traction field* $\mathbf{T}(\mathbf{n},\mathbf{x},t)$ that represents this distribution in a particular configuration of the body at a given time t. It is not a vector field because it depends not only on the position $\mathbf{x}$ of a particular material point, but also on the local orientation of the surface element as defined by its normal vector $\mathbf{n}$.

Any differential area dS with normal vector $\mathbf{n}$ of a given internal surface area S, bounding a portion of the body, experiences a contact force $d\mathbf{F}_C$ arising from the contact between both portions of the body on each side of S, and it is given by

$$d\mathbf{F}_C = \mathbf{T}^{(\mathbf{n})}\, dS$$

where $\mathbf{T}^{(\mathbf{n})}$ is the *surface traction*, also called *stress vector*, *traction*, or *traction vector*. The stress vector is a frame-indifferent vector.

The total contact force on the particular internal surface S is then expressed as the sum (surface integral) of the contact forces on all differential surfaces dS:

$$\mathbf{F}_C = \int_S \mathbf{T}^{(\mathbf{n})}\, dS$$

In continuum mechanics a body is considered stress-free if the only forces present are those inter-atomic forces (ionic, metallic, and van der Waals forces) required to hold the body together and to keep its shape in the absence of all external influences, including gravitational attraction. Stresses generated during manufacture of the body to a specific configuration are also excluded when considering stresses in a body. Therefore, the stresses considered in continuum mechanics are only those produced by deformation of the body, *sc.* only relative changes in stress are considered, not the absolute values of stress.

Body forces are forces originating from sources outside of the body that act on the volume (or mass) of the body. Saying that body forces are due to outside sources implies that the interaction between different parts of the body (internal forces) are manifested through the contact forces alone. These forces arise from the presence of the body in force fields, *e.g.* gravitational field (gravitational forces) or electromagnetic field (electromagnetic forces), or from inertial forces when bodies are in motion. As the mass of a continuous body is assumed to be continuously distributed, any force originating from the mass is also continuously distributed. Thus, body forces are specified by vector fields which are assumed to be continuous over the entire volume of the body, *i.e.* acting on every point in it. Body forces are represented by a body force density $\mathbf{b}(\mathbf{x},t)$ (per unit of mass), which is a frame-indifferent vector field.

In the case of gravitational forces, the intensity of the force depends on, or is proportional to, the mass density $\rho(\mathbf{x},t)$ of the material, and it is specified in terms of force per unit mass (b_i) or per unit volume (p_i). These two specifications are related through the material density by the equation $\rho b_i = p_i$. Similarly, the intensity of electromagnetic forces depends upon the strength (electric charge) of the electromagnetic field.

The total body force applied to a continuous body is expressed as

$$\mathbf{F}_B = \int_V \mathbf{b} dm = \int_V \rho \mathbf{b} dV$$

Body forces and contact forces acting on the body lead to corresponding moments of force (torques) relative to a given point. Thus, the total applied torque $\mathcal{M}$ about the origin is given by

$$\mathcal{M} = \mathbf{M}_B + \mathbf{M}_C$$

In certain situations, not commonly considered in the analysis of the mechanical behavior of materials, it becomes necessary to include two other types of forces: these are *body moments* and *couple stresses* (surface couples, contact torques). Body moments, or body couples, are moments per unit volume or per unit mass applied to the volume of the body. Couple stresses are moments per unit area applied on a surface. Both are important in the analysis of stress for a polarized dielectric solid under the action of an electric field, materials where the molecular structure is taken into consideration (*e.g.* bones), solids under the action of an external magnetic field, and the dislocation theory of metals.

Materials that exhibit body couples and couple stresses in addition to moments produced exclusively by forces are called *polar materials*. *Non-polar materials* are then those materials with only moments of forces. In the classical branches of continuum mechanics the development of the theory of stresses is based on non-polar materials.

Thus, the sum of all applied forces and torques (with respect to the origin of the coordinate system) in the body can be given by

$$\mathcal{F} = \int_V \mathbf{a} dm = \int_S \mathbf{T} dS + \int_V \rho \mathbf{b} dV$$
$$\mathcal{M} = \int_S \mathbf{r} \times \mathbf{T} dS + \int_V \mathbf{r} \times \rho \mathbf{b} dV$$

Kinematics: Deformation and Motion

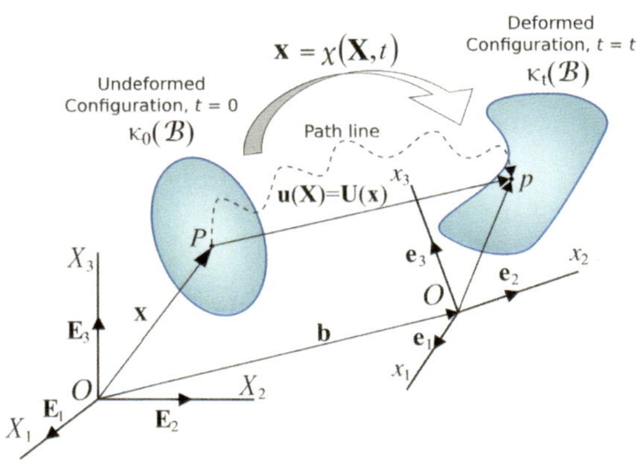

Figure: Motion of a continuum body

A change in the configuration of a continuum body results in a displacement. The displacement of a body has two components: a rigid-body displacement and a deformation. A rigid-body displacement consists of a simultaneous translation and rotation of the body without changing its shape or size. Deformation implies the change in shape and/or size of the body from an initial or undeformed configuration $\kappa_0(\mathcal{B})$ to a current or deformed configuration $\kappa_t(\mathcal{B})$.

The motion of a continuum body is a continuous time sequence of displacements. Thus, the material body will occupy different configurations at different times so that a particle occupies a series of points in space which describe a pathline.

There is continuity during deformation or motion of a continuum body in the sense that:

- The material points forming a closed curve at any instant will always form a closed curve at any subsequent time.
- The material points forming a closed surface at any instant will always form a closed surface at any subsequent time and the matter within the closed surface will always remain within.

It is convenient to identify a reference configuration or initial condition which all subsequent configurations are referenced from. The reference configuration need not be one that the body will ever occupy. Often, the configuration at $t = 0$ is considered the reference configuration, $\kappa_0(\mathcal{B})$. The components X_i of the position vector $\mathbf{X}$ of a particle, taken with respect to the reference configuration, are called the material or reference coordinates.

When analyzing the deformation or motion of solids, or the flow of fluids, it is necessary to describe the sequence or evolution of configurations throughout time. One description for motion is made in terms of the material or referential coordinates, called material description or Lagrangian description.

Lagrangian Description

In the Lagrangian description the position and physical properties of the particles are described in terms of the material or referential coordinates and time. In this case the reference configuration is the configuration at $t = 0$. An observer standing in the frame of reference observes the changes in the position and physical properties as the material body moves in space as time progresses. The results obtained are independent of the choice of initial time and reference configuration, $\kappa_0(\mathcal{B})$. This description is normally used in solid mechanics.

In the Lagrangian description, the motion of a continuum body is expressed by the mapping function $\chi(\cdot)$,

$$\mathbf{x} = \chi(\mathbf{X}, t)$$

which is a mapping of the initial configuration $\kappa_0(\mathcal{B})$ onto the current configuration $\kappa_t(\mathcal{B})$, giving a geometrical correspondence between them, i.e. giving the position vector $\mathbf{x} = x_i\mathbf{e}_i$ that a particle X, with a position vector $\mathbf{X}$ in the undeformed or reference configuration $\kappa_0(\mathcal{B})$, will occupy in the current or deformed configuration $\kappa_t(\mathcal{B})$ at time t. The components x_i are called the spatial coordinates.

Physical and kinematic properties $P_{ij\ldots}$, i.e. thermodynamic properties and flow velocity, which describe or characterize features of the material body, are expressed as continuous functions of position and time, i.e. $P_{ij\ldots} = P_{ij\ldots}(\mathbf{X},t)$.

The material derivative of any property $P_{ij\ldots}$ of a continuum, which may be a scalar, vector, or tensor, is the time rate of change of that property for a specific group of particles of the moving continuum body. The material derivative is also known as the *substantial derivative*, or *comoving derivative*, or *convective derivative*. It can be thought as the rate at which the property changes when measured by an observer traveling with that group of particles.

In the Lagrangian description, the material derivative of $P_{ij\ldots}$ is simply the partial derivative with respect to time, and the position vector $\mathbf{X}$ is held constant as it does not change with time. Thus, we have

$$\frac{d}{dt}[P_{ij\ldots}(\mathbf{X},t)] = \frac{\partial}{\partial t}[P_{ij\ldots}(\mathbf{X},t)]$$

The instantaneous position $\mathbf{x}$ is a property of a particle, and its material derivative is the *instantaneous flow velocity* $\mathbf{v}$ of the particle. Therefore, the flow velocity field of the continuum is given by

$$\mathbf{v} = \dot{\mathbf{x}} = \frac{d\mathbf{x}}{dt} = \frac{\partial \chi(\mathbf{X},t)}{\partial t}$$

Similarly, the acceleration field is given by

$$\mathbf{a} = \dot{\mathbf{v}} = \ddot{\mathbf{x}} = \frac{d^2\mathbf{x}}{dt^2} = \frac{\partial^2 \chi(\mathbf{X},t)}{\partial t^2}.$$

Continuity in the Lagrangian description is expressed by the spatial and temporal continuity of the mapping from the reference configuration to the current configuration of the material points. All physical quantities characterizing the continuum are described this way. In this sense, the function $\chi(\cdot)$ and $P_{ij\ldots}(\cdot)$ are single-valued and continuous, with continuous derivatives with respect to space and time to whatever order is required, usually to the second or third.

Eulerian Description

Continuity allows for the inverse of $\chi(\cdot)$ to trace backwards where the particle currently located at $\mathbf{x}$ was located in the initial or referenced configuration $\kappa_0(\mathcal{B})$. In this case the description of motion is made in terms of the spatial coordinates, in which case is called the spatial description or Eulerian description, i.e. the current configuration is taken as the reference configuration.

The Eulerian description, introduced by d'Alembert, focuses on the current configuration $\kappa_t(\mathcal{B})$, giving attention to what is occurring at a fixed point in space as time progresses, instead of giving attention to individual particles as they move through space and time. This approach is conveniently applied in the study of fluid flow where the kinematic property of greatest interest is the rate at which change is taking place rather than the shape of the body of fluid at a reference time.

Mathematically, the motion of a continuum using the Eulerian description is expressed by the mapping function:

$$\mathbf{X} = \chi^{-1}(\mathbf{x}, t)$$

which provides a tracing of the particle which now occupies the position $\mathbf{x}$ in the current configuration $\kappa_t(\mathcal{B})$ to its original position $\mathbf{X}$ in the initial configuration $\kappa_0(\mathcal{B})$.

A necessary and sufficient condition for this inverse function to exist is that the determinant of the Jacobian Matrix, often referred to simply as the Jacobian, should be different from zero. Thus,

$$J = \left|\frac{\partial \chi_i}{\partial X_J}\right| = \left|\frac{\partial x_i}{\partial X_J}\right| \neq 0$$

In the Eulerian description, the physical properties $P_{ij\ldots}$ are expressed as

$$P_{ij\ldots} = P_{ij\ldots}(\mathbf{X}, t) = P_{ij\ldots}[\chi^{-1}(\mathbf{x}, t), t] = p_{ij\ldots}(\mathbf{x}, t)$$

where the functional form of $P_{ij\ldots}$ in the Lagrangian description is not the same as the form of $P_{ij\ldots}$ in the Eulerian description.

The material derivative of $p_{ij\ldots}(\mathbf{x}, t)$, using the chain rule, is then

$$\frac{d}{dt}[p_{ij\ldots}(\mathbf{x}, t)] = \frac{\partial}{\partial t}[p_{ij\ldots}(\mathbf{x}, t)] + \frac{\partial}{\partial x_k}[p_{ij\ldots}(\mathbf{x}, t)]\frac{dx_k}{dt}$$

The first term on the right-hand side of this equation gives the *local rate of change* of the property $p_{ij\ldots}(\mathbf{x}, t)$ occurring at position $\mathbf{x}$. The second term of the right-hand side is the *convective rate of change* and expresses the contribution of the particle changing position in space (motion).

Continuity in the Eulerian description is expressed by the spatial and temporal continuity and continuous differentiability of the flow velocity field. All physical quantities are defined this way at each instant of time, in the current configuration, as a function of the vector position $\mathbf{x}$.

Displacement Field

The vector joining the positions of a particle P in the undeformed configuration and deformed configuration is called the displacement vector $\mathbf{u}(\mathbf{X}, t) = u_i \mathbf{e}_i$, in the Lagrangian description, or $\mathbf{U}(\mathbf{x}, t) = U_J \mathbf{E}_J$, in the Eulerian description.

A *displacement field* is a vector field of all displacement vectors for all particles in the body, which relates the deformed configuration with the undeformed configuration. It is convenient to do the analysis of deformation or motion of a continuum body in terms of the displacement field, In general, the displacement field is expressed in terms of the material coordinates as

$$\mathbf{u}(\mathbf{X}, t) = \mathbf{b} + \mathbf{x}(\mathbf{X}, t) - \mathbf{X} \qquad \text{or} \qquad u_i = \alpha_{iJ} b_J + x_i - \alpha_{iJ} X_J$$

or in terms of the spatial coordinates as

$$\mathbf{U}(\mathbf{x}, t) = \mathbf{b} + \mathbf{x} - \mathbf{X}(\mathbf{x}, t) \qquad \text{or} \qquad U_J = b_J + \alpha_{Ji} x_i - X_J$$

where α_{Ji} are the direction cosines between the material and spatial coordinate systems with unit vectors $\mathbf{E}_J$ and $\mathbf{e}_i$, respectively. Thus

$$\mathbf{E}_J \cdot \mathbf{e}_i = \alpha_{Ji} = \alpha_{iJ}$$

and the relationship between u_i and U_J is then given by

$$u_i = \alpha_{iJ} U_J \qquad \text{or} \qquad U_J = \alpha_{Ji} u_i$$

Knowing that

$$\mathbf{e}_i = \alpha_{iJ} \mathbf{E}_J$$

then

$$\mathbf{u}(\mathbf{X},t) = u_i \mathbf{e}_i = u_i(\alpha_{iJ} \mathbf{E}_J) = U_J \mathbf{E}_J = \mathbf{U}(\mathbf{x},t)$$

It is common to superimpose the coordinate systems for the undeformed and deformed configurations, which results in $\mathbf{b} = 0$, and the direction cosines become Kronecker deltas, i.e.

$$\mathbf{E}_J \cdot \mathbf{e}_i = \delta_{Ji} = \delta_{iJ}$$

Thus, we have

$$\mathbf{u}(\mathbf{X},t) = \mathbf{x}(\mathbf{X},t) - \mathbf{X} \qquad \text{or} \qquad u_i = x_i - \delta_{iJ} X_J$$

or in terms of the spatial coordinates as

$$\mathbf{U}(\mathbf{x},t) = \mathbf{x} - \mathbf{X}(\mathbf{x},t) \qquad \text{or} \qquad U_J = \delta_{Ji} x_i - X_J$$

Governing Equations

Continuum mechanics deals with the behavior of materials that can be approximated as continuous for certain length and time scales. The equations that govern the mechanics of such materials include the balance laws for mass, momentum, and energy. Kinematic relations and constitutive equations are needed to complete the system of governing equations. Physical restrictions on the form of the constitutive relations can be applied by requiring that the second law of thermodynamics be satisfied under all conditions. In the continuum mechanics of solids, the second law of thermodynamics is satisfied if the Clausius–Duhem form of the entropy inequality is satisfied.

The balance laws express the idea that the rate of change of a quantity (mass, momentum, energy) in a volume must arise from three causes:

1. The physical quantity itself flows through the surface that bounds the volume,
2. There is a source of the physical quantity on the surface of the volume, or/and,
3. There is a source of the physical quantity inside the volume.

Let Ω be the body (an open subset of Euclidean space) and let $\partial\Omega$ be its surface (the boundary of Ω).

Let the motion of material points in the body be described by the map:

$$\mathbf{x} = \chi(\mathbf{X}) = \mathbf{x}(\mathbf{X})$$

where $\mathbf{X}$ is the position of a point in the initial configuration and $\mathbf{x}$ is the location of the same point in the deformed configuration.

The deformation gradient is given by

$$\mathbf{F} = \frac{\partial \mathbf{x}}{\partial \mathbf{X}} = \nabla \mathbf{x} .$$

Balance Laws

Let $f(\mathbf{x},t)$ be a physical quantity that is flowing through the body. Let $g(\mathbf{x},t)$ be sources on the surface of the body and let $h(\mathbf{x},t)$ be sources inside the body. Let $\mathbf{n}(\mathbf{x},t)$ be the outward unit normal to the surface $\partial\Omega$. Let $\mathbf{v}(\mathbf{x},t)$ be the flow velocity of the physical particles that carry the physical quantity that is flowing. Also, let the speed at which the bounding surface $\partial\Omega$ is moving be u_n (in the direction $\mathbf{n}$).

Then, balance laws can be expressed in the general form

$$\frac{d}{dt}\left[\int_{\Omega} f(\mathbf{x},t)\mathrm{dV}\right] = \int_{\partial\Omega} f(\mathbf{x},t)[u_n(\mathbf{x},t) - \mathbf{v}(\mathbf{x},t)\cdot\mathbf{n}(\mathbf{x},t)]\mathrm{dA} + \int_{\partial\Omega} g(\mathbf{x},t)\mathrm{dA} + \int_{\Omega} h(\mathbf{x},t)\mathrm{dV} .$$

Note that the functions $f(\mathbf{x},t)$, $g(\mathbf{x},t)$, and $h(\mathbf{x},t)$ can be scalar valued, vector valued, or tensor valued - depending on the physical quantity that the balance equation deals with. If there are internal boundaries in the body, jump discontinuities also need to be specified in the balance laws.

If we take the Eulerian point of view, it can be shown that the balance laws of mass, momentum, and energy for a solid can be written as (assuming the source term is zero for the mass and angular momentum equations)

$$\dot{\rho} + \rho\nabla\cdot\mathbf{v} = 0 \qquad \textit{Balance of Mass}$$

$$\rho\mathbf{v} - \nabla\cdot\boldsymbol{\sigma} - \rho\mathbf{b} = 0 \qquad \textit{BalanceofLinearMomentum(Cauchy's first law of motion)}$$

$$\boldsymbol{\sigma} = \boldsymbol{\sigma}^T \qquad \textit{BalanceofAngularMomentum(Cauchy's second law of motion)}$$

$$\rho\dot{e} - \boldsymbol{\sigma} : (\nabla\mathbf{v}) + \nabla\cdot\mathbf{q} - \rho s = 0 \qquad \textit{Balance of Energy.}$$

In the above equations $\rho(\mathbf{x},t)$ is the mass density (current), $\dot{\rho}$ is the material time derivative of ρ, $\mathbf{v}(\mathbf{x},t)$ is the particle velocity, $\mathbf{v}$ is the material time derivative of $\mathbf{v}$, $\boldsymbol{\sigma}(\mathbf{x},t)$ is the Cauchy stress tensor, $\mathbf{b}(\mathbf{x},t)$ is the body force density, $e(\mathbf{x},t)$ is the internal energy per unit mass, $\dot{e}$ is the material time derivative of e, $\mathbf{q}(\mathbf{x},t)$ is the heat flux vector, and $s(\mathbf{x},t)$ is an energy source per unit mass.

With respect to the reference configuration (the Lagrangian point of view), the balance laws can be written as

$$\rho\,\det(\mathbf{F}) - \rho_o = 0 \qquad \text{Balance of Mass}$$

$$\rho_o\,\ddot{\mathbf{x}} - \nabla_\circ \cdot \mathbf{P}^T - \rho_o\,\mathbf{b} = 0 \qquad \text{Balance of Linear Momentum}$$

$$\mathbf{F}\cdot\mathbf{P}^T = \mathbf{P}\cdot\mathbf{F}^T \qquad \text{Balance of Angular Momentum}$$

$$\rho_o\,\dot{e} - \mathbf{P}^T : \dot{\mathbf{F}} + \nabla_\circ \cdot \mathbf{q} - \rho_o\,s = 0 \qquad \text{Balance of Energy.}$$

In the above, P is the first Piola-Kirchhoff stress tensor, and ρ_o is the mass density in the reference configuration. The first Piola-Kirchhoff stress tensor is related to the Cauchy stress tensor by

$$\mathbf{P} = J\,\boldsymbol{\sigma}\cdot\mathbf{F}^{-T} \text{ where } J = \det(\mathbf{F})$$

We can alternatively define the nominal stress tensor $\boldsymbol{N}$ which is the transpose of the first Piola-Kirchhoff stress tensor such that

$$\boldsymbol{N} = \mathbf{P}^T = \boldsymbol{J}\,\mathbf{F}^{-1}\cdot\boldsymbol{\sigma}\,.$$

Then the balance laws become

$$\rho\,\det(\mathbf{F}) - \rho_o = 0 \qquad \textit{Balance of Mass}$$

$$\rho_o\,\ddot{\mathbf{x}} - \nabla_\circ \cdot \mathbf{N} - \rho_o\,\mathbf{b} = 0 \qquad \textit{Balance of Linear Momentum}$$

$$\mathbf{F}\cdot\mathbf{N} = \mathbf{N}^T\cdot\mathbf{F}^T \qquad \textit{Balance of Angular Momentum}$$

$$\rho_o\,\dot{\mathbf{e}} - \mathbf{N} : \dot{\mathbf{F}} + \nabla_\circ \cdot \mathbf{q} - \rho_o\,\mathbf{s} = 0 \qquad \textit{Balance of Energy.}$$

The operators in the above equations are defined as such that

$$\nabla\mathbf{v} = \sum_{i,j=1}^{3} \frac{\partial \boldsymbol{v}_i}{\partial \boldsymbol{x}_j}\mathbf{e}_i \otimes \mathbf{e}_j = \boldsymbol{v}_{i,j}\mathbf{e}_i \otimes \mathbf{e}_j \ ; \ \nabla\cdot\mathbf{v} = \sum_{i=1}^{3} \frac{\partial \boldsymbol{v}_i}{\partial \boldsymbol{x}_i} = \boldsymbol{v}_{i,i} \ ; \ \nabla\cdot\mathbf{S} = \sum_{i,j=1}^{3} \frac{\partial \mathbf{S}_{ij}}{\partial \boldsymbol{x}_j}\mathbf{e}_i = \sigma_{ij,j}\,\mathbf{e}_i\,.$$

where $\mathbf{v}$ is a vector field, $\boldsymbol{S}$ is a second-order tensor field, and $\mathbf{e}_i$ are the components of an orthonormal basis in the current configuration. Also,

$$\nabla_\circ\mathbf{v} = \sum_{i,j=1}^{3} \frac{\partial \boldsymbol{v}_i}{\partial \boldsymbol{X}_j}\mathbf{E}_i \otimes \mathbf{E}_j = \boldsymbol{v}_{i,j}\mathbf{E}_i \otimes \mathbf{E}_j \ ; \ \nabla_\circ\cdot\mathbf{v} = \sum_{i=1}^{3} \frac{\partial \boldsymbol{v}_i}{\partial \boldsymbol{X}_i} = \boldsymbol{v}_{i,i} \ ; \ \nabla_\circ\cdot\mathbf{S} = \sum_{i,j=1}^{3} \frac{\partial \mathbf{S}_{ij}}{\partial \boldsymbol{X}_j}\mathbf{E}_i = \boldsymbol{S}_{ij,j}\,\mathbf{E}_i$$

where $\mathbf{v}$ is a vector field, $\boldsymbol{S}$ is a second-order tensor field, and $\mathbf{E}_i$ are the components of an orthonormal basis in the reference configuration.

The inner product is defined as

$$\mathbf{A} : \mathbf{B} = \sum_{i,j=1}^{3} \boldsymbol{A}_{ij}\,\boldsymbol{B}_{ij} = \mathrm{trace}(\mathbf{A}\mathbf{B}^T)\,.$$

Clausius–Duhem Inequality

The Clausius–Duhem inequality can be used to express the second law of thermodynamics for elastic-plastic materials. This inequality is a statement concerning the irreversibility of natural processes, especially when energy dissipation is involved.

Just like in the balance laws in the previous section, we assume that there is a flux of a quantity, a source of the quantity, and an internal density of the quantity per unit mass. The quantity of interest in this case is the entropy. Thus, we assume that there is an entropy flux, an entropy source, an internal mass density $\tilde{n}$ and an internal specific entropy (i.e. entropy per unit mass) η in the region of interest.

Let Ω be such a region and let $\partial\Omega$ be its boundary. Then the second law of thermodynamics states that the rate of increase of η in this region is greater than or equal to the sum of that supplied to Ω (as a flux or from internal sources) and the change of the internal entropy density $\rho\eta$ due to material flowing in and out of the region.

Let $\partial\Omega$ move with a flow velocity u_n and let particles inside Ω have velocities $\mathbf{v}$. Let $\mathbf{n}$ be the unit outward normal to the surface $\partial\Omega$. Let $\tilde{n}$ be the density of matter in the region, $\bar{q}$ be the entropy flux at the surface, and r be the entropy source per unit mass. Then the entropy inequality may be written as

$$\frac{\mathbf{d}}{\mathbf{dt}}\left(\int_{\Omega}\rho\eta dV\right) \geq \int_{\partial\Omega}\rho\eta\,(\boldsymbol{u}_n - \mathbf{v}\cdot\mathbf{n})dA + \int_{\partial\Omega}\bar{\boldsymbol{q}}\,dA + \int_{\Omega}\rho\boldsymbol{r}\,dV.$$

The scalar entropy flux can be related to the vector flux at the surface by the relation $\bar{q}$=-$\psi(\mathbf{x})\cdot\mathbf{n}$. Under the assumption of incrementally isothermal conditions, we have

$$\psi(x)=\frac{\mathbf{q}(x)}{T}\ ;\ r=\frac{s}{T}$$

where $\mathbf{q}$ is the heat flux vector, s is an energy source per unit mass, and T is the absolute temperature of a material point at $\mathbf{x}$ at time t.

We then have the Clausius–Duhem inequality in integral form:

$$\frac{d}{dt}\left(\int_{\Omega}\rho\eta\ dV\right)^{3}\int_{\partial\Omega}\rho\eta\ (u_n\text{-}\mathbf{v}\cdot\mathbf{n})\ dA\text{-}\int_{\partial\Omega}\frac{\mathbf{q}\cdot\mathbf{n}}{T}\ dA\text{+}\int_{\Omega}\frac{\rho\, s}{T}\ dV.$$

We can show that the entropy inequality may be written in differential form as

$$\rho\dot{\eta} \geq -\nabla\cdot\left(\frac{\mathbf{q}}{\boldsymbol{T}}\right)+\frac{\rho\boldsymbol{s}}{\boldsymbol{T}}.$$

In terms of the Cauchy stress and the internal energy, the Clausius–Duhem inequality may be written as

$$\rho\ (\dot{e}\text{-}T\,\dot{\eta}) - \boldsymbol{\sigma}:\nabla\mathbf{v} \leq -\frac{\mathrm{q}\cdot\nabla\mathrm{T}}{\mathrm{T}}.$$

Kinematics

Kinematics is the study of the motion of points, objects, and groups of objects without considering the causes of its motion.

Kinematics is the branch of classical mechanics that describes the motion of points, objects and systems of groups of objects, without reference to the causes of motion (i.e., forces). The study of kinematics is often referred to as the "geometry of motion."

Objects are in motion all around us. Everything from a tennis match to a space-probe flyby of the planet Neptune involves motion. When you are resting, your heart moves blood through your veins. Even in inanimate objects there is continuous motion in the vibrations of atoms and molecules. Interesting questions about motion can arise: how long will it take for a space probe to travel to Mars? Where will a football land if thrown at a certain angle? An understanding of motion, however, is also key to understanding other concepts in physics. An understanding of acceleration, for example, is crucial to the study of force.

To describe motion, kinematics studies the trajectories of points, lines and other geometric objects, as well as their differential properties (such as velocity and acceleration). Kinematics is used in astrophysics to describe the motion of celestial bodies and systems; and in mechanical engineering, robotics and biomechanics to describe the motion of systems composed of joined parts (such as an engine, a robotic arm, or the skeleton of the human body).

A formal study of physics begins with kinematics. The word "kinematics" comes from a Greek word "kinesis" meaning motion, and is related to other English words such as "cinema" (movies) and "kinesiology" (the study of human motion). Kinematic analysis is the process of measuring the kinematic quantities used to describe motion. The study of kinematics can be abstracted into purely mathematical expressions, which can be used to calculate various aspects of motion such as velocity, acceleration, displacement, time, and trajectory.

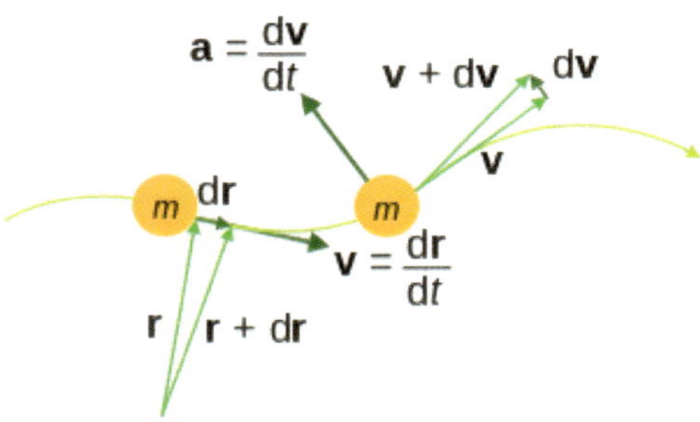

Kinematics of a particle trajectory: Kinematic equations can be used to calculate the trajectory of particles or objects. The physical quantities relevant to the motion of a particle include: mass m, position r, velocity v, acceleration a.

Reference Frames and Displacement

In order to describe an object's motion, you need to specify its position relative to a convenient reference frame.

In order to describe the motion of an object, you must first describe its position — where it is at any particular time. More precisely, you need to specify its position relative to a convenient reference frame. Earth is often used as a reference frame, and we often describe the position of objects related to its position to or from Earth. Mathematically, the position of an object is generally represented by the variable x.

Frame of Reference

There are two choices you have to make in order to define a position variable x. You have to decide where to put $x = 0$ and which direction will be positive. This is referred to as choosing a coordinate system, or choosing a frame of reference. As long as you are consistent, any frame is equally valid. But you don't want to change coordinate systems in the middle of a calculation. Imagine sitting in a train in a station when suddenly you notice that the station is moving backward. Most people would say that they just failed to notice that the train was moving — it only *seemed* like the station was moving. But this shows that there is a *third* arbitrary choice that goes into choosing a coordinate system: valid frames of reference can differ from each other by moving relative to one another. It might seem strange to use a coordinate system moving relative to the earth — but, for instance, the frame of reference moving along with a train might be far more convenient for describing things happening inside the train. Frames of reference are particularly important when describing an object's displacement.

Displacement

Displacement is the change in position of an object relative to its reference frame. For example, if a car moves from a house to a grocery store, its displacement is the relative distance of the grocery store to the reference frame, or the house. The word "displacement" implies that an object has moved or has been displaced. Displacement is the change in position of an object and can be represented mathematically as follows:

$$\Delta x = x_f - x_0$$

where Δx is displacement, x_f is the final position, and x_0 is the initial position.

shows the importance of using a frame of reference when describing the displacement of a passenger on an airplane.

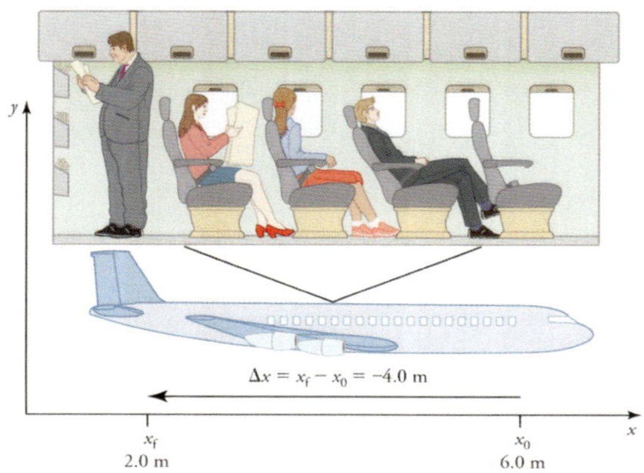

Displacement in Terms of Frame of Reference: A passenger moves from his seat to the back of the plane. His location relative to the airplane is given by x. The -4.0m displacement of the passenger relative to the plane is represented by an arrow toward the rear of the plane. Notice that the arrow representing his displacement is twice as long as the arrow representing the displacement of the professor (he moves twice as far).

Scalars and Vectors

A vector is any quantity that has both magnitude and direction, whereas a scalar has only magnitude.

What is the difference between distance and displacement? Whereas displacement is defined by both direction and magnitude, distance is defined by magnitude alone. Displacement is an example of a vector quantity. Distance is an example of a scalar quantity. A vector is any quantity with both magnitude and direction. Other examples of vectors include a velocity of 90 km/h east and a force of 500 newtons straight down.

In mathematics, physics, and engineering, a vector is a geometric object that has a magnitude (or length) and direction and can be added to other vectors according to vector algebra. The direction of a vector in one-dimensional motion is given simply by a plus (+) or minus (–) sign. A vector is frequently represented by a line segment with a definite direction, or graphically as an arrow, connecting an initial point A with a terminal point B, as shown in.

$$\overrightarrow{AB}$$

Vector representation: A vector is frequently represented by a line segment with a definite direction, or graphically as an arrow, connecting an initial point A with a terminal point B.

Some physical quantities, like distance, either have no direction or no specified direction. In physics, a scalar is a simple physical quantity that is not changed by coordinate system rotations or translations. It is any quantity that can be expressed by a single number and has a magnitude, but no direction. For example, a 20°C temperature, the 250 kilocalories (250 Calories) of energy in a candy bar, a 90 km/h speed limit, a person's 1.8 m height, and a distance of 2.0 m are all scalars, or quantities with no specified direction. Note, however, that a scalar can be negative, such as a –20°C temperature. In this case, the minus sign indicates a point on a scale rather than a direction. Scalars are never represented by arrows. (A comparison of scalars vs. vectors is shown in figure).

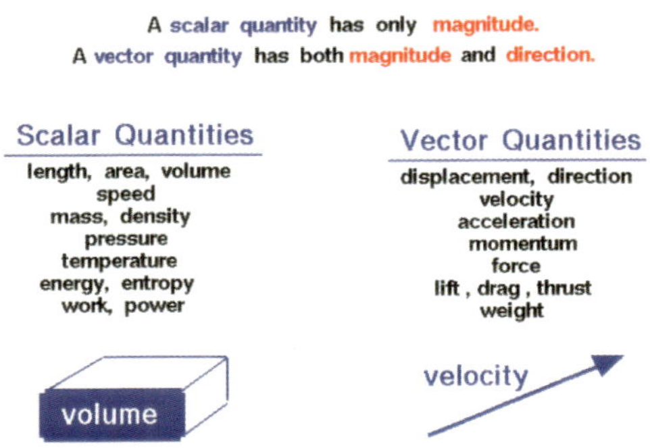

Scalars vs. Vectors: A brief list of quantities that are either scalars or vectors.

Particle Trajectories under Constant Acceleration

For the case of constant acceleration, the differential equation can be integrated as the acceleration vector A of a point *P* is constant in magnitude and direction. Such a point is said to undergo *uniformly accelerated motion*. In this case, the velocity V(t) and then the trajectory P(t) of the particle can be obtained by integrating the acceleration equation A with respect to time.

Assuming that the initial conditions of the position, $\mathbf{P}_0$, and velocity $\mathbf{V}_0$ at time $t=0$ are known, the first integration yields the velocity of the particle as a function of time.

$$V(t)=V_0+\int_0^t A d\tau=V_0+At.$$

A second integration yields its path (trajectory),

$$P(t)=P_0+\int_0^t V(\tau)d\tau=\mathbf{P}_0+\int_0^t (V_0+A\tau)\, d\tau=\mathbf{P}_0+\mathbf{V}_0 t+\tfrac{1}{2}\mathbf{A}t^2.$$

Additional relations between displacement, velocity, acceleration, and time can be derived. Since the acceleration is constant,

$\mathbf{A}=\dfrac{\Delta\mathbf{V}}{\Delta t}=\dfrac{\mathbf{V}-\mathbf{V}_0}{t}$ can be substituted into the above equation to give:

$$\mathbf{P}(t)=\mathbf{P}_0+\left(\frac{\mathbf{V}+\mathbf{V}_0}{2}\right)t.$$

A relationship between velocity, position and acceleration without explicit time dependence can be had by solving the average acceleration for time and substituting and simplifying,

$$t=\frac{\mathbf{V}-\mathbf{V}_0}{\mathbf{A}}$$

$$(\mathbf{P}-\mathbf{P}_0)^\circ\mathbf{A}=(\mathbf{V}-\mathbf{V}_0)\circ\frac{\mathbf{V}+\mathbf{V}_0}{2},$$

where $\circ$ denotes the dot product, which is appropriate as the products are scalars rather than vectors.

$$2(\mathbf{P}-\mathbf{P}_0)^\circ\mathbf{A}=|\mathbf{V}|^2-|\mathbf{V}_0|^2.$$

The dot can be replaced by the cosine of the angle α between the vectors and the vectors by their magnitudes, in which case:

$$2|\mathbf{P}-\mathbf{P}_0|\cdot|\mathbf{A}|\cdot\cos\alpha=|\mathbf{V}|^2-|\mathbf{V}_0|^2.$$

In the case of acceleration always in the direction of the motion, the angle between the vectors (α) is 0, so $\cos 0 \quad 1$, and

$$|\mathbf{V}|^2=|\mathbf{V}_0|^2+2|\mathbf{A}|\cdot|\mathbf{P}-\mathbf{P}_0|.$$

This can be simplified using the notation for the magnitudes of the vectors $||\mathbf{A}|| = a, ||\mathbf{V}|| = v, ||\mathbf{P} - \mathbf{P}_0|| = \Delta x$ where Δx can be any curvaceous path taken as the constant tangential acceleration is applied along that path, so

$$v^2 = v_0^2 + 2a \cdot \Delta x.$$

This reduces the parametric equations of motion of the particle to a cartesian relationship of speed versus position. This relation is useful when time is unknown. We also know that $\Delta x = \int (v)dt$ or Δx is the area under a v, t graph

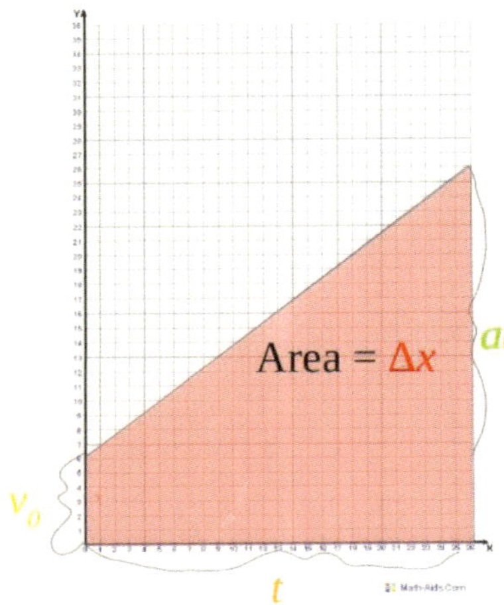

Velocity Time physics graph

We can take Δx by adding the top area and the bottom area. The bottom area is a rectangle, and the area of a rectangle is the $A \cdot B$ where A is the width and B is the height. In this case $A = t$ and $B = v_0$ (note that the A here is different from the acceleration a). This means that the bottom area is tv_0. Now let's find the top area (a triangle). The area of a trangle is $\frac{1}{2}BH$ where B is the base and H is the height. In this case, $B = t$ & $H = at$ or $A = \frac{1}{2}BH = \frac{1}{2}att = \frac{1}{2}at^2 = \frac{at^2}{2}$. Adding tv_0 and $\frac{at^2}{2}$ results in the equation Δx results in the equation $\Delta x = tv_0 + \frac{at^2}{2}$. This equation is very useful when the final velocity v is unknown.

Particle Trajectories in Cylindrical-polar Coordinates

It is often convenient to formulate the trajectory of a particle P(t) = (X(t), Y(t) and Z(t)) using polar coordinates in the X–Y plane. In this case, its velocity and acceleration take a convenient form.

Recall that the trajectory of a particle P is defined by its coordinate vector P measured in a fixed reference frame F. As the particle moves, its coordinate vector P(t) traces its trajectory, which is a curve in space, given by:

$$\mathbf{P}(t) = X(t)\hat{i} + Y(t)\hat{j} + Z(t)\hat{k},$$

where i, j, and k are the unit vectors along the X, Y and Z axes of the reference frame F, respectively.

Consider a particle P that moves only on the surface of a circular cylinder R(t)=constant, it is possible to align the Z axis of the fixed frame F with the axis of the cylinder. Then, the angle θ around this axis in the $X-Y$ plane can be used to define the trajectory as,

$$\mathbf{P}(t) = R\cos\theta(t)\hat{i} + R\sin\theta(t)\hat{j} + Z(t)\hat{k}.$$

The cylindrical coordinates for P(t) can be simplified by introducing the radial and tangential unit vectors,

$$\mathbf{e}_r = \cos\theta(\mathrm{t})\hat{i} + \sin\theta(\mathrm{t})\hat{j}, \quad \mathbf{e}_\theta = -\sin\theta(\mathrm{t})\hat{i} + \cos\theta(\mathrm{t})\hat{j}.$$

and their time derivatives from elementary calculus:

$$\frac{d}{dt}\mathbf{e}_r = \dot{\mathbf{e}}_r = \dot{\theta}\mathbf{e}_\theta$$
$$\frac{d}{dt}\dot{\mathbf{e}}_r = \ddot{\mathbf{e}}_r = \ddot{\theta}\mathbf{e}_\theta - \dot{\theta}\mathbf{e}_r$$
$$\frac{d}{dt}\mathbf{e}_\theta = \dot{\mathbf{e}}_\theta = -\dot{\theta}\mathbf{e}_r$$
$$\frac{d}{dt}\dot{\mathbf{e}}_\theta = \ddot{\mathbf{e}}_\theta = -\ddot{\theta}\mathbf{e}_r - \dot{\theta}^2\mathbf{e}_\theta.$$

Using this notation, P(t) takes the form,

$$\mathbf{P}(t) = R\mathbf{e}_r + Z(t)\hat{k},$$

where R is constant in the case of the particle moving only on the surface of a cylinder of radius R.

In general, the trajectory P(t) is not constrained to lie on a circular cylinder, so the radius R varies with time and the trajectory of the particle in cylindrical-polar coordinates becomes:

$$\mathbf{P}(t) = R(t)\mathbf{e}_r + Z(t)\hat{k}.$$

Where R, theta, and Z might be continuously differentiable functions of time and the function notation is dropped for simplicity. The velocity vector V_P is the time derivative of the trajectory P(t), which yields:

$$\mathbf{V}_P = \frac{d}{dt}(R\mathbf{e}_r + Z\hat{k}) = \dot{R}\mathbf{e}_r + R\dot{\mathbf{e}}_r + \dot{Z}\hat{k} = \dot{R}\mathbf{e}_r + R\dot{\theta}\mathbf{e}_\theta + \dot{Z}\hat{k}.$$

Similarly, the acceleration A_P, which is the time derivative of the velocity V_P, is given by:

$$\mathbf{A}_P = \frac{d}{dt}(\dot{R}\mathbf{e}_r + R\dot{\theta}\mathbf{e}_\theta + \dot{Z}\hat{k}) = (\ddot{R} - R\dot{\theta}^2)\mathbf{e}_r + (R\ddot{\theta} + 2\dot{R}\dot{\theta})\mathbf{e}_\theta + \ddot{Z}\hat{k}.$$

The term $-R\dot{\theta}^2\mathbf{e}_r$ acts toward the center of curvature of the path at that point on the path, is commonly called the centripetal acceleration. The term $2\dot{R}\dot{\theta}\mathbf{e}_\theta$ is called the Coriolis acceleration.

Constant Radius

If the trajectory of the particle is constrained to lie on a cylinder, then the radius R is constant and the velocity and acceleration vectors simplify. The velocity of V_P is the time derivative of the trajectory P(t),

$$\mathbf{V}_P = \frac{d}{dt}(R\mathbf{e}_r + Z\hat{k}) = R\dot{\theta}\mathbf{e}_\theta + \dot{Z}\hat{k}.$$

The acceleration vector becomes:

$$\mathbf{A}_P = \frac{d}{dt}(R\dot{\theta}\mathbf{e}_\theta + \dot{Z}\hat{k}) = -R\dot{\theta}^2\mathbf{e}_r + R\ddot{\theta}\mathbf{e}_\theta + \ddot{Z}\hat{k}.$$

Planar Circular Trajectories

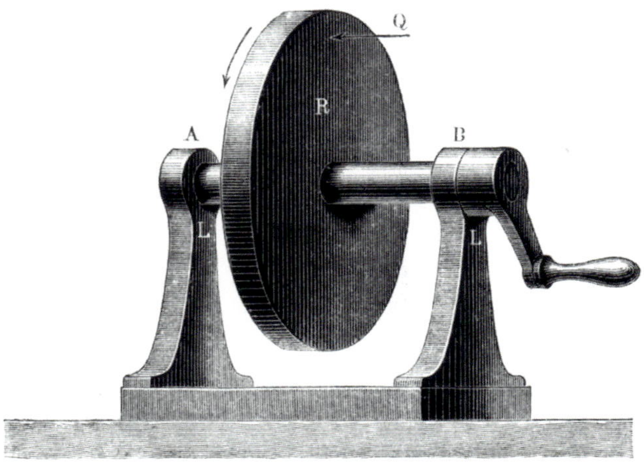

Each particle on the wheel travels in a planar circular trajectory

A special case of a particle trajectory on a circular cylinder occurs when there is no movement along the Z axis:

$$\mathbf{P}(t) = R\mathbf{e}_r + Z_0\hat{k},$$

where R and Z_0 are constants. In this case, the velocity V_P is given by:

$$\mathbf{V}_P = \frac{d}{dt}(R\mathbf{e}_r + Z_0\hat{k}) = R\dot{\theta}\mathbf{e}_\theta = R\omega\mathbf{e}_\theta,$$

where

$$\omega = \dot{\theta},$$

is the angular velocity of the unit vector e_θ around the z axis of the cylinder.

The acceleration A_P of the particle P is now given by:

$$\mathbf{A}_P = \frac{d}{dt}(R\dot{\theta}\mathbf{e}_\theta) = -R\dot{\theta}^2\mathbf{e}_r + R\ddot{\theta}\mathbf{e}_\theta.$$

The components

$$a_r = -R\dot{\theta}^2, \quad a_\theta = R\ddot{\theta},$$

are called, respectively, the *radial* and *tangential components* of acceleration.

The notation for angular velocity and angular acceleration is often defined as

$$\omega = \dot{\theta}, \quad \alpha = \ddot{\theta},$$

so the radial and tangential acceleration components for circular trajectories are also written as

$$a_r = -R\omega^2, \quad a_\theta = R\alpha.$$

Point Trajectories in a Body Moving in the Plane

The movement of components of a mechanical system are analyzed by attaching a reference frame to each part and determining how the various reference frames move relative to each other. If the structural stiffness of the parts are sufficient, then their deformation can be neglected and rigid transformations can be used to define this relative movement. This reduces the description of the motion of the various parts of a complicated mechanical system to a problem of describing the geometry of each part and geometric association of each part relative to other parts.

Geometry is the study of the properties of figures that remain the same while the space is transformed in various ways—more technically, it is the study of invariants under a set of transformations. These transformations can cause the displacement of the triangle in the plane, while leaving the vertex angle and the distances between vertices unchanged. Kinematics is often described as applied geometry, where the movement of a mechanical system is described using the rigid transformations of Euclidean geometry.

The coordinates of points in a plane are two-dimensional vectors in R^2 (two dimensional space). Rigid transformations are those that preserve the distance between any two points. The set of rigid transformations in an *n*-dimensional space is called the special Euclidean group on R^n, and denoted *SE(n)*.

Displacements and Motion

The position of one component of a mechanical system relative to another is defined by introducing a reference frame, say *M*, on one that moves relative to a fixed frame, *F*, on the other. The rigid transformation, or displacement, of *M* relative to *F* defines the relative position of the two components. A displacement consists of the combination of a rotation and a translation.

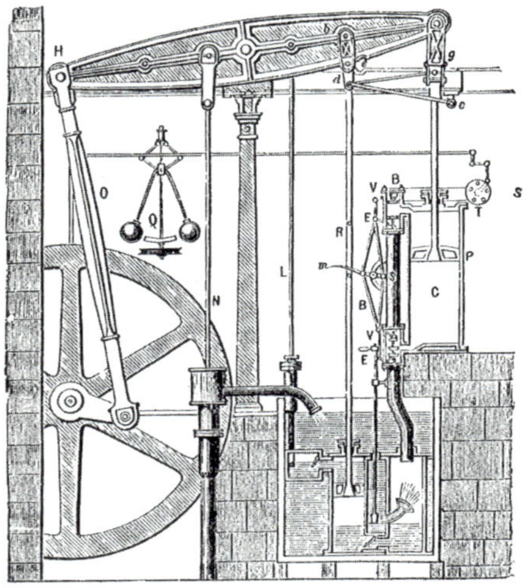

The movement of each of the components of the Boulton & Watt Steam Engine is modeled by a continuous set of rigid displacements.

The set of all displacements of *M* relative to *F* is called the configuration space of *M*. A smooth curve from one position to another in this configuration space is a continuous set of displacements, called the motion of *M* relative to *F*. The motion of a body consists of a continuous set of rotations and translations.

Matrix Representation

The combination of a rotation and translation in the plane R^2 can be represented by a certain type of 3x3 matrix known as a homogeneous transform. The 3x3 homogeneous transform is constructed from a 2x2 rotation matrix A(φ) and the 2x1 translation vector $d=(d_x, d_y)$, as:

$$[T(\phi,\mathbf{d})] = \begin{bmatrix} A(\phi) & \mathbf{d} \\ 0 & 1 \end{bmatrix} = \begin{bmatrix} \cos\phi & -\sin\phi & d_x \\ \sin\phi & \cos\phi & d_y \\ 0 & 0 & 1 \end{bmatrix}.$$

These homogeneous transforms perform rigid transformations on the points in the plane z=1, that is on points with coordinates p=(x, y, 1).

In particular, let p define the coordinates of points in a reference frame *M* coincident with a fixed frame *F*. Then, when the origin of *M* is displaced by the translation vector d relative to the origin of *F* and rotated by the angle φ relative to the x-axis of *F*, the new coordinates in *F* of points in *M* are given by:

$$\mathbf{P} = [T(\phi,\mathbf{d})]\mathbf{p} = \begin{bmatrix} \cos\phi & -\sin\phi & d_x \\ \sin\phi & \cos\phi & d_y \\ 0 & 0 & 1 \end{bmatrix} \begin{Bmatrix} x \\ y \\ 1 \end{Bmatrix}.$$

Homogeneous transforms represent affine transformations. This formulation is necessary because a translation is not a linear transformation of R^2. However, using projective geometry, so that R^2 is considered a subset of R^3, translations become affine linear transformations.

Pure Translation

If a rigid body moves so that its reference frame *M* does not rotate (∅=0) relative to the fixed frame *F*, the motion is called pure translation. In this case, the trajectory of every point in the body is an offset of the trajectory d(t) of the origin of *M*, that is:

$$\mathbf{P}(t) = [T(0, \mathbf{d}(t))]\mathbf{p} = \mathbf{d}(t) + \mathbf{p}.$$

Thus, for bodies in pure translation, the velocity and acceleration of every point *P* in the body are given by:

$$\mathbf{V}_P = \dot{\mathbf{P}}(t) = \dot{\mathbf{d}}(t) = \mathbf{V}_O, \quad \mathbf{A}_P = \ddot{\mathbf{P}}(t) = \ddot{\mathbf{d}}(t) = \mathbf{A}_O,$$

where the dot denotes the derivative with respect to time and V_O and A_O are the velocity and acceleration, respectively, of the origin of the moving frame *M*. Recall the coordinate vector p in *M* is constant, so its derivative is zero.

Rotation of a Body around a Fixed Axis

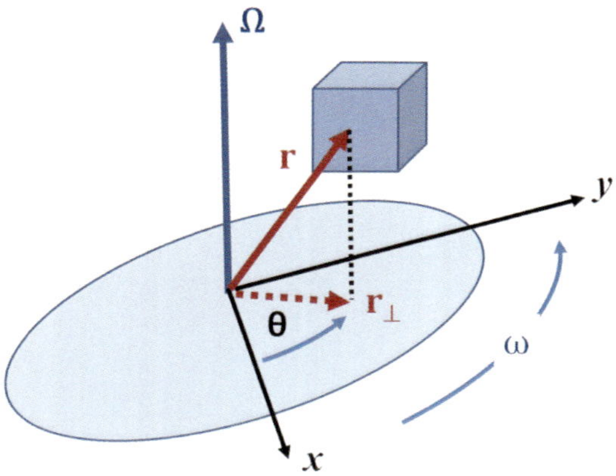

Figure: The angular velocity vector Ω points up for counterclockwise rotation and down for clockwise rotation, as specified by the right-hand rule. Angular position $\theta(t)$ changes with time at a rate $\omega(t) = d\theta/dt$.

Rotational or angular kinematics is the description of the rotation of an object. The description of rotation requires some method for describing orientation. Common descriptions include Euler angles and the kinematics of turns induced by algebraic products.

In what follows, attention is restricted to simple rotation about an axis of fixed orientation. The *z*-axis has been chosen for convenience.

Position

This allows the description of a rotation as the angular position of a planar reference frame *M* relative to a fixed *F* about this shared *z*-axis. Coordinates p = (*x*, *y*) in *M* are related to coordinates P = (X, Y) in *F* by the matrix equation:

$$\mathbf{P}(t) = [A(t)]\mathbf{p},$$

where

$$[A(t)] = \begin{bmatrix} \cos\theta(t) & -\sin\theta(t) \\ \sin\theta(t) & \cos\theta(t) \end{bmatrix},$$

is the rotation matrix that defines the angular position of *M* relative to *F* as a function of time.

Velocity

If the point p does not move in *M*, its velocity in *F* is given by

$$\mathbf{V}_P = \dot{\mathbf{P}} = [\dot{A}(t)]\mathbf{p}.$$

It is convenient to eliminate the coordinates p and write this as an operation on the trajectory P(t),

$$\mathbf{V}_P = [\dot{A}(t)][A(t)^{-1}]\mathbf{P} = [\Omega]\mathbf{P},$$

where the matrix

$$[\Omega] = \begin{bmatrix} 0 & -\omega \\ \omega & 0 \end{bmatrix},$$

is known as the angular velocity matrix of *M* relative to *F*. The parameter ω is the time derivative of the angle θ, that is:

$$\omega = \frac{d\theta}{dt}.$$

Acceleration

The acceleration of P(t) in *F* is obtained as the time derivative of the velocity,

$$\mathbf{A}_P = \ddot{P}(t) = [\dot{\Omega}]\mathbf{P} + [\Omega]\dot{\mathbf{P}},$$

which becomes

$$\mathbf{A}_P = [\dot{\Omega}]\mathbf{P} + [\Omega][\Omega]\mathbf{P},$$

where

$$[\dot{\Omega}] = \begin{bmatrix} 0 & -\alpha \\ \alpha & 0 \end{bmatrix},$$

is the angular acceleration matrix of *M* on *F*, and

$$\alpha = \frac{d^2\theta}{dt^2}.$$

The description of rotation then involves these three quantities:

- Angular position: the oriented distance from a selected origin on the rotational axis to a point of an object is a vector r (t) locating the point. The vector r(t) has some projection (or, equivalently, some component) $r_\perp(t)$ on a plane perpendicular to the axis of rotation. Then the *angular position* of that point is the angle θ from a reference axis (typically the positive x-axis) to the vector $r_\perp(t)$ in a known rotation sense (typically given by the right-hand rule).
- Angular velocity: the angular velocity ω is the rate at which the angular position θ changes with respect to time t:

  $$\omega = \frac{d\theta}{dt}$$

  The angular velocity is represented by a vector Ω pointing along the axis of rotation with magnitude ω and sense determined by the direction of rotation as given by the right-hand rule.
- Angular acceleration: the magnitude of the angular acceleration α is the rate at which the angular velocity ω changes with respect to time t:

  $$\alpha = \frac{d\omega}{dt}$$

The equations of translational kinematics can easily be extended to planar rotational kinematics for constant angular acceleration with simple variable exchanges:

$$\omega_f = \omega_i + \alpha t$$
$$\theta_f - \theta_i = \omega_i t + \tfrac{1}{2}\alpha t^2$$
$$\theta_f - \theta_i = \tfrac{1}{2}(\omega_f + \omega_i)t$$
$$\omega_f^2 = \omega_i^2 + 2\alpha(\theta_f - \theta_i).$$

Here θ_i and θ_f are, respectively, the initial and final angular positions, ω_i and ω_f are, respectively, the initial and final angular velocities, and α is the constant angular acceleration. Although position in space and velocity in space are both true vectors (in terms of their properties under rotation), as is angular velocity, angle itself is not a true vector.

Point Trajectories in Body Moving in Three Dimensions

Important formulas in kinematics define the velocity and acceleration of points in a moving body as they trace trajectories in three-dimensional space. This is particularly important for the center of mass of a body, which is used to derive equations of motion using either Newton's second law or Lagrange's equations.

Position

In order to define these formulas, the movement of a component *B* of a mechanical system is defined by the set of rotations [A(t)] and translations d(t) assembled into the homogeneous transformation [T(t)]=[A(t), d(t)]. If p is the coordinates of a point *P* in *B* measured in the moving reference frame *M*, then the trajectory of this point traced in *F* is given by:

$$\mathbf{P}(t)=[T(t)]\mathbf{p}=\begin{Bmatrix}\mathbf{P}\\1\end{Bmatrix}=\begin{bmatrix}A(t) & \mathbf{d}(t)\\0 & 1\end{bmatrix}\begin{Bmatrix}\mathbf{p}\\1\end{Bmatrix}.$$

This notation does not distinguish between P = (X, Y, Z, 1), and P = (X, Y, Z), which is hopefully clear in context.

This equation for the trajectory of *P* can be inverted to compute the coordinate vector p in *M* as:

$$\mathbf{p}=[T(t)]^{-1}\mathbf{P}(t)=\begin{Bmatrix}\mathbf{p}\\1\end{Bmatrix}=\begin{bmatrix}A(t)^T & -A(t)^T\mathbf{d}(t)\\0 & 1\end{bmatrix}\begin{Bmatrix}\mathbf{P}(t)\\1\end{Bmatrix}.$$

This expression uses the fact that the transpose of a rotation matrix is also its inverse, that is:

$$[A(t)]^T[A(t)]=I.$$

Velocity

The velocity of the point *P* along its trajectory P(t) is obtained as the time derivative of this position vector,

$$\mathbf{V}_P=[\dot{T}(t)]\mathbf{p}=\begin{Bmatrix}\mathbf{V}_P\\0\end{Bmatrix}=\begin{bmatrix}A(t) & \mathbf{d}(t)\\0 & 1\end{bmatrix}\begin{Bmatrix}\mathbf{p}\\1\end{Bmatrix}=\begin{bmatrix}\dot{A}(t) & \dot{\mathbf{d}}(t)\\0 & 0\end{bmatrix}\begin{Bmatrix}\mathbf{p}\\1\end{Bmatrix}.$$

The dot denotes the derivative with respect to time; because p is constant, its derivative is zero.

This formula can be modified to obtain the velocity of *P* by operating on its trajectory P(t) measured in the fixed frame *F*. Substituting the inverse transform for p into the velocity equation yields:

$$\begin{aligned}\mathbf{V}_P=[\dot{T}(t)][T(t)]^{-1}\mathbf{P}(t)=\begin{Bmatrix}\mathbf{V}_P\\0\end{Bmatrix}&=\begin{bmatrix}\dot{A} & \dot{\mathbf{d}}\\0 & 0\end{bmatrix}\begin{bmatrix}A & \mathbf{d}\\0 & 1\end{bmatrix}^{-1}\begin{Bmatrix}\mathbf{P}(t)\\1\end{Bmatrix}\\&=\begin{bmatrix}\dot{A} & \dot{\mathbf{d}}\\0 & 0\end{bmatrix}A^{-1}\begin{bmatrix}1 & -\mathbf{d}\\0 & A\end{bmatrix}\begin{Bmatrix}\mathbf{P}(t)\\1\end{Bmatrix}\\&=\begin{bmatrix}\dot{A}A^{-1} & -\dot{A}A^{-1}\mathbf{d}+\dot{\mathbf{d}}\\0 & 0\end{bmatrix}\begin{Bmatrix}\mathbf{P}(t)\\1\end{Bmatrix}\\&=\begin{bmatrix}\dot{A}A^{T} & -\dot{A}A^{T}\mathbf{d}+\dot{\mathbf{d}}\\0 & 0\end{bmatrix}\begin{Bmatrix}\mathbf{P}(t)\\1\end{Bmatrix}\\\mathbf{V}_P&=[S]\mathbf{P}.\end{aligned}$$

The matrix [S] is given by:

$$[S] = \begin{bmatrix} \Omega & -\Omega\mathbf{d} + \dot{\mathbf{d}} \\ 0 & 0 \end{bmatrix}$$

where

$$[\Omega] = \dot{A}A^T,$$

is the angular velocity matrix.

Multiplying by the operator [S], the formula for the velocity V_P takes the form:

$$\mathbf{V}_P = [\Omega](\mathbf{P} - \mathbf{d}) + \dot{\mathbf{d}} = \omega \times \mathbf{R}_{P/O} + \mathbf{V}_O,$$

where the vector ω is the angular velocity vector obtained from the components of the matrix [Ω]; the vector

$$\mathbf{R}_{P/O} = \mathbf{P} - \mathbf{d},$$

is the position of *P* relative to the origin *O* of the moving frame *M*; and

$$\mathbf{V}_O = \dot{\mathbf{d}}$$

is the velocity of the origin *O*.

Acceleration

The acceleration of a point *P* in a moving body *B* is obtained as the time derivative of its velocity vector:

$$\mathbf{A}_P = \frac{d}{dt}\mathbf{V}_P = \frac{d}{dt}\big([S]\mathbf{P}\big) = [\dot{S}]\mathbf{P} + [S]\dot{\mathbf{P}} = [\dot{S}]\mathbf{P} + [S][S]\mathbf{P}.$$

This equation can be expanded firstly by computing

$$[\dot{S}] = \begin{bmatrix} \dot{\Omega} & -\dot{\Omega}\mathbf{d} - \Omega\dot{\mathbf{d}} + \ddot{\mathbf{d}} \\ 0 & 0 \end{bmatrix} = \begin{bmatrix} \dot{\Omega} & -\dot{\Omega}\mathbf{d} - \Omega\mathbf{V}_O + \mathbf{A}_O \\ 0 & 0 \end{bmatrix}$$

and

$$[S]^2 = \begin{bmatrix} \Omega & -\Omega\mathbf{d} + \mathbf{V}_O \\ 0 & 0 \end{bmatrix}^2 = \begin{bmatrix} \Omega^2 & -\Omega^2\mathbf{d} + \Omega\mathbf{V}_O \\ 0 & 0 \end{bmatrix}.$$

The formula for the acceleration A_P can now be obtained as:

$$\mathbf{A}_P = \dot{\Omega}(\mathbf{P}-\mathbf{d}) + \mathbf{A}_O + \Omega^2(\mathbf{P}-\mathbf{d}),$$

or

$$\mathbf{A}_P = \alpha \times \mathbf{R}_{P/O} + \omega \times \omega \times \mathbf{R}_{P/O} + \mathbf{A}_O,$$

where α is the angular acceleration vector obtained from the derivative of the angular velocity matrix;

$$\mathbf{R}_{P/O} = \mathbf{P} - \mathbf{d},$$

is the relative position vector (the position of *P* relative to the origin *O* of the moving frame *M*); and

$$\mathbf{A}_O = \ddot{\mathbf{d}}$$

is the acceleration of the origin of the moving frame *M*.

Kinematic Constraints

Kinematic constraints are constraints on the movement of components of a mechanical system. Kinematic constraints can be considered to have two basic forms, (i) constraints that arise from hinges, sliders and cam joints that define the construction of the system, called holonomic constraints, and (ii) constraints imposed on the velocity of the system such as the knife-edge constraint of ice-skates on a flat plane, or rolling without slipping of a disc or sphere in contact with a plane, which are called non-holonomic constraints. The following are some common examples.

Kinematic Coupling

A kinematic coupling exactly constrains all 6 degrees of freedom.

Rolling without Slipping

An object that rolls against a surface without slipping obeys the condition that the velocity of its center of mass is equal to the cross product of its angular velocity with a vector from the point of contact to the center of mass:

$$\mathbf{v}_G(t) = \mathbf{\Omega} \times \mathbf{r}_{G/O}.$$

For the case of an object that does not tip or turn, this reduces to $v = r\omega$.

Inextensible Cord

This is the case where bodies are connected by an idealized cord that remains in tension and cannot change length. The constraint is that the sum of lengths of all segments of the cord is the total length, and accordingly the time derivative of this sum is zero. A dynamic problem of this type is the pendulum. Another example is a drum turned by the pull of gravity upon a falling weight attached to the rim by the inextensible cord. An *equilibrium* problem (i.e. not kinematic) of this type is the catenary.

Kinematic Pairs

Reuleaux called the ideal connections between components that form a machine kinematic pairs. He distinguished between higher pairs which were said to have line contact between the two links and lower pairs that have area contact between the links. J. Phillips shows that there are many ways to construct pairs that do not fit this simple classification.

Lower Pair

A lower pair is an ideal joint, or holonomic constraint, that maintains contact between a point, line or plane in a moving solid (three-dimensional) body to a corresponding point line or plane in the fixed solid body. There are the following cases:

- A revolute pair, or hinged joint, requires a line, or axis, in the moving body to remain co-linear with a line in the fixed body, and a plane perpendicular to this line in the moving body maintain contact with a similar perpendicular plane in the fixed body. This imposes five constraints on the relative movement of the links, which therefore has one degree of freedom, which is pure rotation about the axis of the hinge.
- A prismatic joint, or slider, requires that a line, or axis, in the moving body remain co-linear with a line in the fixed body, and a plane parallel to this line in the moving body maintain contact with a similar parallel plane in the fixed body. This imposes five constraints on the relative movement of the links, which therefore has one degree of freedom. This degree of freedom is the distance of the slide along the line.
- A cylindrical joint requires that a line, or axis, in the moving body remain co-linear with a line in the fixed body. It is a combination of a revolute joint and a sliding joint. This joint has two degrees of freedom. The position of the moving body is defined by both the rotation about and slide along the axis.
- A spherical joint, or ball joint, requires that a point in the moving body maintain contact with a point in the fixed body. This joint has three degrees of freedom.
- A planar joint requires that a plane in the moving body maintain contact with a plane in fixed body. This joint has three degrees of freedom.

Higher Pairs

Generally speaking, a higher pair is a constraint that requires a curve or surface in the moving body to maintain contact with a curve or surface in the fixed body. For example, the contact between a cam and its follower is a higher pair called a *cam joint*. Similarly, the contact between the involute curves that form the meshing teeth of two gears are cam joints.

Kinematic Chains

Rigid bodies ("links") connected by kinematic pairs ("joints") are known as *kinematic chains*. Mechanisms and robots are examples of kinematic chains. The degree of freedom of a kinematic chain is computed from the number of links and the number and type of joints using the mobility formula. This formula can also be used to enumerate the topologies of kinematic chains that have a given degree of freedom, which is known as *type synthesis* in machine design.

Illustration of a four-bar linkage

Examples

The planar one degree-of-freedom linkages assembled from *N* links and *j* hinged or sliding joints are:

- N=2, j=1 : a two-bar linkage that is the lever;
- N=4, j=4 : the four-bar linkage;
- N=6, j=7 : a six-bar linkage. This must have two links ("ternary links") that support three joints. There are two distinct topologies that depend on how the two ternary linkages are connected. In the Watt topology, the two ternary links have a common joint; in the Stephenson topology, the two ternary links do not have a common joint and are connected by binary links.
- N=8, j=10 : eight-bar linkage with 16 different topologies;
- N=10, j=13 : ten-bar linkage with 230 different topologies;
- N=12, j=16 : twelve-bar linkage with 6,856 topologies.

Statics

Statics, in physics, the subdivision of mechanics that is concerned with the forces that act on bodies at rest under equilibrium conditions. Its foundations were laid more than 2,200 years ago by the ancient Greek mathematician Archimedes and others while studying the force-amplifying properties of simple machines such as the lever and the axle. The methods and results of the science of statics have proved especially useful in designing buildings, bridges, and dams, as well as cranes and other similar mechanical devices. To be able to calculate the dimensions of such structures and machines, architects and engineers must first determine the forces that act on their interconnected parts. Statics provides the analytical and graphical procedures needed to identify and describe these unknown forces.

Statics assumes that the bodies with which it deals are perfectly rigid. It also holds that the sum of all the forces acting on a body at rest has to be zero (i.e., the forces involved balance one another) and that there must be no tendency for the forces to turn the body about any axis. These three conditions are independent of one another, and their expression in mathematical form comprises the equations of equilibrium. There are three equations, and so only three unknown forces can be calculated. If more than three unknown forces exist, it means that there are more components in the structure or machine than are required to support the applied loads or that there are more restraints than are needed to keep the body from moving. Such unnecessary components or restraints are termed redundant (e.g., a table with four legs has one redundant leg) and the system of forces is said to be statically indeterminate. The number of equations available in statics is limited because of a neglect of the deformations of loaded bodies, a direct consequence of the underlying premise that any solid body under consideration is ideally rigid and immutable as to shape and size under all conditions.

You encounter a great number and variety of examples of forces every day, such as when you press a button, turn a doorknob, or run your hands through your hair. Motion and changes in shape are critical to the functionality of man-made objects as well as objects the nature.

Statics is an essential prerequisite for many branches of engineering, such as mechanical, civil, aeronautical, and bioengineering, which address the various consequences of forces.

Vectors

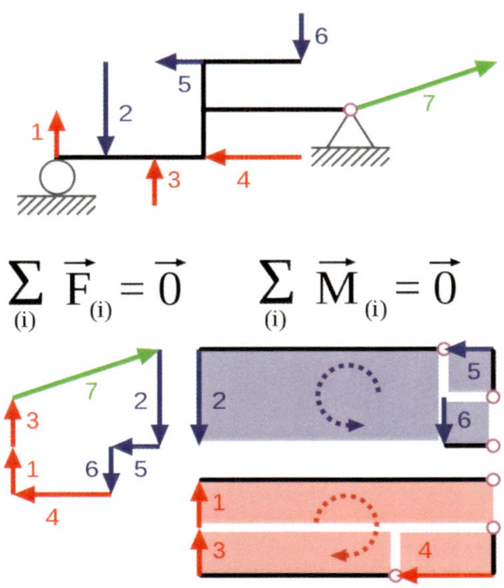

Example of a beam in static equilibrium. The sum of force and moment is zero.

A scalar is a quantity which only has a magnitude, such as mass or temperature. A vector has a magnitude and a direction. There are several notations to identify a vector, including:

- A bold faced character **V**
- An underlined character $\underline{V}$
- A character with an arrow over it $\vec{V}$.

Vectors are added using the parallelogram law or the triangle law. Vectors contain components in orthogonal bases. Unit vectors i, j, and k are, by convention, along the x, y, and z axes, respectively.

Force

Force is the action of one body on another. A force is either a push or a pull. A force tends to move a body in the direction of its action. The action of a force is characterized by its magnitude, by the direction of its action, and by its point of application. Thus, force is a vector quantity, because its effect depends on the direction as well as on the magnitude of the action.

Forces are classified as either contact or body forces. A contact force is produced by direct physical contact; an example is the force exerted on a body by a supporting surface. A body force is generated by virtue of the position of a body within a force field such as a gravitational, electric, or magnetic field and is independent of contact with any other body. An example of a body force is the weight of a body in the Earth's gravitational field.

Moment of a Force

In addition to the tendency to move a body in the direction of its application, a force can also tend to rotate a body about an axis. The axis may be any line which neither intersects nor is parallel to the line of action of the force. This rotational tendency is known as the *moment* (M) of the force. Moment is also referred to as *torque*.

Moment about a Point

The magnitude of the moment of a force at a point O, is equal to the perpendicular distance from O to the line of action of F, multiplied by the magnitude of the force: $M = F \cdot d$, where

F = the force applied.

d = the perpendicular distance from the axis to the line of action of the force. This perpendicular distance is called the moment arm.

The direction of the moment is given by the right hand rule, where counter clockwise (CCW) is out of the page, and clockwise (CW) is into the page. The moment direction may be accounted for by using a stated sign convention, such as a plus sign (+) for counterclockwise moments and a minus sign (–) for clockwise moments, or vice versa. Moments can be added together as vectors.

In vector format, the moment can be defined as the cross product between the radius vector, r (the vector from point O to the line of action), and the force vector, F:

$$\mathbf{M}_O = \mathbf{r} \times \mathbf{F}$$

Varignon's Theorem

Varignon's theorem states that the moment of a force about any point is equal to the sum of the moments of the components of the force about the same point.

Equilibrium Equations

The static equilibrium of a particle is an important concept in statics. A particle is in equilibrium only if the resultant of all forces acting on the particle is equal to zero. In a rectangular coordinate system the equilibrium equations can be represented by three scalar equations, where the sums of forces in all three directions are equal to zero. An engineering application of this concept is determining the tensions of up to three cables under load, for example the forces exerted on each cable of a hoist lifting an object or of guy wires restraining a hot air balloon to the ground.

Moment of Inertia

In classical mechanics, moment of inertia, also called mass moment,rotational inertia, polar moment of inertia of mass, or the angular mass, (SI units $kg \cdot m^2$) is a measure of an object's resistance to changes to its rotation. It is the inertia of a rotating body with respect to its rotation. The moment of inertia plays much the same role in rotational dynamics as mass does in linear dynamics, describing the relationship between angular momentum and angular velocity, torque and angular acceleration, and several other quantities. The symbols I and J are usually used to refer to the moment of inertia or polar moment of inertia.

While a simple scalar treatment of the moment of inertia suffices for many situations, a more advanced tensor treatment allows the analysis of such complicated systems as spinning tops and gyroscopic motion.

The concept was introduced by Leonhard Euler in his 1765 book *Theoria motus corporum solidorum seu rigidorum*; he discussed the moment of inertia and many related concepts, such as the principal axis of inertia.

Solids

Statics is used in the analysis of structures, for instance in architectural and structural engineering. Strength of materials is a related field of mechanics that relies heavily on the application of static equilibrium. A key concept is the center of gravity of a body at rest: it represents an imaginary point at which all the mass of a body resides. The position of the point relative to the foundations on which a body lies determines its stability in response to external forces. If the center of gravity exists outside the foundations, then the body is unstable because there is a torque acting: any small disturbance will cause the body to fall or topple. If the center of gravity exists within the foundations, the body is stable since no net torque acts on the body. If the center of gravity coincides with the foundations, then the body is said to be metastable.

Fluids

Hydrostatics, also known as fluid statics, is the study of fluids at rest (i.e. in static equilibrium). The characteristic of any fluid at rest is that the force exerted on any particle of the fluid is the same at all points at the same depth (or altitude) within the fluid. If the net force is greater than zero the fluid will move in the direction of the resulting force. This concept was first formulated in a slightly extended form by French mathematician and philosopher Blaise Pascal in 1647 and became known as Pascal's Law. It has many important applications in hydraulics. Archimedes, Abū Rayhān al-Bīrūnī, Al-Khazini and Galileo Galilei were also major figures in the development of hydrostatics.

Newton's Laws of Motion

Sir Isaac Newton's three laws of motion describe the motion of massive bodies and how they interact. While Newton's laws may seem obvious to us today, more than three centuries ago they were considered revolutionary.

Newton was one of the most influential scientists of all time. His ideas became the basis for modern physics. He built upon ideas put forth from the works of previous scientists including Galileo and Aristotle and was able to prove some ideas that had only been theories in the past. He studied optics, astronomy and math — he invented calculus. (German mathematician Gottfried Leibniz is also credited with developing it independently at about the same time.)

Newton is perhaps best known for his work in studying gravity and the motion of planets. Urged on by astronomer Edmond Halley after admitting he had lost his proof of elliptical orbits a few years prior, Newton published his laws in 1687, in his seminal work "Philosophiæ Naturalis Principia Mathematica" (Mathematical Principles of Natural Philosophy) in which he formalized the description of how massive bodies move under the influence of external forces.

In formulating his three laws, Newton simplified his treatment of massive bodies by considering them to be mathematical points with no size or rotation. This allowed him to ignore factors such as friction, air resistance, temperature, material properties, etc., and concentrate on phenomena that can be described solely in terms of mass, length and time. Consequently, the three laws cannot be used to describe precisely the behavior of large rigid or deformable objects; however, in many cases they provide suitably accurate approximations.

Newton's laws pertain to the motion of massive bodies in an *inertial reference frame*, sometimes called a *Newtonian reference frame*, although Newton himself never described such a reference frame. An inertial reference frame can be described as a 3-dimensional coordinate system that is either stationary or in uniform linear motion., i.e., it is not accelerating or rotating. He found that motion within such an inertial reference frame could be described by three simple laws.

Newton's first Law

The first law states that if the net force (the vector sum of all forces acting on an object) is zero, then the velocity of the object is constant. Velocity is a vector quantity which expresses both the object's speed and the direction of its motion; therefore, the statement that the object's velocity is constant is a statement that both its speed and the direction of its motion are constant.

The first law can be stated mathematically when the mass is a non-zero constant, as,

$$\sum \mathbf{F} = 0 \Leftrightarrow \frac{d\mathbf{v}}{dt} = 0.$$

Consequently,

- An object that is at rest will stay at rest unless a force acts upon it.
- An object that is in motion will not change its velocity unless a force acts upon it.

This is known as *uniform motion*. An object *continues* to do whatever it happens to be doing unless a force is exerted upon it. If it is at rest, it continues in a state of rest (demonstrated when a tablecloth is skilfully whipped from under dishes on a tabletop and the dishes remain in their initial state of rest). If an object is moving, it continues to move without turning or changing its speed. This is evident in space probes that continuously move in outer space. Changes in motion must be imposed against the tendency of an object to retain its state of motion. In the absence of net forces, a moving object tends to move along a straight line path indefinitely.

Newton placed the first law of motion to establish frames of reference for which the other laws are applicable. The first law of motion postulates the existence of at least one frame of reference called a Newtonian or inertial reference frame, relative to which the motion of a particle not subject to forces is a straight line at a constant speed. Newton's first law is often referred to as the *law of inertia*. Thus, a condition necessary for the uniform motion of a particle relative to an inertial reference frame is that the total net force acting on it is zero. In this sense, the first law can be restated as:

In every material universe, the motion of a particle in a preferential reference frame Φ is determined by the action of forces whose total vanished for all times when and only when the velocity of the particle is constant in Φ. That is, a particle initially at rest or in uniform motion in the preferential frame Φ continues in that state unless compelled by forces to change it.

Newton's first and second laws are valid only in an inertial reference frame. Any reference frame that is in uniform motion with respect to an inertial frame is also an inertial frame, i.e. Galilean invariance or the principle of Newtonian relativity.

Newton's Second Law

The second law states that the rate of change of momentum of a body is directly proportional to the force applied, and this change in momentum takes place in the direction of the applied force.

$$\mathbf{F} = \frac{d\mathbf{p}}{dt} = \frac{d(m\mathbf{v})}{dt}.$$

The second law can also be stated in terms of an object's acceleration. Since Newton's second law is valid only for constant-mass systems, m can be taken outside the differentiation operator by the constant factor rule in differentiation. Thus,

$$\mathbf{F} = m\frac{d\mathbf{v}}{dt} = m\mathbf{a},$$

where F is the net force applied, m is the mass of the body, and a is the body's acceleration. Thus, the net force applied to a body produces a proportional acceleration. In other words, if a body is accelerating, then there is a force on it. An application of this notation is the derivation of G Subscript C.

Consistent with the first law, the time derivative of the momentum is non-zero when the momentum changes direction, even if there is no change in its magnitude; such is the case with uniform circular motion. The relationship also implies the conservation of momentum: when the net force on the body is zero, the momentum of the body is constant. Any net force is equal to the rate of change of the momentum.

Any mass that is gained or lost by the system will cause a change in momentum that is not the result of an external force. A different equation is necessary for variable-mass systems.

Newton's second law is an approximation that is increasingly worse at high speeds because of relativistic effects.

Impulse

An impulse J occurs when a force F acts over an interval of time Δt, and it is given by

$$\mathbf{J} = \int_{\Delta t} \mathbf{F} \mathrm{d}t.$$

Since force is the time derivative of momentum, it follows that

$$\mathbf{J} = \Delta\mathbf{p} = m\Delta\mathbf{v}.$$

This relation between impulse and momentum is closer to Newton's wording of the second law.

Impulse is a concept frequently used in the analysis of collisions and impacts.

Variable-mass Systems

Variable-mass systems, like a rocket burning fuel and ejecting spent gases, are not closed and cannot be directly treated by making mass a function of time in the second law; that is, the following formula is wrong:

$$\mathbf{F}_{\text{net}} = \frac{\mathrm{d}}{\mathrm{d}t}\big[m(t)\mathbf{v}(t)\big] = m(t)\frac{\mathrm{d}\mathbf{v}}{\mathrm{d}t} + \mathbf{v}(t)\frac{\mathrm{d}m}{\mathrm{d}t}. \qquad \text{(wrong)}$$

The falsehood of this formula can be seen by noting that it does not respect Galilean invariance: a variable-mass object with F = 0 in one frame will be seen to have F ≠ 0 in another frame. The correct equation of motion for a body whose mass m varies with time by either ejecting or accreting mass is obtained by applying the second law to the entire, constant-mass system consisting of the body and its ejected/accreted mass; the result is

$$\mathbf{F} + \mathbf{u}\frac{\mathrm{d}m}{\mathrm{d}t} = m\frac{\mathrm{d}\mathbf{v}}{\mathrm{d}t}$$

where u is the velocity of the escaping or incoming mass relative to the body. From this equation one can derive the equation of motion for a varying mass system, for example, the Tsiolkovsky rocket equation. Under some conventions, the quantity $u\,dm/dt$ on the left-hand side, which represents the advection of momentum, is defined as a force (the force exerted on the body by the changing mass, such as rocket exhaust) and is included in the quantity F. Then, by substituting the definition of acceleration, the equation becomes $F = ma$.

Newton's Third Law

An illustration of Newton's third law in which two skaters push against each other.

The first skater on the left exerts a normal force N_{12} on the second skater directed towards the right, and the second skater exerts a normal force N_{21} on the first skater directed towards the left. The magnitudes of both forces are equal, but they have opposite directions, as dictated by Newton's third law.

The third law states that all forces between two objects exist in equal magnitude and opposite direction: if one object *A* exerts a force F_A on a second object *B*, then *B* simultaneously exerts a force F_B on *A*, and the two forces are equal in magnitude and opposite in direction: $F_A = -F_B$. The third law means that all forces are *interactions* between different bodies, or different regions within one body, and thus that there is no such thing as a force that is not accompanied by an equal and opposite force. In some situations, the magnitude and direction of the forces are determined entirely by one of the two bodies, say Body *A*; the force exerted by Body *A* on Body *B* is called the "action", and the force exerted by Body *B* on Body *A* is called the "reaction". This law is sometimes referred to as the *action-reaction law*, with F_A called the "action" and F_B the "reaction". In other situations the magnitude and directions of the forces are determined jointly by both bodies and it isn't necessary to identify one force as the "action" and the other as the "reaction". The action and the reaction are simultaneous, and it does not matter which is called the *action* and which is called *reaction*; both forces are part of a single interaction, and neither force exists without the other.

The two forces in Newton's third law are of the same type (e.g., if the road exerts a forward frictional force on an accelerating car's tires, then it is also a frictional force that Newton's third law predicts for the tires pushing backward on the road).

From a conceptual standpoint, Newton's third law is seen when a person walks: they push against the floor, and the floor pushes against the person. Similarly, the tires of a car push against the road while the road pushes back on the tires—the tires and road simultaneously push against each other. In swimming, a person interacts with the water, pushing the water backward, while the water simultaneously pushes the person forward—both the person and the water push against each other. The reaction forces account for the motion in these examples. These forces depend on friction; a person or car on ice, for example, may be unable to exert the action force to produce the needed reaction force.

Importance and Range of Validity

Newton's laws were verified by experiment and observation for over 200 years, and they are excellent approximations at the scales and speeds of everyday life. Newton's laws of motion, together with his law of universal gravitation and the mathematical techniques of calculus, provided for the first time a unified quantitative explanation for a wide range of physical phenomena.

These three laws hold to a good approximation for macroscopic objects under everyday conditions. However, Newton's laws (combined with universal gravitation and classical electrodynamics) are inappropriate for use in certain circumstances, most notably at very small scales, very high speeds (in special relativity, the Lorentz factor must be included in the expression for momentum along with the rest mass and velocity) or very strong gravitational fields. Therefore, the laws cannot be used to explain phenomena such as conduction of electricity in a semiconductor, optical properties of substances, errors in non-relativistically corrected GPS systems and superconductivity. Explanation of these phenomena requires more sophisticated physical theories, including general relativity and quantum field theory.

In quantum mechanics, concepts such as force, momentum, and position are defined by linear operators that operate on the quantum state; at speeds that are much lower than the speed of light, Newton's laws are just as exact for these operators as they are for classical objects. At speeds comparable to the speed of light, the second law holds in the original form $F = dp/dt$, where F and p are four-vectors.

Relationship to the Conservation Laws

In modern physics, the laws of conservation of momentum, energy, and angular momentum are of more general validity than Newton's laws, since they apply to both light and matter, and to both classical and non-classical physics.

This can be stated simply, "Momentum, energy and angular momentum cannot be created or destroyed."

Because force is the time derivative of momentum, the concept of force is redundant and subordinate to the conservation of momentum, and is not used in fundamental theories (e.g., quantum mechanics, quantum electrodynamics, general relativity, etc.). The standard model explains in detail how the three fundamental forces known as gauge forces originate out of exchange by virtual particles. Other forces, such as gravity and fermionic degeneracy pressure, also arise from the momentum conservation. Indeed, the conservation of 4-momentum in inertial motion via curved space-time results in what we call gravitational force in general relativity theory. The application of the space derivative (which is a momentum operator in quantum mechanics) to the overlapping wave functions of a pair of fermions (particles with half-integer spin) results in shifts of maxima of compound wavefunction away from each other, which is observable as the "repulsion" of the fermions.

Newton stated the third law within a world-view that assumed instantaneous action at a distance between material particles. However, he was prepared for philosophical criticism of this action at a distance, and it was in this context that he stated the famous phrase "I feign no hypotheses". In modern physics, action at a distance has been completely eliminated, except for subtle effects involving quantum entanglement. (In particular, this refers to Bell's theorem – that no local model can reproduce the predictions of quantum theory.) Despite only being an approximation, in modern engineering and all practical applications involving the motion of vehicles and satellites, the concept of action at a distance is used extensively.

The discovery of the second law of thermodynamics by Carnot in the 19th century showed that not every physical quantity is conserved over time, thus disproving the validity of inducing the opposite metaphysical view from Newton's laws. Hence, a "steady-state" worldview based solely on Newton's laws and the conservation laws does not take entropy into account.

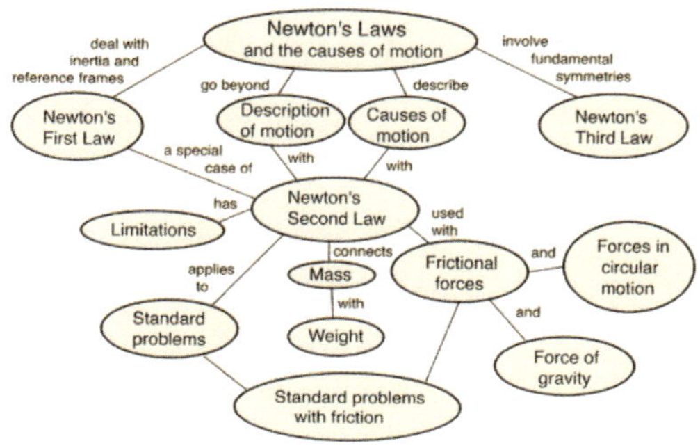

D'Alembert's Principle

D'Alembert's principle, alternative form of Newton's second law of motion, stated by the 18th-century French polymath Jean le Rond d'Alembert. In effect, the principle reduces a problem in dynamics to a problem in statics. The second law states that the force F acting on a body is equal to the product of the mass m and acceleration a of the body, or $F = ma$; in d'Alembert's form, the force F plus the negative of the mass m times acceleration a of the body is equal to zero: $F - ma = 0$. In other words, the body is in equilibrium under the action of the real force F and the fictitious force $-ma$. The fictitious force is also called an inertial force and a reversed effective force.

Because unknown forces are more easily determined on bodies in equilibrium than on moving bodies, the force and stress analysis of machine components can usually be simplified by using inertial forces. When developing the formulas for the stresses in a rotating disk, for example, it is convenient to assume that a representative element in the disk is in equilibrium under the action of a system of radial and tangential forces produced by the stresses and an outward-acting inertial (centrifugal) force.

The standard does not have any significant bearing for irreversible relocations, for example, sliding contact, and more broad detail of the irreversibility is required.

Write the d'Alembert's equation.

$$\sum_i (F_i - m_i a_i)\delta r_i = 0$$

Where, integer is I, total applied force is F_i, mass for the given integer is m_i, acceleration of integer is a_i, and virtual displacement of integer is δr_i.

General Case with Changing Masses

The general statement of d'Alembert's principle mentions "the time derivatives of the momenta of the system". The momentum of the i-th mass is the product of its mass and velocity:

$$\mathbf{p}_i = m_i \mathbf{v}_i$$

and its time derivative is

$$\dot{\mathbf{p}}_i = \dot{m}_i \mathbf{v}_i + m_i \dot{\mathbf{v}}_i .$$

In many applications, the masses are constant and this equation reduces to

$$\dot{\mathbf{p}}_i = m_i \dot{\mathbf{v}}_i = m_i \mathbf{a}_i ,$$

which appears in the formula given above. However, some applications involve changing masses (for example, chains being rolled up or being unrolled) and in those cases both terms $\dot{m}_i \mathbf{v}_i$ and $m_i \dot{\mathbf{v}}_i$ have to remain present, giving

$$\sum_i (\mathbf{F}_i - m_i \mathbf{a}_i - \dot{m}_i \mathbf{v}_i) \cdot \delta \mathbf{r}_i = 0.$$

Derivation for Special Cases

To date, nobody has shown that D'Alembert's principle is equivalent to Newton's Second Law. This is true only for some very special cases e.g. rigid body constraints. However, an approximate solution to this problem does exist.

Consider Newton's law for a system of particles, i. The total force on each particle is

$$\mathbf{F}_i^{(T)} = m_i \mathbf{a}_i ,$$

where

$\mathbf{F}_i^{(T)}$ are the total forces acting on the system's particles,

$m_i \mathbf{a}_i$ are the inertial forces that result from the total forces.

Moving the inertial forces to the left gives an expression that can be considered to represent quasi-static equilibrium, but which is really just a small algebraic manipulation of Newton's law:

$$\mathbf{F}_i^{(T)} = m_i \mathbf{a}_i = 0.$$

Considering the virtual work, δW, , done by the total and inertial forces together through an arbitrary virtual displacement, $\delta \mathbf{r}_i$, of the system leads to a zero identity, since the forces involved sum to zero for each particle.

$$\delta W = \sum_i \mathbf{F}_i^{(T)} \cdot \delta \mathbf{r}_i - \sum_i m_i \mathbf{a}_i \cdot \delta \mathbf{r}_i = 0$$

The original vector equation could be recovered by recognizing that the work expression must hold for arbitrary displacements. Separating the total forces into applied forces, $\mathbf{F}_i$, and constraint forces, $\mathbf{C}_i$, yields

$$\delta W = \sum_i \mathbf{F}_i \cdot \delta \mathbf{r}_i + \sum_i \mathbf{C}_i \cdot \delta \mathbf{r}_i - \sum_i m_i \mathbf{a}_i \cdot \delta \mathbf{r}_i = 0.$$

If arbitrary virtual displacements are assumed to be in directions that are orthogonal to the constraint forces (which is not usually the case, so this derivation works only for special cases), the constraint forces do no work. Such displacements are said to be *consistent* with the constraints. This leads to the formulation of *d'Alembert's principle*, which states that the difference of applied forces and inertial forces for a dynamic system does no virtual work:.

$$\delta W = \sum_i (\mathbf{F}_i - m_i \mathbf{a}_i) \cdot \delta \mathbf{r}_i = 0.$$

There is also a corresponding principle for static systems called the principle of virtual work for applied forces.

D'Alembert's Principle of Inertial Forces

D'Alembert showed that one can transform an accelerating rigid body into an equivalent static system by adding the so-called "inertial force" and "inertial torque" or moment. The inertial force must act through the center of mass and the inertial torque can act anywhere. The system can then be analyzed exactly as a static system subjected to this "inertial force and moment" and the external forces. The advantage is that, in the equivalent static system one can take moments about any point (not just the center of mass). This often leads to simpler calculations because any force (in turn) can be eliminated from the moment equations by choosing the appropriate point about which to apply the moment equation (sum of moments = zero). Even in the course of Fundamentals of Dynamics and Kinematics of machines, this principle helps in analyzing the forces that act on a link of a mechanism when it is in motion.

Example for 1D Motion of a Rigid Body

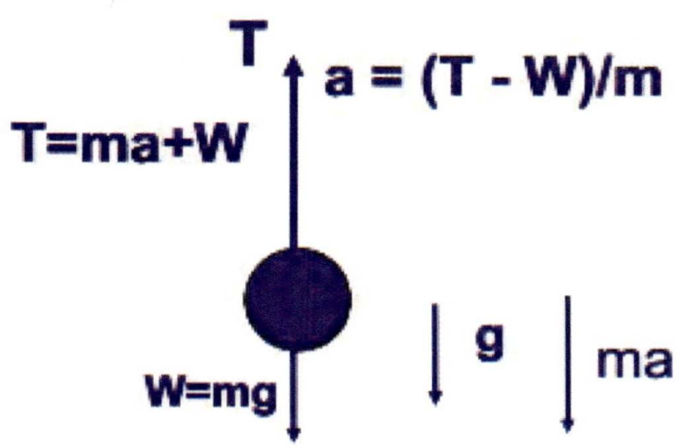

Free body diagram of a wire pulling on a mass with weight W, showing the d'Alembert inertia "force" ma.

To illustrate the concept of *d'Alembert's principle*, let's use a simple model with a weight W, suspended from a wire. The weight is subjected to a gravitational force, $W = mg$, , and a tension force T in the wire. The mass accelerates upward with an acceleration a. Newton's Second Law becomes $T - W = ma$ or $T = W + ma$. As an observer with feet planted firmly on the ground, we see that the force T accelerates the weight, W, but, if we are moving with the wire we don't see the acceleration, we feel it. The tension in the wire seems to counteract an acceleration "force" ma or $(W / g)a$.

Example for Plane 2D Motion of a Rigid Body

For a planar rigid body, moving in the plane of the body (the x–y plane), and subjected to forces and torques causing rotation only in this plane, the inertial force is;

$$\mathbf{F}_i = -m\ddot{\mathbf{r}}_c$$

where $\mathbf{r}_c$ is the position vector of the centre of mass of the body, and m is the mass of the body. The inertial torque (or moment) is;

$$T_i = -I\ddot{\theta}$$

where I is the moment of inertia of the body. If, in addition to the external forces and torques acting on the body, the inertia force acting through the center of mass is added and the inertial torque is added (acting around the centre of mass is as good as anywhere) the system is equivalent to one in static equilibrium. Thus the equations of static equilibrium;

$$\sum F_x = 0$$
$$\sum F_y = 0$$
$$\sum T = 0$$

hold. The important thing is that $\sum T$ is the sum of torques (or moments, including the inertial moment and the moment of the inertial force) taken about *any* point. The direct application of Newton's laws requires that the angular acceleration equation be applied *only* about the center of mass.

Dynamic Equilibrium

D'Alembert's form of the principle of virtual work states that a system of rigid bodies is in dynamic equilibrium when the virtual work of the sum of the applied forces and the inertial forces is zero for any virtual displacement of the system. Thus, dynamic equilibrium of a system of n rigid bodies with m generalized coordinates requires that is to be

$$\delta W = (Q_1 + Q_1^*)\delta q_1 + \ldots + (Q_m + Q_m^*)\delta q_m = 0,$$

for any set of virtual displacements δq_j. This condition yields m equations,

$$Q_j + Q_j^* = 0, \quad j = 1, \ldots, m,$$

which can also be written as

$$\frac{d}{dt}\frac{\partial T}{\partial \dot{q}_j} - \frac{\partial T}{\partial q_j} = Q_j, \quad j = 1,\ldots,m.$$

The result is a set of m equations of motion that define the dynamics of the rigid body system.

Momentum

Momentum is a physics term; it refers to the quantity of motion that an object has. A sports team that is *on the move* has the momentum. If an object is in motion (*on the move*) then it has momentum.

Momentum can be defined as "mass in motion." All objects have mass; so if an object is moving, then it has momentum - it has its mass in motion. The amount of momentum that an object has is dependent upon two variables: how much *stuff* is moving and how fast the *stuff* is moving. Momentum depends upon the variables mass and velocity. In terms of an equation, the momentum of an object is equal to the mass of the object times the velocity of the object.

Momentum = mass • velocity

In physics, the symbol for the quantity momentum is the lower case p. Thus, the above equation can be rewritten as

$p = m \bullet v$

where m is the mass and v is the velocity. The equation illustrates that momentum is directly proportional to an object's mass and directly proportional to the object's velocity.

The units for momentum would be mass units times velocity units. The standard metric unit of momentum is the kg•m/s. While the kg•m/s is the standard metric unit of momentum, there are a variety of other units that are acceptable (though not conventional) units of momentum. Examples include kg•mi/hr, kg•km/hr, and g•cm/s. In each of these examples, a mass unit is multiplied by a velocity unit to provide a momentum unit. This is consistent with the equation for momentum.

Momentum as a Vector Quantity

Momentum is a vector quantity. As discussed in an earlier unit, a vector quantity is a quantity that is fully described by both magnitude and direction. To fully describe the momentum of a 5-kg bowling ball moving westward at 2 m/s, you must include information about both the magnitude and the direction of the bowling ball. It is not enough to say that the ball has 10 kg•m/s of momentum; the momentum of the ball is not fully described until information about its direction is given. The direction of the momentum vector is the same as the direction of the velocity of the ball. In a previous unit, it was said that the direction of the velocity vector is the same as the direction that an object is moving. If the bowling ball is moving westward, then its momentum can be fully described by saying that it is 10 kg•m/s, westward. As a vector quantity, the momentum of an object is fully described by both magnitude and direction.

Newtonian

Momentum is a vector quantity: it has both magnitude and direction. Since momentum has a direction, it can be used to predict the resulting direction and speed of motion of objects after they collide. Below, the basic properties of momentum are described in one dimension. The vector equations are almost identical to the scalar equations.

Single Particle

The momentum of a particle is conventionally represented by the letter p. It is the product of two quantities, the particle's mass (represented by the letter m) and its velocity (v):

$$p = mv.$$

The unit of momentum is the product of the units of mass and velocity. In SI units, if the mass is in kilograms and the velocity is in meters per second then the momentum is in kilogram meters per second (kg·m/s). In cgs units, if the mass is in grams and the velocity in centimeters per second, then the momentum is in gram centimeters per second (g·cm/s).

Being a vector, momentum has magnitude and direction. For example, a 1 kg model airplane, traveling due north at 1 m/s in straight and level flight, has a momentum of 1 kg·m/s due north measured with reference to the ground.

Many Particles

The momentum of a system of particles is the sum of their momenta. If two particles have respective masses m_1 and m_2, and velocities v_1 and v_2, the total momentum is

$$\begin{aligned} p &= p_1 + p_2 \\ &= m_1 v_1 + m_2 v_2 . \end{aligned}$$

The momenta of more than two particles can be added more generally with the following:

$$p = \sum_i m_i v_i .$$

A system of particles has a center of mass, a point determined by the weighted sum of their positions:

$$r_{\text{cm}} = \frac{m_1 r_1 + m_2 r_2 + \cdots}{m_1 + m_2 + \cdots} = \frac{\sum_i m_i r_i}{\sum_i m_i} .$$

If all the particles are moving, the center of mass will generally be moving as well (unless the system is in pure rotation around it). If the center of mass is moving at velocity v_{cm}, the momentum is:

$$p = m v_{\text{cm}} .$$

This is known as Euler's first law.

Relation to Force

If the net force applied to a particle is a constant F, and is applied for a time interval Δt, the momentum of the particle changes by an amount

$$\Delta p = F\Delta t.$$

In differential form, this is Newton's second law; the rate of change of the momentum of a particle is equal to the instantaneous force F acting on it,

$$F = \frac{dp}{dt}.$$

If the net force experienced by a particle changes as a function of time, $F(t)$, the change in momentum (or impulse J) between times t_1 and t_2 is

$$\Delta p = J = \int_{t_1}^{t_2} F(t)dt.$$

Impulse is measured in the derived units of the newton second (1 N·s = 1 kg·m/s) or dyne second (1 dyne·s = 1 g·m/s)

Under the assumption of constant mass m, it is equivalent to write

$$F = \frac{d(mv)}{dt} = m\frac{dv}{dt} = ma,$$

hence the net force is equal to the mass of the particle, times its acceleration.

Example: A model airplane of mass 1 kg accelerates from rest to a velocity of 6 m/s due north in 2 s. The net force required to produce this acceleration is 3 newtons due north. The change in momentum is 6 kg·m/s. The rate of change of momentum is 3 (kg·m/s)/s = 3 N.

Conservation

In a closed system (one that does not exchange any matter with its surroundings and is not acted on by external forces) the total momentum is constant. This fact, known as the *law of conservation of momentum*, is implied by Newton's laws of motion. Suppose, for example, that two particles interact. Because of the third law, the forces between them are equal and opposite. If the particles are numbered 1 and 2, the second law states that $F_1 = \frac{dp_1}{dt}$ and $F_2 = \frac{dp_2}{dt}$. Therefore,

$$\frac{dp_1}{dt} = -\frac{dp_2}{dt},$$

with the negative sign indicating that the forces oppose. Equivalently,

$$\frac{d}{dt}(p_1 + p_2) = 0.$$

If the velocities of the particles are u_1 and u_2 before the interaction, and afterwards they are v_1 and v_2, then

$$m_1 u_1 + m_2 u_2 = m_1 v_1 + m_2 v_2.$$

This law holds no matter how complicated the force is between particles. Similarly, if there are several particles, the momentum exchanged between each pair of particles adds up to zero, so the total change in momentum is zero. This conservation law applies to all interactions, including collisions and separations caused by explosive forces. It can also be generalized to situations where Newton's laws do not hold, for example in the theory of relativity and in electrodynamics.

Dependence on Reference Frame

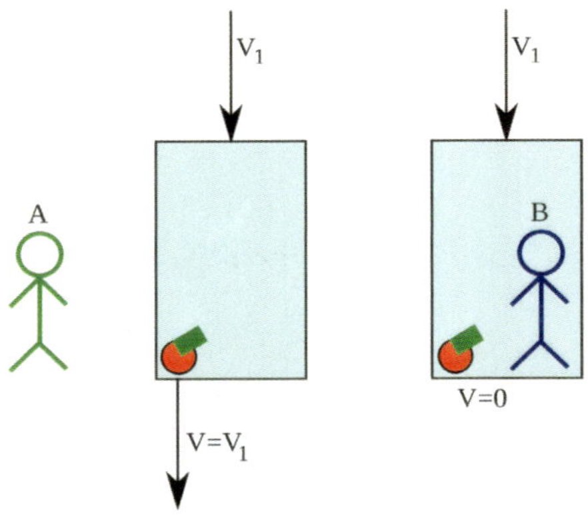

Newton's apple in Einstein's elevator. In person A's frame of reference, the apple has non-zero velocity and momentum. In the elevator's and person B's frames of reference, it has zero velocity and momentum.

Momentum is a measurable quantity, and the measurement depends on the motion of the observer. For example: if an apple is sitting in a glass elevator that is descending, an outside observer, looking into the elevator, sees the apple moving, so, to that observer, the apple has a non-zero momentum. To someone inside the elevator, the apple does not move, so, it has zero momentum. The two observers each have a frame of reference, in which, they observe motions, and, if the elevator is descending steadily, they will see behavior that is consistent with those same physical laws.

Suppose a particle has position x in a stationary frame of reference. From the point of view of another frame of reference, moving at a uniform speed u, the position (represented by a primed coordinate) changes with time as:

$$x' = x - ut.$$

This is called a Galilean transformation. If the particle is moving at speed $\frac{dx}{dt} = v$ in the first frame of reference, in the second, it is moving at speed:

$$v' = \frac{dx'}{dt} = v - u.$$

Since u does not change, the accelerations are the same:

$$a' = \frac{dv'}{dt} = a.$$

Thus, momentum is conserved in both reference frames. Moreover, as long as the force has the same form, in both frames, Newton's second law is unchanged. Forces such as Newtonian gravity, which depend only on the scalar distance between objects, satisfy this criterion. This independence of reference frame is called Newtonian relativity or Galilean invariance.

A change of reference frame, can, often, simplify calculations of motion. For example, in a collision of two particles, a reference frame can be chosen, where, one particle begins at rest. Another, commonly used reference frame, is the center of mass frame – one that is moving with the center of mass. In this frame, the total momentum is zero.

Application to Collisions

By itself, the law of conservation of momentum is not enough to determine the motion of particles after a collision. Another property of the motion, kinetic energy, must be known. This is not necessarily conserved. If it is conserved, the collision is called an *elastic collision*; if not, it is an *inelastic collision*.

Elastic Collisions

An elastic collision is one in which no kinetic energy is absorbed in the collision. Perfectly elastic "collisions" can occur when the objects do not touch each other, as for example in atomic or nuclear scattering where electric repulsion keeps them apart. A slingshot maneuver of a satellite around a planet can also be viewed as a perfectly elastic collision. A collision between two pool balls is a good example of an *almost* totally elastic collision, due to their high rigidity, but when bodies come in contact there is always some dissipation.

A head-on elastic collision between two bodies can be represented by velocities in one dimension, along a line passing through the bodies. If the velocities are u_1 and u_2 before the collision and v_1 and v_2 after, the equations expressing conservation of momentum and kinetic energy are:

$$m_1u_1 + m_2u_2 = m_1v_1 + m_2v_2$$
$$\tfrac{1}{2}m_1u_1^2 + \tfrac{1}{2}m_2u_2^2 = \tfrac{1}{2}m_1v_1^2 + \tfrac{1}{2}m_2v_2^2.$$

A change of reference frame can simplify analysis of a collision. For example, suppose there are two bodies of equal mass m, one stationary and one approaching the other at a speed v. The center of mass is moving at speed $\frac{v}{2}$ and both bodies are moving towards it at speed $\frac{v}{2}$. Because of the symmetry, after the collision both must be moving away from the center of mass at the same speed. Adding the speed of the center of mass to both, we find that the body that was moving is now stopped and the other is moving away at speed v. The bodies have exchanged their velocities. Regardless of the velocities of the bodies, a switch to the center of mass frame leads us to the same conclusion. Therefore, the final velocities are given by;

$$v_1 = u_2$$
$$v_2 = u_1$$

In general, when the initial velocities are known, the final velocities are given by

$$v_1 = \left(\frac{m_1 - m_2}{m_1 + m_2}\right) u_1 + \left(\frac{2m_2}{m_1 + m_2}\right) u_2$$
$$v_2 = \left(\frac{m_2 - m_1}{m_1 + m_2}\right) u_2 + \left(\frac{2m_1}{m_1 + m_2}\right) u_1 .$$

If one body has much greater mass than the other, its velocity will be little affected by a collision while the other body will experience a large change.

Inelastic Collisions

In an inelastic collision, some of the kinetic energy of the colliding bodies is converted into other forms of energy (such as heat or sound). Examples include traffic collisions, in which the effect of lost kinetic energy can be seen in the damage to the vehicles; electrons losing some of their energy to atoms (as in the Franck–Hertz experiment); and particle accelerators in which the kinetic energy is converted into mass in the form of new particles.

In a perfectly inelastic collision (such as a bug hitting a windshield), both bodies have the same motion afterwards. If one body is motionless to begin with, the equation for conservation of momentum is:

$$m_1 u_1 = (m_1 + m_2) v,$$

so

$$v = \frac{m_1}{m_1 + m_2} u_1 .$$

In a frame of reference moving at the speed v), the objects are brought to rest by the collision and 100% of the kinetic energy is converted to other forms of energy.

One measure of the inelasticity of the collision is the coefficient of restitution C_R, defined as the ratio of relative velocity of separation to relative velocity of approach. In applying this measure to a ball bouncing from a solid surface, this can be easily measured using the following formula:

$$C_R = \sqrt{\frac{\text{bounce height}}{\text{drop height}}} .$$

The momentum and energy equations also apply to the motions of objects that begin together and then move apart. For example, an explosion is the result of a chain reaction that transforms potential energy stored in chemical, mechanical, or nuclear form into kinetic energy, acoustic energy, and electromagnetic radiation. Rockets also make use of conservation of momentum: propellant is thrust outward, gaining momentum, and an equal and opposite momentum is imparted to the rocket.

Multiple Dimensions

Real motion has both direction and velocity and must be represented by a vector. In a coordinate system with x, y, z axes, velocity has components v_x in the x-direction, v_y in the y-direction, v_z in the z-direction. The vector is represented by a boldface symbol:

$$\mathbf{v} = (v_x, v_y, v_z).$$

Similarly, the momentum is a vector quantity and is represented by a boldface symbol:

$$\mathbf{p} = (p_x, p_y, p_z).$$

The equations in the previous sections, work in vector form if the scalars p and v are replaced by vectors p and v. Each vector equation represents three scalar equations. For example,

$$\mathbf{p} = m\mathbf{v}$$

represents three equations:

$$p_x = mv_x$$
$$p_y = mv_y$$
$$p_z = mv_z.$$

The kinetic energy equations are exceptions to the above replacement rule. The equations are still one-dimensional, but each scalar represents the magnitude of the vector, for example,

$$v^2 = v_x^2 + v_y^2 + v_z^2.$$

Each vector equation represents three scalar equations. Often coordinates can be chosen so that only two components are needed, as in the figure. Each component can be obtained separately and the results combined to produce a vector result.

A simple construction involving the center of mass frame can be used to show that if a stationary elastic sphere is struck by a moving sphere, the two will head off at right angles after the collision.

Objects of Variable Mass

The concept of momentum plays a fundamental role in explaining the behavior of variable-mass objects such as a rocket ejecting fuel or a star accreting gas. In analyzing such an object, one treats the object's mass as a function that varies with time: $m(t)$. The momentum of the object at time t is therefore $p(t) = m(t)v(t)$. One might then try to invoke Newton's second law of motion by saying that the external force F on the object is related to its momentum $p(t)$ by $F=\frac{dp}{dt}$, but this is incorrect, as is the related expression found by applying the product rule to $\frac{d(mv)}{dt}$:

$$F=m(t)\frac{dv}{dt}+v(t)\frac{dm}{dt}. \text{ (incorrect)}$$

This equation does not correctly describe the motion of variable-mass objects. The correct equation is

$$F=m(t)\frac{dv}{dt}-u\frac{dm}{dt},$$

where u is the velocity of the ejected/accreted mass *as seen in the object's rest frame*. This is distinct from v, which is the velocity of the object itself as seen in an inertial frame.

This equation is derived by keeping track of both the momentum of the object as well as the momentum of the ejected/accreted mass (dm). When considered together, the object and the mass (dm) constitute a closed system in which total momentum is conserved:

$$P(t+dt)=(m-dm)(v+dv)+dm(v-u)=mv+mdv-udm=P(t)+mdv-udm.$$

Relativistic

Lorentz Invariance

Newtonian physics assumes that absolute time and space exist outside of any observer; this gives rise to Galilean invariance. It also results in a prediction that the speed of light can vary from one reference frame to another. This is contrary to observation. In the special theory of relativity, Einstein keeps the postulate that the equations of motion do not depend on the reference frame, but assumes that the speed of light c is invariant. As a result, position and time in two reference frames are related by the Lorentz transformation instead of the Galilean transformation.

Consider, for example, a reference frame moving relative to another at velocity v in the x direction. The Galilean transformation gives the coordinates of the moving frame as

$$t'=t$$
$$x'=x-vt$$

while the Lorentz transformation gives

$$t'=\gamma\left(t-\frac{vx}{c^2}\right)$$
$$x'=\gamma(x-vt)$$

where γ is the Lorentz factor:

$$\gamma = \frac{1}{\sqrt{1 - v^2 / c^2}}.$$

Newton's second law, with mass fixed, is not invariant under a Lorentz transformation. However, it can be made invariant by making the *inertial mass m* of an object a function of velocity:

$$m = \gamma m_0;$$

m_0 is the object's invariant mass.

The modified momentum,

$$\mathbf{p} = \gamma m_0 \mathbf{v},$$

obeys Newton's second law:

$$\mathbf{F} = \frac{d\mathbf{p}}{dt}.$$

Within the domain of classical mechanics, relativistic momentum closely approximates Newtonian momentum: at low velocity, γm_0v is approximately equal to m_0v, the Newtonian expression for momentum.

Four-vector Formulation

In the theory of special relativity, physical quantities are expressed in terms of four-vectors that include time as a fourth coordinate along with the three space coordinates. These vectors are generally represented by capital letters, for example R for position. The expression for the *four-momentum* depends on how the coordinates are expressed. Time may be given in its normal units or multiplied by the speed of light so that all the components of the four-vector have dimensions of length. If the latter scaling is used, an interval of proper time, τ, defined by

$$c^2 d\tau^2 = c^2 dt^2 - dx^2 - dy^2 - dz^2,$$

is invariant under Lorentz transformations (in this expression and in what follows the (+ − − −) metric signature has been used, different authors use different conventions). Mathematically this invariance can be ensured in one of two ways: by treating the four-vectors as Euclidean vectors and multiplying time by $\sqrt{-1}$; or by keeping time a real quantity and embedding the vectors in a Minkowski space. In a Minkowski space, the scalar product of two four-vectors U = (U_0, U_1, U_2, U_3) and V = (V_0, V_1, V_2, V_3) is defined as

$$\mathbf{U} \cdot \mathbf{V} = U_0 V_0 - U_1 V_1 - U_2 V_2 - U_3 V_3.$$

In all the coordinate systems, the (contravariant) relativistic four-velocity is defined by

$$\mathbf{U} \equiv \frac{d\mathbf{R}}{d\tau} = \gamma \frac{d\mathbf{R}}{dt},$$

and the (contravariant) four-momentum is

$$\mathbf{P} = m_0 \mathbf{U},$$

where m_0 is the invariant mass. If R = (ct,x,y,z) (in Minkowski space), then

$$\mathbf{P} = \gamma m_0 (c, \mathbf{v}) = (mc, \mathbf{p}).$$

Using Einstein's mass-energy equivalence, $E = mc^2$, this can be rewritten as

$$\mathbf{P} = \left(\frac{E}{c}, \mathbf{p}\right).$$

Thus, conservation of four-momentum is Lorentz-invariant and implies conservation of both mass and energy.

The magnitude of the momentum four-vector is equal to $m_0 c$:

$$||\mathbf{P}||^2 = \mathbf{P} \cdot \mathbf{P} = \gamma^2 m_0^2 (c^2 - v^2) = (m_0 c)^2,$$

and is invariant across all reference frames.

The relativistic energy–momentum relationship holds even for massless particles such as photons; by setting $m_0 = 0$ it follows that

$$E = pc.$$

In a game of relativistic "billiards", if a stationary particle is hit by a moving particle in an elastic collision, the paths formed by the two afterwards will form an acute angle. This is unlike the non-relativistic case where they travel at right angles.

The four-momentum of a planar wave can be related to a wave four-vector:

$$\mathbf{P} = \left(\frac{E}{c}, \vec{\mathbf{p}}\right) = \hbar \mathbf{K} = \hbar \left(\frac{\omega}{c}, \vec{\mathbf{k}}\right)$$

For a particle, the relationship between temporal components, $E = \hbar\,\omega$, is the Planck–Einstein relation, and the relation between spatial components, $p = \hbar\,k$, describes a de Broglie matter wave.

Generalized

Newton's laws can be difficult to apply to many kinds of motion because the motion is limited by *constraints*. For example, a bead on an abacus is constrained to move along its wire and a pendulum bob is constrained to swing at a fixed distance from the pivot. Many such constraints can be incorporated by changing the normal Cartesian coordinates to a set of *generalized coordinates* that may be fewer in number. Refined mathematical methods have been developed for solving mechanics problems in generalized coordinates. They introduce a *generalized momentum*, also known as the *canonical* or *conjugate momentum*, that extends the concepts of both linear momentum and angular momentum. To distinguish it from generalized momentum, the product of mass and velocity is also referred to as *mechanical, kinetic* or *kinematic momentum*. The two main methods are described below.

Lagrangian Mechanics

In Lagrangian mechanics, a Lagrangian is defined as the difference between the kinetic energy T and the potential energy V:

$$\mathcal{L} = T - V.$$

If the generalized coordinates are represented as a vector q = $(q_1, q_2, \dots, q_N)$ and time differentiation is represented by a dot over the variable, then the equations of motion (known as the Lagrange or Euler–Lagrange equations) are a set of N equations:

$$\frac{d}{dt}\left(\frac{\partial \mathcal{L}}{\partial \dot{q}_j}\right) - \frac{\partial \mathcal{L}}{\partial q_j} = 0.$$

If a coordinate q_i is not a Cartesian coordinate, the associated generalized momentum component p_i does not necessarily have the dimensions of linear momentum. Even if q_i is a Cartesian coordinate, p_i will not be the same as the mechanical momentum if the potential depends on velocity. Some sources represent the kinematic momentum by the symbol Π.

In this mathematical framework, a generalized momentum is associated with the generalized coordinates. Its components are defined as

$$p_j = \frac{\partial \mathcal{L}}{\partial \dot{q}_j}.$$

Each component p_j is said to be the *conjugate momentum* for the coordinate q_j.

Now if a given coordinate q_i does not appear in the Lagrangian (although its time derivative might appear), then

$$p_j = \text{constant}.$$

This is the generalization of the conservation of momentum.

Even if the generalized coordinates are just the ordinary spatial coordinates, the conjugate momenta are not necessarily the ordinary momentum coordinates. An example is found in the section on electromagnetism.

Hamiltonian Mechanics

In Hamiltonian mechanics, the Lagrangian (a function of generalized coordinates and their derivatives) is replaced by a Hamiltonian that is a function of generalized coordinates and momentum.

The Hamiltonian is defined as:

$$\mathcal{H}(\mathbf{q},\mathbf{p},t)=\mathbf{p}\cdot\dot{\mathbf{q}}-\mathcal{L}\left(\mathbf{q},\dot{\mathbf{q}},t\right),$$

where the momentum is obtained by differentiating the Lagrangian as above. The Hamiltonian equations of motion are

$$\begin{aligned}\dot{q}_i &= \frac{\partial\mathcal{H}}{\partial p_i}\\ -\dot{p}_i &= \frac{\partial\mathcal{H}}{\partial q_i}\\ -\frac{\partial\mathcal{L}}{\partial t} &= \frac{d\mathcal{H}}{dt}.\end{aligned}$$

As in Lagrangian mechanics, if a generalized coordinate does not appear in the Hamiltonian, its conjugate momentum component is conserved.

Symmetry and Conservation

Conservation of momentum is a mathematical consequence of the homogeneity (shift symmetry) of space (position in space is the canonical conjugate quantity to momentum). That is, conservation of momentum is a consequence of the fact that the laws of physics do not depend on position; this is a special case of Noether's theorem.

Electromagnetic

Particle in a Field

In Maxwell's equations, the forces between particles are mediated by electric and magnetic fields. The electromagnetic force (*Lorentz force*) on a particle with charge q due to a combination of electric field E and magnetic field B is:

$$\mathbf{F}=q(\mathbf{E}+\mathbf{v}\times\mathbf{B}).$$

It has an electric potential φ(r, t) and magnetic vector potential A(r, t). In the non-relativistic regime, its generalized momentum is:

$$\mathbf{P}=m\mathbf{v}+q\mathbf{A},$$

while in relativistic mechanics this becomes

$$\mathbf{P}=\gamma m\mathbf{v}+q\mathbf{A}.$$

Conservation

In Newtonian mechanics, the law of conservation of momentum can be derived from the law of action and reaction, which states that every force has a reciprocating equal and opposite force.

Under some circumstances, moving charged particles can exert forces on each other in non-opposite directions. Nevertheless, the combined momentum of the particles and the electromagnetic field is conserved.

Vacuum

The Lorentz force imparts a momentum to the particle, so by Newton's second law the particle must impart a momentum to the electromagnetic fields.

In a vacuum, the momentum per unit volume is:

$$\mathbf{g} = \frac{1}{\mu_0 c^2} \mathbf{E} \times \mathbf{B},$$

where μ_0 is the vacuum permeability and c is the speed of light. The momentum density is proportional to the Poynting vector S which gives the directional rate of energy transfer per unit area:

$$\mathbf{g} = \frac{\mathbf{S}}{c^2}.$$

If momentum is to be conserved over the volume V over a region Q, changes in the momentum of matter through the Lorentz force must be balanced by changes in the momentum of the electromagnetic field and outflow of momentum. If P_{mech} is the momentum of all the particles in Q, and the particles are treated as a continuum, then Newton's second law gives:

$$\frac{d\mathbf{P}_{\text{mech}}}{dt} = \iiint_Q (\rho\mathbf{E} + \mathbf{J} \times \mathbf{B}) dV.$$

The electromagnetic momentum is

$$\mathbf{P}_{\text{field}} = \frac{1}{\mu_0 c^2} \iiint_Q \mathbf{E} \times \mathbf{B} dV,$$

and the equation for conservation of each component i of the momentum is

$$\frac{d}{dt}(\mathbf{P}_{\text{mech}} + \mathbf{P}_{\text{field}})_i = \iint_\sigma \left(\sum_j T_{ij} n_j \right) d\Sigma.$$

The term on the right is an integral over the surface area Σ of the surface σ representing momentum flow into and out of the volume, and n_j is a component of the surface normal of S. The quantity T_{ij} is called the Maxwell stress tensor, defined as

$$T_{ij} \equiv \epsilon_0 \left(E_i E_j - \frac{1}{2} \delta_{ij} E^2 \right) + \frac{1}{\mu_0} \left(B_i B_j - \frac{1}{2} \delta_{ij} B^2 \right).$$

Media

The above results are for the *microscopic* Maxwell equations, applicable to electromagnetic forces in a vacuum (or on a very small scale in media). It is more difficult to define momentum density in media because the division into electromagnetic and mechanical is arbitrary. The definition of electromagnetic momentum density is modified to

$$\mathbf{g} = \frac{1}{c^2}\mathbf{E}\times\mathbf{H} = \frac{\mathbf{S}}{c^2},$$

where the H-field H is related to the B-field and the magnetization M by:

$$\mathbf{B} = \mu_o(\mathbf{H}+\mathbf{M}).$$

The electromagnetic stress tensor depends on the properties of the media.

Quantum Mechanical

In quantum mechanics, momentum is defined as a self-adjoint operator on the wave function. The Heisenberg uncertainty principle defines limits on how accurately the momentum and position of a single observable system can be known at once. In quantum mechanics, position and momentum are conjugate variables.

For a single particle described in the position basis the momentum operator can be written as

$$\mathbf{p} = \frac{\hbar}{i}\nabla = -i\hbar\nabla,$$

where ∇ is the gradient operator, $\hbar$ is the reduced Planck constant, and i is the imaginary unit. This is a commonly encountered form of the momentum operator, though the momentum operator in other bases can take other forms. For example, in momentum space the momentum operator is represented as

$$\mathbf{p}\psi(p) = p\psi(p),$$

where the operator p acting on a wave function $\psi(p)$ yields that wave function multiplied by the value p, in an analogous fashion to the way that the position operator acting on a wave function $\psi(x)$ yields that wave function multiplied by the value x.

For both massive and massless objects, relativistic momentum is related to the phase constant β by

$$p = \hbar\beta$$

Electromagnetic radiation (including visible light, ultraviolet light, and radio waves) is carried by photons. Even though photons (the particle aspect of light) have no mass, they still carry momentum. This leads to applications such as the solar sail. The calculation of the momentum of light within dielectric media is somewhat controversial.

In Deformable Bodies and Fluids

Conservation in a Continuum

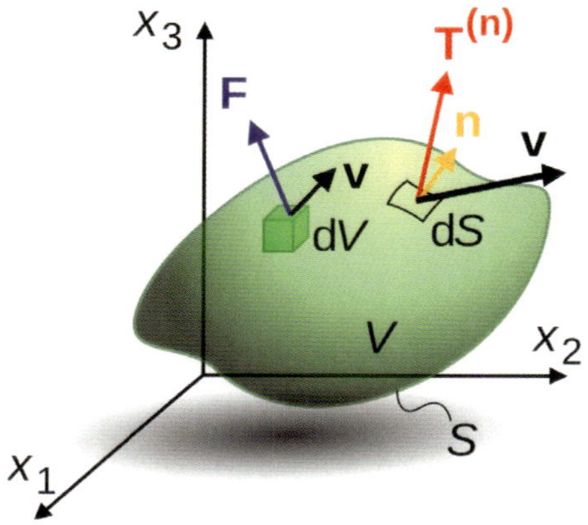

Motion of a material body

In fields such as fluid dynamics and solid mechanics, it is not feasible to follow the motion of individual atoms or molecules. Instead, the materials must be approximated by a continuum in which there is a particle or fluid parcel at each point that is assigned the average of the properties of atoms in a small region nearby. In particular, it has a density ρ and velocity v that depend on time t and position r. The momentum per unit volume is ρv.

Consider a column of water in hydrostatic equilibrium. All the forces on the water are in balance and the water is motionless. On any given drop of water, two forces are balanced. The first is gravity, which acts directly on each atom and molecule inside. The gravitational force per unit volume is ρg, where g is the gravitational acceleration. The second force is the sum of all the forces exerted on its surface by the surrounding water. The force from below is greater than the force from above by just the amount needed to balance gravity. The normal force per unit area is the pressure p. The average force per unit volume inside the droplet is the gradient of the pressure, so the force balance equation is

$$-\nabla p + \rho \mathbf{g} = 0.$$

If the forces are not balanced, the droplet accelerates. This acceleration is not simply the partial derivative $\frac{\partial \mathrm{v}}{\partial t}$ because the fluid in a given volume changes with time. Instead, the material derivative is needed:

$$\frac{D}{Dt} \equiv \frac{\partial}{\partial t} + \mathbf{v} \cdot \nabla.$$

Applied to any physical quantity, the material derivative includes the rate of change at a point and the changes due to advection as fluid is carried past the point. Per unit volume, the rate of change in momentum is equal to $p\frac{D\mathrm{v}}{Dt}$. This is equal to the net force on the droplet.

Forces that can change the momentum of a droplet include the gradient of the pressure and gravity, as above. In addition, surface forces can deform the droplet. In the simplest case, a shear stress

τ, exerted by a force parallel to the surface of the droplet, is proportional to the rate of deformation or strain rate. Such a shear stress occurs if the fluid has a velocity gradient because the fluid is moving faster on one side than another. If the speed in the x direction varies with z, the tangential force in direction x per unit area normal to the z direction is

$$\sigma_{zx} = -\mu \frac{\partial v_x}{\partial z},$$

where μ is the viscosity. This is also a flux, or flow per unit area, of x-momentum through the surface.

Including the effect of viscosity, the momentum balance equations for the incompressible flow of a Newtonian fluid are

$$\rho \frac{D\mathbf{v}}{Dt} = -\nabla p + \mu \nabla^2 \mathbf{v} + \rho \mathbf{g}.$$

These are known as the Navier–Stokes equations.

The momentum balance equations can be extended to more general materials, including solids. For each surface with normal in direction i and force in direction j, there is a stress component σ_{ij}. The nine components make up the Cauchy stress tensor σ, which includes both pressure and shear. The local conservation of momentum is expressed by the Cauchy momentum equation:

$$\rho \frac{D\mathbf{v}}{Dt} = \nabla \cdot \boldsymbol{\sigma} + \mathbf{f},$$

where f is the body force.

The Cauchy momentum equation is broadly applicable to deformations of solids and liquids. The relationship between the stresses and the strain rate depends on the properties of the material.

Acoustic Waves

A disturbance in a medium gives rise to oscillations, or waves, that propagate away from their source. In a fluid, small changes in pressure p can often be described by the acoustic wave equation:

$$\frac{\partial^2 p}{\partial t^2} = c^2 \nabla^2 p,$$

where c is the speed of sound. In a solid, similar equations can be obtained for propagation of pressure (P-waves) and shear (S-waves).

The flux, or transport per unit area, of a momentum component ρv_j by a velocity v_i is equal to $\rho\, v_j v_j$. In the linear approximation that leads to the above acoustic equation, the time average of this flux is zero. However, nonlinear effects can give rise to a nonzero average. It is possible for momentum flux to occur even though the wave itself does not have a mean momentum.

Angular Momentum

Angular momentum, property characterizing the rotary inertia of an object or system of objects in motion about an axis that may or may not pass through the object or system. The Earth has orbital angular momentum by reason of its annual revolution about the Sun and spin angular momentum because of its daily rotation about its axis. Angular momentum is a vector quantity, requiring the specification of both a magnitude and a direction for its complete description. The magnitude of the angular momentum of an orbiting object is equal to its linear momentum (product of its mass m and linear velocity v) times the perpendicular distance r from the centre of rotation to a line drawn in the direction of its instantaneous motion and passing through the object's centre of gravity, or simply mvr. For a spinning object, on the other hand, the angular momentum must be considered as the summation of the quantity mvr for all the particles composing the object. Angular momentum may be formulated equivalently as the product of I, the moment of inertia, and ω, the angular velocity, of a rotating body or system, or simply $I\omega$. The direction of the angular-momentum vector is that of the axis of rotation of the given object and is designated as positive in the direction that a right-hand screw would advance if turned similarly. Appropriate MKS or SI units for angular momentum are kilogram metres squared per second (kg-m^2/sec).

For a given object or system isolated from external forces, the total angular momentum is a constant, a fact that is known as the law of conservation of angular momentum. A rigid spinning object, for example, continues to spin at a constant rate and with a fixed orientation unless influenced by the application of an external torque. (The rate of change of the angular momentum is, in fact, equal to the applied torque.) A figure skater spins faster, or has a greater angular velocity ω, when the arms are drawn inward, because this action reduces the moment of inertia I while the product $I\omega$, the skater's angular momentum, remains constant. Because of the conservation of direction as well as magnitude, a spinning gyrocompass in an airplane remains fixed in its orientation, independent of the motion of the airplane.

For the extension of the conception of orbital and spin angular momentum to analogous properties of subatomic particles such as electrons.

Angular momentum and its conservation is a useful quantity to consider when analysing rotational dynamics problems.

The angular momentum of an object may be calculated directly, if its moment of inertia and angular velocity are known, by the formula:

$$\underline{L} = I\underline{\omega}.$$

If the object is a point particle of mass m , rotating with instantaneous speed v about an axis and at a fixed distance r from that axis (so that the particle is in a circular orbit), then $I=mr^2$ and $\underline{\omega}$ has magnitude $\omega = \frac{v}{r}$. Hence $\underline{L}$ has magnitude:

$$L = mr^2 \frac{v}{r} = mrv$$

in this special case.

More generally, one can define the instantaneous angular momentum $\underline{L}$ of a point particle of mass m about an arbitrary point P, when the particle has velocity $\underline{v}$ and is located at position $\underline{r}$ relative to P. The magnitude of $\underline{L}$ is:

$$L = mrv \sin\theta,$$

where θ is the angle between vectors $\underline{r}$ and $\underline{v}$. This agrees with the result for the particle in a circular orbit since in that case θ=90°.

Angular Momentum of a Particle

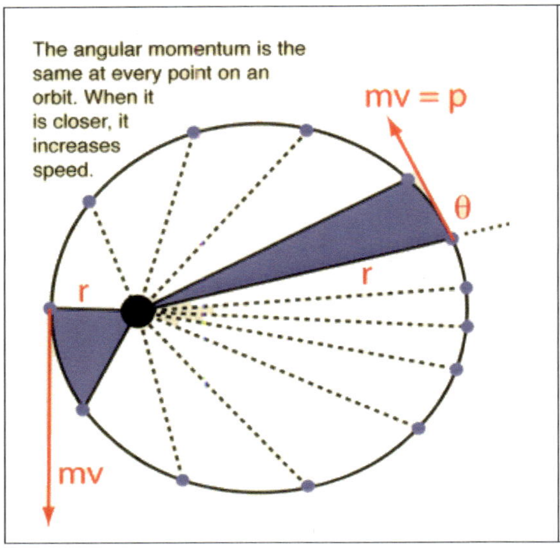

The angular momentum of a particle of mass m with respect to a chosen origin is given by

$$L = mvr \sin\theta$$

or more formally by the vector product

$$L = r \times p$$

The direction is given by the right hand rule which would give L the direction out of the diagram. For an orbit, angular momentum is conserved, and this leads to one of Kepler's laws. For a circular orbit, L becomes

$$L = mvr$$

References

- Knudsen, Jens M.; Hjorth, Poul (2012). Elements of Newtonian Mechanics (illustrated ed.). Springer Science & Business Media. p. 30. ISBN 978-3-642-97599-8
- Jesseph, Douglas M. (1998). "Leibniz on the Foundations of the Calculus: The Question of the Reality of Infinitesimal Magnitudes". Perspectives on Science. 6.1&2: 6–40. Retrieved 31 December 2011
- Galili, I.; Tseitlin, M. (2003). "Newton's First Law: Text, Translations, Interpretations and Physics Education". Science & Education. 12 (1): 45–73. Bibcode:2003Sc&Ed..12...45G. doi:10.1023/A:1022632600805
- Doggett, LeRoy E. (1997), "Celestial Mechanics", in Lankford, John, History of Astronomy: An Encyclopedia, New York: Taylor & Francis, pp. 131–140, ISBN 9780815303220
- C Hellingman (1992). "Newton's third law revisited". Phys. Educ. 27 (2): 112–115. Bibcode:1992PhyEd..27..112H. doi:10.1088/0031-9120/27/2/011
- Carl Nave (2010). "Elastic and inelastic collisions". Hyperphysics. Archived from the original on 18 August 2012. Retrieved 2 August 2012
- Aydin Sayili (1987). «Ibn Sīnā and Buridan on the Motion of the Projectile». Annals of the New York Academy of Sciences. 500 (1): 477–482. Bibcode:1987NYASA.500..477S. doi:10.1111/j.1749-6632.1987.tb37219.x

2

Rotational and Circular Motion

The motion of an object with respect to a central point in a circular path is called circular motion. The circular motion of an object around a center of rotation or axis of rotation is called rotational motion. This chapter discusses the fundamentals of rotational and circular motion such as rotating reference frame, centrifugal force and centripetal force.

Rotational Motion

Rotational motion deals with changes in angle (θ). An object will always rotate—that is, spin or move in a circle—around an axis of rotation. The earth spins around on the axis that runs from the North Pole to the South Pole, creating day and night; figure skaters spin around on the axis from their head to their feet, creating dizzying gold medals; Star Fox spins around on the axis from his spaceship's nose to stern, creating DO A BARREL ROLLs.

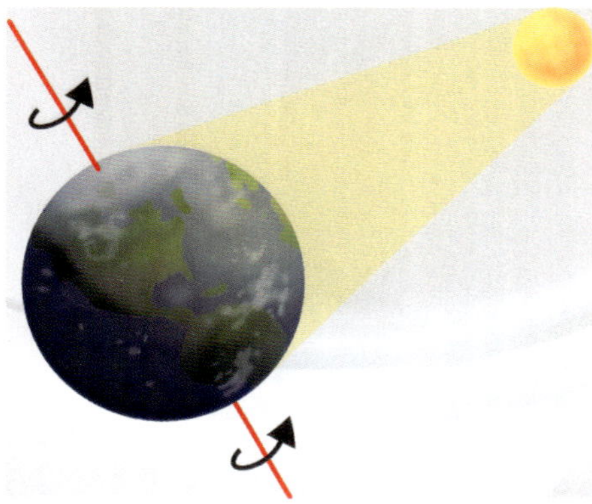

The axis of rotation, however, can also be outside the rotating object. The sun, for example, is on the axis that the earth rotates around in its orbit.

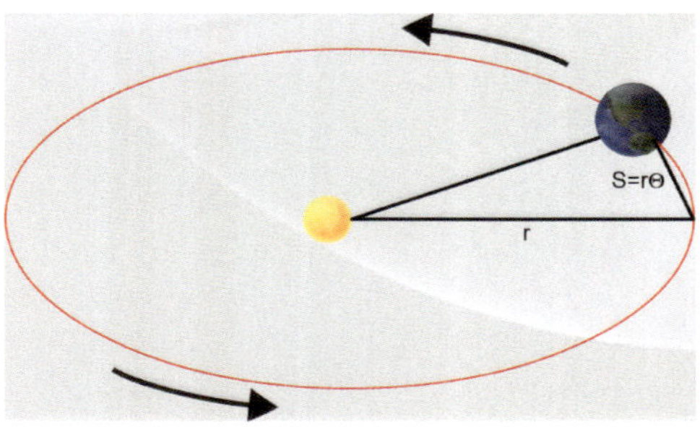

An angle is just a way to measure how much exactly an object has spun, or how far around the circle an object has gotten. Of course, we could also measure the distance around the circle in terms of position (say, along the circumference), but angle is a very convenient shorthand, and handy if we're more concerned about the details of the rotation than the translation.

Since angle and position give us the same information, we can convert between the two if we know the radius of the circle. An object that rotates through an angle θ will move along an arc of length:

$$s = r\theta$$

Here, r is the radius of the circle the object travels along. Both s and r are measured in meters, so angle is technically dimensionless. As a bookkeeping tool, we measure angle in radians. This is similar to the degrees you're probably familiar with—90° is a right angle or a quarter circle, 180° is a change in direction or a half circle, etc.—but on a different scale. Whereas 1° is equal to $\left(\frac{1}{360}\right)^{th}$ of a circle (since a full rotation is 360°), 1 rad is equal to $\left(\frac{1}{2\pi}\right)^{th}$ of a circle: radians break a circle down into 2π parts, instead of 360. Unfortunately, "2π heelflip" doesn't have quite the same ring as a "360 heelflip".

Here's the radian equivalent of some common angles:

Degree	Radian
0°	0 rad
30°	$\frac{\pi}{6}$ rad
45°	$\frac{\pi}{4}$ rad
60°	$\frac{\pi}{3}$ rad
90°	$\frac{\pi}{2}$ rad
180°	π rad
270°	$\frac{3\pi}{2}$ rad
360°	2 π rad

You can translate radians into degrees by multiplying by $\frac{360^\circ}{2\pi rad}$.

Rotation around a Fixed Axis

Rotation about a fixed axis is a special case of rotational motion. It is very common to analyze problems that involve this type of rotation – for example, a wheel.

The figure illustrates rotational motion of a rigid body about a fixed axis at point *O*. This type of motion occurs in a plane perpendicular to the axis of rotation.

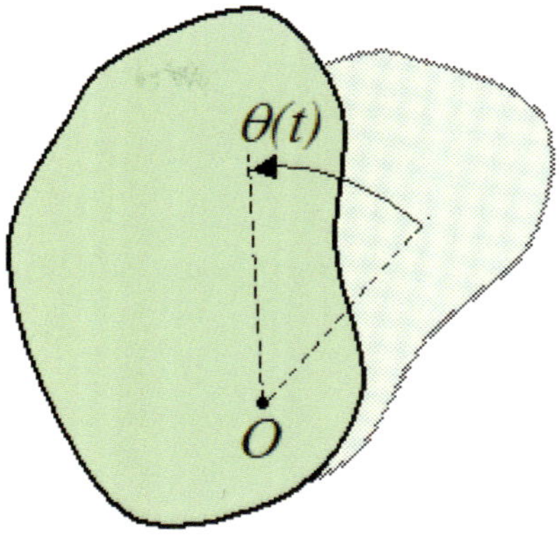

In the figure, the angle is defined as the angular position of the body, as a function of time *t*. This angle can be measured in any unit one desires, such as radians, or degrees.

Every point in the rigid body rotates by the same angle *(t)*.

Given the angular position of the rigid body we can calculate the angular displacement, angular velocity, and angular acceleration. These are important quantities to consider when evaluating the rotational kinematics of a problem.

A common assumption, which applies to numerous problems involving rotation about a fixed axis, is that angular acceleration is constant. With angular acceleration as constant we can derive equations for the angular position, angular displacement, and angular velocity of a rigid body experiencing rotation about a fixed axis.

The easiest way to derive these equations is by using Calculus. The derivations that follow are of the exact same form as the equations derived for rectilinear motion, with constant acceleration.

The angular acceleration is given by

$$\frac{d^2\theta}{dt^2} = \alpha$$

where α is the angular acceleration, which we define as constant.

Integrate the above equation with respect to time, to obtain angular velocity. This gives us

$$w(t) = \int \alpha \, dt = C_1 + \alpha t$$

where $w(t)$ is the angular velocity and C_1 is a constant.

Integrate the above equation with respect to time, to obtain angular position. This gives us

$$\theta(t) = \int w(t)dt = C_2 + C_1 t + \frac{1}{2}\alpha t^2$$

where $\theta(t)$ is the angular position and C_2 is a constant.

The constants C_1 and C_2 are determined by the initial conditions at time $t = 0$. The initial conditions are:

At time $t = 0$ the angular position is θ_1.

At time $t = 0$ the angular velocity is w_1.

Substituting these two initial conditions into the above two equations we get

$$w(0) = w_1 = C_1$$

$$\theta(0) = \theta_1 = C_2$$

Therefore $C_1 = w_1$ and $C_2 = \theta_1$.

This gives us

$$\theta(t) = \theta_1 + w_1 t + \frac{1}{2}\alpha t^2$$
$$w(t) = w_1 + \alpha t$$

For convenience, set $\theta(t) = \theta_2$ and $w(t) = w_2$. As a result

$$\theta_2 = \theta_1 + w_1 t + \frac{1}{2}\alpha t^2 \qquad (1) - \text{angular position equation}$$

$$w_2 = w_1 + \alpha t \qquad (2) - \text{angular velocity equation}$$

Angular displacement is defined as $\Delta\theta = \theta_2 - \theta_1$. Therefore, equation (1) becomes

$$\Delta\theta = w_1 t + \frac{1}{2}\alpha t^2 \qquad (3) - \text{angular displacement equation}$$

If we wish to find an equation that doesn't involve time t we can combine equations (2) and (3) to eliminate time as a variable. This gives us

$$w_2^2 = w_1^2 + 2\alpha(\Delta\theta) \qquad (4) - 2^{\text{nd}} \text{ angular velocity equation}$$

Equations (1), (2), (3), and (4) fully describe the rotational motion of rigid bodies (or particles) rotating about a fixed axis, where angular acceleration α is constant.

For the cases where angular acceleration is not constant, new expressions have to be derived for the angular position, angular displacement, and angular velocity. If the angular acceleration is known as a function of time, we can use Calculus to find the angular position, angular displacement, and angular velocity, in the same manner as before.

Alternatively, if we are given the angular position $\theta(t)$ as a function of time, we determine the angular velocity by differentiating $\theta(t)$ once, and we determine the angular acceleration by differentiating $\theta(t)$ twice.

For example, let's say the angular position $\theta(t)$ of a particle is given by:

$$\theta(t) = \sin(3t) + 3t^2$$

Thus, the angular velocity $w(t)$ is given by:

$$w(t) = \frac{d\theta}{dt} = 3\cos(3t) + 6t$$

The angular acceleration $\alpha(t)$ is given by:

$$\alpha(t) = \frac{d^2\theta}{dt^2} = -9\sin(3t) + 6\cdot$$

Angular Speed and Angular Velocity

Angular velocity is the change in angular displacement per unit time. The symbol for angular velocity is ω and the units are typically rad s^{-1}. Angular speed is the magnitude of angular velocity.

$$\bar{\omega} = \frac{\Delta\theta}{\Delta t} = \frac{\theta_2 - \theta_1}{t_2 - t_1}.$$

The instantaneous angular velocity is given by

$$\omega(t) = \frac{d\theta}{dt}.$$

Using the formula for angular position and letting $v = \frac{ds}{dt}$, we have also

$$\omega = \frac{d\theta}{dt} = \frac{v}{r},$$

where v is the translational speed of the particle.

Angular velocity and frequency are related by

$$\omega = 2\pi f.$$

Angular Acceleration

A changing angular velocity indicates the presence of an angular acceleration in rigid body, typically measured in rad s^{-2}. The average angular acceleration $\bar{\alpha}$ over a time interval Δt is given by

$$\bar{\alpha} = \frac{\Delta\omega}{\Delta t} = \frac{\omega_2 - \omega_1}{t_2 - t_1}.$$

The instantaneous acceleration $\alpha(t)$ is given by

$$\alpha(t) = \frac{d\omega}{dt} = \frac{d^2\theta}{dt^2}.$$

Thus, the angular acceleration is the rate of change of the angular velocity, just as acceleration is the rate of change of velocity.

The translational acceleration of a point on the object rotating is given by

$$a = r\alpha,$$

where r is the radius or distance from the axis of rotation. This is also the tangential component of acceleration: it is tangential to the direction of motion of the point. If this component is 0, the motion is uniform circular motion, and the velocity changes in direction only.

The radial acceleration (perpendicular to direction of motion) is given by

$$a_R = \frac{v^2}{r} = \omega^2 r.$$

It is directed towards the center of the rotational motion, and is often called the *centripetal acceleration.*

The angular acceleration is caused by the torque, which can have a positive or negative value in accordance with the convention of positive and negative angular frequency. The ratio of torque and angular acceleration (how difficult it is to start, stop, or otherwise change rotation) is given by the moment of inertia: $T = I\alpha$.

Equations of Kinematics

When the angular acceleration is constant, the five quantities angular displacement θ, initial angular velocity ω_i, final angular velocity ω_f, angular acceleration α, and time t can be related by four equations of kinematics:

$$\omega_f = \omega_i + \alpha t$$

$$\theta = \omega_i t + \frac{1}{2}\alpha t^2$$

$$\omega_f^2 = \omega_i^2 + 2\alpha\theta$$

$$\theta = \frac{1}{2}(\omega_f + \omega_i)t$$

Dynamics

Moment of Inertia

The moment of inertia of an object, symbolized by I, is a measure of the object's resistance to changes to its rotation. The moment of inertia is measured in kilogram metre^2 (kg m^2). It depends on the object's mass: increasing the mass of an object increases the moment of inertia. It also depends on the distribution of the mass: distributing the mass further from the centre of rotation increases the moment of inertia by a greater degree. For a single particle of mass m a distance r from the axis of rotation, the moment of inertia is given by

$$I = mr^2.$$

Torque

Torque τ is the twisting effect of a force F applied to a rotating object which is at position r from its axis of rotation. Mathematically,

$$\tau = r \times F,$$

where $\times$ denotes the cross product. A net torque acting upon an object will produce an angular acceleration of the object according to

$$\tau = I\alpha,$$

just as $F = ma$ in linear dynamics.

The work done by a torque acting on an object equals the magnitude of the torque times the angle through which the torque is applied:

$$W = \tau\theta.$$

The power of a torque is equal to the work done by the torque per unit time, hence:

$$P = \tau\omega.$$

Angular Momentum

The angular momentum L is a measure of the difficulty of bringing a rotating object to rest. It is given by

$$L = r \times p.$$

Angular momentum is related to angular velocity by

$$L = I\omega,$$

just as $p = mv$ in linear dynamics.

The equivalent of linear momentum in rotational motion is angular momentum. The greater the angular momentum of the spinning object such as a top, the greater its tendency to continue to spin.

The Angular Momentum of a rotating body is proportional to its mass and to how rapidly it is turning. In addition the angular momentum depends on how the mass is distributed relative to the axis of rotation: the further away the mass is located from the axis of rotation, the greater the angular momentum . A flat disk such as a record turntable has less angular momentum than a hollow cylinder of the same mass and velocity of rotation.

Like linear momentum, angular momentum is vector quantity, and its conservation implies that the direction of the spin axis tends to remain unchanged. For this reason the spinning top remains upright whereas a stationary one falls over immediately.

The angular momentum equation can be used to relate the moment of the resultant force on a body about an axis (sometimes called torque), and the rate of rotation about that axis.

Torque and angular momentum are related according to:

$$\tau = \frac{d\mathbf{L}}{dt},$$

just as F = dp/dt in linear dynamics. In the absence of an external torque, the angular momentum of a body remains constant. The conservation of angular momentum is notably demonstrated in figure skating: when pulling the arms closer to the body during a spin, the moment of inertia is decreased, and so the angular velocity is increased.

Kinetic Energy

The kinetic energy K_{rot} due to the rotation of the body is given by

$$K_{rot} = \frac{1}{2} I\omega^2,$$

just as $K_{trans} = \frac{1}{2}mv^2$ in linear dynamics.

Kinetic energy is the energy of motion. The amount of translational kinetic energy found in two variables: the mass of the object (m) and the speed of the object (v) as show in the equation above. Kinetic energy must always be either zero or a positive value. While velocity can have either a positive or negative value, velocity squared will always be positive.

Vector Expression

The above development is a special case of general rotational motion. In the general case, angular displacement, angular velocity, angular acceleration and torque are considered to be vectors.

An angular displacement is considered to be a vector, pointing along the axis, of magnitude equal to that of Äθ . A right-hand rule is used to find which way it points along the axis; if the fingers of the right hand are curled to point in the way that the object has rotated, then the thumb of the right hand points in the direction of the vector.

The angular velocity vector also points along the axis of rotation in the same way as the angular displacements it causes. If a disk spins counterclockwise as seen from above, its angular velocity vector points upwards. Similarly, the angular acceleration vector points along the axis of rotation in the same direction that the angular velocity would point if the angular acceleration were maintained for a long time.

The torque vector points along the axis around which the torque tends to cause rotation. To maintain rotation around a fixed axis, the total torque vector has to be along the axis, so that it only changes the magnitude and not the direction of the angular velocity vector. In the case of a hinge, only the component of the torque vector along the axis has effect on the rotation, other forces and torques are compensated by the structure.

Velocity and Acceleration of a Particle Traveling around a Circle

The figure below shows a particle *P* traveling around a circle of radius *R*, with fixed center at *O*.

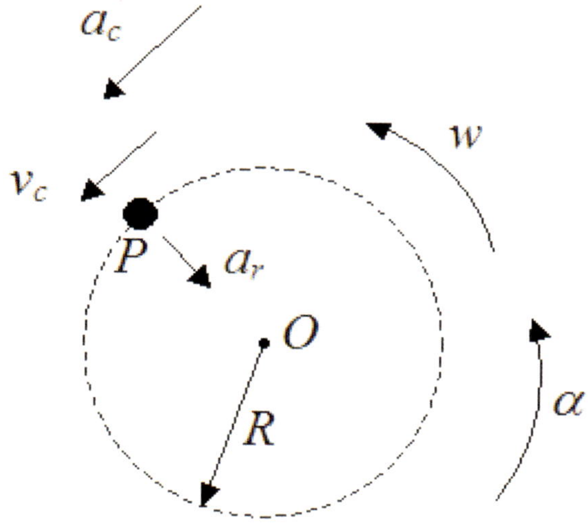

Where:

w is the angular velocity associated with the rotational motion of the particle *P*. This quantity is measured in rad/s

α is the angular acceleration associated with the rotational motion of the particle *P*. This quantity is measured in rad/s^2

a_r is the radial acceleration of the particle *P*, pointing towards the center of the circle (this is commonly called centripetal acceleration)

v_c is the circumferential velocity of the particle *P* (this is tangent to the circle)

a_c is the circumferential acceleration of the particle *P* (this is tangent to the circle)

We can use the equations for curvilinear motion to derive expressions for a_r , v_c , and a_c. Since *R* is constant we get

$$a_r = Rw^2$$
$$v_c = Rw$$
$$a_c = R\alpha$$

Note that there is no radial component of velocity pointing towards, or away from the center of the circle. This is because the radius *R* is constant. If *R* were to change with time there would be a radial component of velocity pointing towards, or away from the center of the circle.

Rotational Motion in Three Dimensions

Rotational motion in three dimensions is mathematically more complicated than planar rotation about a fixed axis, since the axis of rotation can change direction. This type of rotation applies to bodies experiencing three-dimensional motion. However, it is generally only necessary (and practical) to account for such rotation when determining the velocity and acceleration of a point on a body that is experiencing three-dimensional motion.

Angular Displacement

Angular displacement is defined as "the angle in radians (degrees, revolutions) through which a point or line has been rotated in a specified sense about a specified axis". It is the angle of the movement of a body in a circular path.

Angular displacement is a vector quantity, which means that angular displacement has a size and a direction associated with it. The direction is important for later mathematical processes, but the definition is a bit confusing. As the object rotates from point "0" to point "1", it rotates about an axis, so the direction of the angular displacement is measured along the axis. A positive value for the direction of the axis is defined by the right hand rule. *Extend your right hand as if to shake hands with someone. Curve your fingers with the base at point "0" and the tips going to point "1". Your thumb points perpendicular to the plane of rotation in the positive direction along the axis of rotation.*

Angular displacement is measured in units of radians. *Two pi radians equals 360 degrees.* The angular displacement is not a length (not measured in meters or feet), so an angular displacement is different than a linear displacement. As the solid object rotates about the axis of rotation, all of the points of the object experience the same angular displacement, but points farther away from the axis move farther than points closer to the axis. On the slide we consider two points; one is located at radius ra on the edge of the disk, and the other is located at radius rb which is less than ra. As the object rotates through the angular displacement phi, the point on the edge of the disk moves distance sa along a circular path. The point at rb also moves in a circular path, but the distance sb is shorter than the distance sa. In general, the length of the circular path s is equal to the radius r times the angular displacement phi, expressed in radians.

for angular displacement phi,

s = phi * r

ra > rb

sa > sb.

Angular Displacement Formula

The angular displacement is the angle through which a body rotates in circular path. It is denoted by θ and expressed in radians or degrees.

$$\theta = \frac{S}{r}$$

Angular displacement formula is used to find the angular displacement θ made by the body in time t. Angular displacement formula in terms of angular velocity is given by

$$\theta = \omega t$$

Where, ω ω is angular velocity,

t is the time taken.

Angular Displacement Problems

Here are some problems given on angular displacement based on the above formula.

Problem 1

A motor bike is moving in circular track covers the whole track of distance 50m having radius of 4m from the center. Find its angular displacement.

Solution

Given: Linear displacement S = 50 m,

Radius of circular path r = 4 m,

The angular displacement is given by

$$\theta = \frac{S}{r}$$

$$= \frac{50}{4}$$

= 12.5 radians.

Problem 2

A body is moving in a circular path with a angular speed of 59 rad/sec. Calculate its angular displacement at time t = 5 s?

Solution

given: Angular speed ω = 59 rad/sec,

Time t = 5 s

The angular displacement is given by

$$\theta = \omega t$$

$$= 59 \times 5$$

= 295 radians.

Three Dimensions

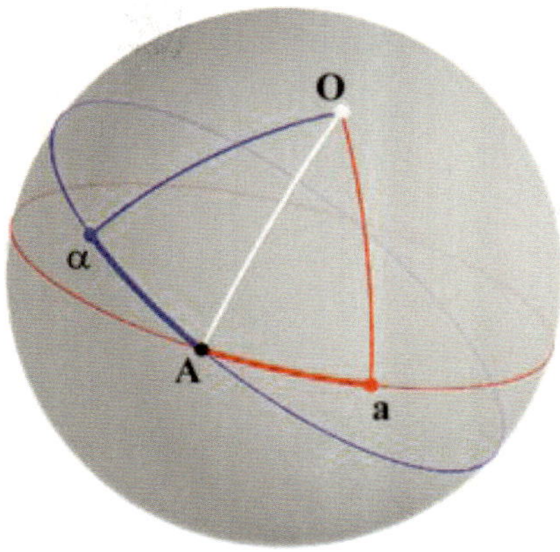

Figure: Euler's rotation theorem. A great circle transforms to another great circle under rotations, leaving always a diameter of the sphere in its original position.

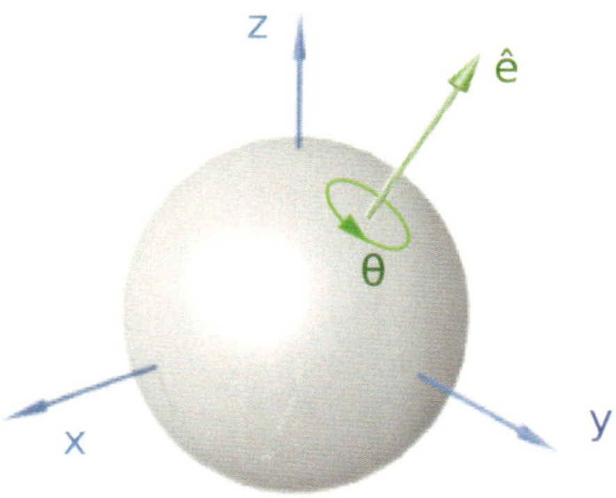

Figure: A rotation represented by an Euler axis and angle.

In three dimensions, angular displacement is an entity with a direction and a magnitude. The direction specifies the axis of rotation, which always exists by virtue of the Euler's rotation theorem; the magnitude specifies the rotation in radians about that axis (using the right-hand rule to determine direction). This entity is called an axis-angle.

Despite having direction and magnitude, angular displacement is not a vector because it does not obey the commutative law for addition. Nevertheless, when dealing with infinitesimal rotations, second order infinitesimals can be discarded and in this case commutativity appears.

Several ways to describe angular displacement exist, like rotation matrices or Euler angles.

Matrix Notation

Given that any frame in the space can be described by a rotation matrix, the displacement among them can also be described by a rotation matrix. Being A_0 and A_f two matrices, the angular displacement matrix between them can be obtained as $\Delta A = A_f . A_0^{-1}$. When this product is performed having a very small difference between both frames we will obtain a matrix close to the identity.

In the limit, we will have an infinitesimal rotation matrix.

Infinitesimal Rotation Matrices

An infinitesimal angular displacement is a infinitesimal rotation matrix:

- As any rotation matrix has a single real eigenvalue, which is +1, this eigenvalue shows the rotation axis.
- Its module can be deduced from the value of the infinitesimal rotation.
- The shape of the matrix is like this:

$$A = \begin{pmatrix} 1 & -d\phi_z(t) & d\phi_y(t) \\ d\phi_z(t) & 1 & -d\phi_x(t) \\ -d\phi_y(t) & d\phi_x(t) & 1 \end{pmatrix}$$

We can introduce here the infinitesimal angular displacement tensor or rotation generator associated:

$$d\Phi(t) = \begin{pmatrix} 0 & -d\phi_z(t) & d\phi_y(t) \\ d\phi_z(t) & 0 & -d\phi_x(t) \\ -d\phi_y(t) & d\phi_x(t) & 0 \end{pmatrix}$$

Such that its associated rotation matrix is $A = I + d\Phi(t)$. When it is divided by the time, this will yield the angular velocity vector.

Generators of Rotations

Suppose we specify an axis of rotation by a unit vector $[x, y, z]$, and suppose we have an infinitely small rotation of angle $\Delta\theta$ about that vector. Expanding the rotation matrix as an infinite addition, and taking the first order approach, the rotation matrix ΔR is represented as:

$$\Delta R = \begin{bmatrix} 1 & 0 & 0 \\ 0 & 1 & 0 \\ 0 & 0 & 1 \end{bmatrix} + \begin{bmatrix} 0 & z & -y \\ -z & 0 & x \\ y & -x & 0 \end{bmatrix} \Delta\theta = \mathbf{I} + \mathbf{A}\Delta\theta.$$

A finite rotation through angle θ about this axis may be seen as a succession of small rotations about the same axis. Approximating $\Delta\theta$ as θ/N where N is a large number, a rotation of θ about the axis may be represented as:

$$R = \left(\mathbf{1} + \frac{\mathbf{A}\theta}{N}\right)^N \approx e^{\mathbf{A}\theta}.$$

It can be seen that Euler's theorem essentially states that all rotations may be represented in this form. The product $\mathbf{A}\theta$ is the "generator" of the particular rotation, being the vector (x,y,z) associated with the matrix A. This shows that the rotation matrix and the axis-angle format are related by the exponential function.

One can derive a simple expression for the generator G. One starts with an arbitrary plane defined by a pair of perpendicular unit vectors a and b. In this plane one can choose an arbitrary vector x with perpendicular y. One then solves for y in terms of x and substituting into an expression for a rotation in a plane yields the rotation matrix R which includes the generator G = $ba^T - ab^T$.

$$x = a\cos(\alpha) + b\sin(\alpha)$$
$$y = -a\sin(\alpha) + b\cos(\alpha)$$
$$\cos(\alpha) = a^T x$$
$$\sin(\alpha) = b^T x$$
$$y = -ab^T x + ba^T x = \left(ba^T - ab^T\right)x$$
$$x' = x\cos(\beta) + y\sin(\beta)$$
$$= \left[I\cos(\beta) + \left(ba^T - ab^T\right)\sin(\beta)\right]x$$
$$R = I\cos(\beta) + \left(ba^T - ab^T\right)\sin(\beta)$$
$$= I\cos(\beta) + G\sin(\beta)$$
$$G = ba^T - ab^T$$

To include vectors outside the plane in the rotation one needs to modify the above expression for R by including two projection operators that partition the space. This modified rotation matrix can be rewritten as an exponential function.

$$P_{ab} = -G^2$$
$$R = I - P_{ab} + \left[I\cos(\beta) + G\sin(\beta)\right]P_{ab} = e^{G\beta}$$

Analysis is often easier in terms of these generators, rather than the full rotation matrix. Analysis in terms of the generators is known as the Lie algebra of the rotation group.

Relationship with Lie Algebras

The matrices in the Lie algebra are not themselves rotations; the skew-symmetric matrices are derivatives, proportional differences of rotations. An actual “differential rotation”, or *infinitesimal rotation matrix* has the form

$$I + A\,d\theta\,,$$

where $d\theta$ is vanishingly small and $A \in \mathrm{so}(n)$, for instance with $A = L_x$,

$$dL_x = \begin{bmatrix} 1 & 0 & 0 \\ 0 & 1 & -d\theta \\ 0 & d\theta & 1 \end{bmatrix}.$$

The computation rules are as usual except that infinitesimals of second order are routinely dropped. With these rules, these matrices do not satisfy all the same properties as ordinary finite rotation

matrices under the usual treatment of infinitesimals. It turns out that *the order in which infinitesimal rotations are applied is irrelevant.*

Exponential Map

Connecting the Lie algebra to the Lie group is the exponential map, which is defined using the standard matrix exponential series for e^A For any skew-symmetric matrix A, exp(A) is always a rotation matrix.

An important practical example is the 3 × 3 case. In rotation group SO(3), it is shown that one can identify every $A \in$ so(3) with an Euler vector $\omega = \theta$ u, where u = (x,y,z) is a unit magnitude vector.

By the properties of the identification $\mathbf{su}(2) \cong 9^3$, u is in the null space of A. Thus, u is left invariant by exp(A) and is hence a rotation axis.

Using Rodrigues' rotation formula on matrix form with $\theta = \theta/_{2} + \theta/_{2}$, together with standard double angle formulae one obtains,

$$\exp(A) = \exp(\theta(u \cdot L)) = \exp\left(\begin{bmatrix} 0 & -z\theta & y\theta \\ z\theta & 0 & -x\theta \\ -y\theta & x\theta & 0 \end{bmatrix}\right) = I + 2\cos\frac{\theta}{2}\sin\frac{\theta}{2}u \cdot L + 2\sin^2\frac{\theta}{2}(u \cdot L)^2,$$

where $c = \cos \theta/_{2}$, $s = \sin \theta/_{2}$.

This is the matrix for a rotation around axis u by the angle θ in half-angle form.

Notice that for infinitesimal angles second order terms can be ignored and remains exp(A) = I + A.

Angular Velocity

Angular velocity, time rate at which an object rotates, or revolves, about an axis, or at which the angular displacement between two bodies changes. In the figure, this displacement is represented by the angle θ between a line on one body and a line on the other.

In engineering, angles or angular displacements are commonly expressed in degrees or revolutions (of 360°), and angular velocities in revolutions per minute (rpm). In mathematics and physics, angles are usually expressed in radians and angular velocities in radians per second. These measures are related through the following conversion factors: 1 degree equals $\pi/180$ (about 0.0175) radian; 1 rpm equals $\pi/30$ (about 0.105) radian per second.

In many situations, an angular velocity—usually symbolized by the Greek letter omega (ω)—is equally well regarded as a frequency, and the choice of terms depends upon the particular aspect of a system being considered. Thus, in electrical engineering, the rotational speed of a generator might be expressed in revolutions per minute, whereas the alternating electric current produced by the generator would be described in terms of its frequency.

The magnitude, or length, of the angular velocity vector is directly proportional to the angular speed, and is measured in the same units as angular speed (radians per second, degrees per second, revolutions per second, or revolutions per minute). The direction of the angular velocity vector is perpendicular to the plane in which the rotation takes place. If the rotation appears clockwise with respect to an observer, then the angular velocity vector points away from the observer. If the rotation appears counterclockwise, then the angular velocity vector points toward the observer.

Example

Consider a car rolling forward along a highway. The angular velocity vectors for all four tires point toward the left along the lines containing the wheel axles. If the car speeds up, the vectors get longer. If the car slows down, the vectors get shorter. If the car stops, the vector lengths become zero. If the car is put into reverse, the vectors reverse their directions, and point toward the right along the lines containing the wheel axles.

Angular Velocity of a Particle

Particle in Two Dimensions

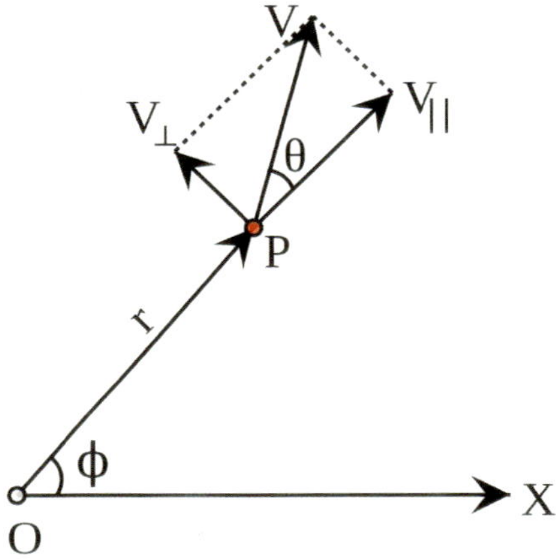

The angular velocity of the particle at *P* with respect to the origin *O* is determined by the perpendicular component of the velocity vector v.

In the simplest case of circular motion at radius *r*, with position given by the angular displacement $\phi(t)$ from the x-axis, the angular velocity is the rate of change of angle with respect to time: $\omega = \frac{d\phi}{dt}$. If ϕ is measured in radians, the distance from the x-axis around the circle to the particle is $\ell = r\phi$, and the linear velocity is $v(t) = \frac{d\ell}{dt} = r\omega(t)$, so that $\omega = \frac{v}{r}$.

In the general case of a particle moving in the plane, the angular velocity is measured relative to a chosen center point, called the origin. The diagram shows the radius vector r from the origin *O* to the particle *P*, with its polar coordinates (r,ϕ). (All variables are functions of time *t*.) The particle has linear velocity splitting as $\mathbf{v} = \mathbf{v}_{\parallel} + \mathbf{v}_{\perp}$, with the radial component $\mathbf{v}_{\parallel}$ parallel to the radius, and the cross-radial (or tangential or circular) component $\mathbf{v}_{\perp}$ perpendicular to the radius. When there is no radial component, the particle moves around the origin in a circle; but when there is no cross-radial component, it moves in a straight line from the origin. Since radial motion leaves the angle unchanged, only the cross-radial component of linear velocity contributes to angular velocity.

The angular velocity ω is the rate of change of angle with respect to time, which can be computed from the cross-radial velocity as:

$$\omega = \frac{d\phi}{dt} = \frac{v_\perp}{r}.$$

Here the cross-radial speed $v_\perp$ is the signed magnitude of $\mathbf{v}_\perp$, positive for counter-clockwise motion, negative for clockwise. Taking polar coordinates for the linear velocity v gives magnitude v (linear speed) and angle θ relative to the radius vector; in these terms, $v_\perp = v\sin(\theta)$, so that

$$\omega = \frac{v\sin(\theta)}{r}.$$

These formulas may be derived from $\mathbf{r} = (x(t), y(t))$, $\mathbf{v} = (x'(t), y'(t))$ and $\phi = \arctan(y(t)/x(t))$, together with the projection formula $v_\perp = \frac{\mathbf{r}^\perp}{r} \cdot \mathbf{v},$, where $\mathbf{r}^\perp = (-y, x)$.

In two dimensions, angular velocity is a number with plus or minus sign indicating orientation, but not pointing in a direction. The sign is conventionally taken to be positive if the radius vector turns counter-clockwise, and negative if clockwise. Angular velocity may be termed a pseudoscalar, a numerical quantity which changes sign under a parity inversion, such as inverting one axis or switching the two axes.

Particle in Three Dimensions

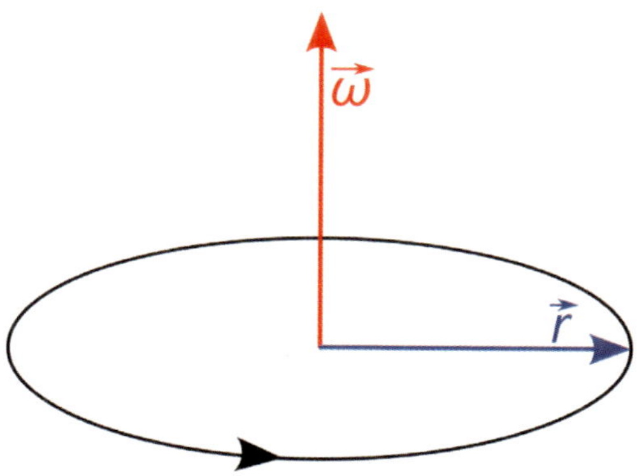

The angular velocity describes the rate of change of angular position, and the instantaneous axis of rotation. In this case (counter-clockwise rotation) the vector points up.

In three-dimensional space, we again have the position vector r of a moving particle, its radius vector from the origin. Angular velocity is a vector (or pseudovector) whose magnitude measures the rate at which the radius sweeps out angle, and whose direction shows the principal axis of rotation. Its up-or-down direction is given by the right-hand rule.

Let the axial vector $\mathbf{u}$ be the unit-length vector perpendicular to the plane of rotation spanned by the radius and velocity vectors, so that the direction of rotation is counter-clockwise looking from the top of $\mathbf{u}$. Taking polar coordinates (r, ϕ) in the plane of rotation, as in the two-dimensional case above, one may define the angular velocity vector as:

$$\omega = \omega\mathbf{u} = \tfrac{d\phi}{dt}\mathbf{u} = \frac{v\sin(\theta)}{r}\mathbf{u},$$

where θ is the angle between r and v. In terms of the cross product, this is:

$$\omega = \frac{\mathbf{r}\times\mathbf{v}}{r^2}.$$

From this, one can recover the tangential velocity as:

$$\mathbf{v}_{\perp} = \omega\times\mathbf{r}.$$

Addition of Angular Velocity Vectors

If a point rotates with angular velocity ω_1 in a coordinate frame F_1 which itself rotates with an angular velocity ω_2 with respect to an external frame F_2, we can define $\omega_1 + \omega_2$ to be the composite angular velocity vector of the point with respect to F_2. This operation coincides with usual addition of vectors, and it gives angular velocity the algebraic structure of a true vector, rather than just a pseudo-vector.

The only non-obvious property of the above addition is commutativity. This can be proven from the fact that the velocity tensor W is skew-symmetric, so that $R = e^{Wt}$ is a rotation matrix which can be expanded as $R = I + W\cdot dt + \frac{1}{2}(W\cdot dt)^2 + \ldots$. The composition of rotations is not commutative, but $(I + W_1\cdot dt)(I + W_2\cdot dt) = (I + W_2\cdot dt)(I + W_1\cdot dt)$ is commutative to first order, and therefore $\omega_1 + \omega_2 = \omega_2 + \omega_1$.

Angular Velocity Vector for a Frame

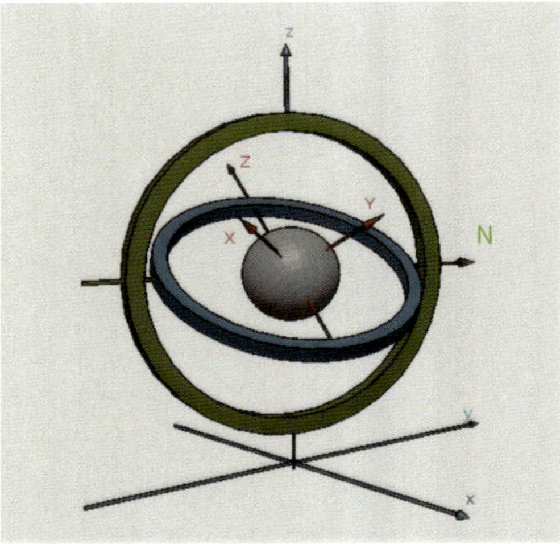

Schematic construction for addition of angular velocity vectors for rotating frames

Given a rotating frame of three unit coordinate vectors, all the three must have the same angular speed at each instant. In such a frame, each vector may be considered as a moving particle with constant scalar radius.

The rotating frame appears in the context of rigid bodies, and special tools have been developed for it: the angular velocity may be described as a vector or equivalently as a tensor.

Consistent with the general definition, the angular velocity of a frame is defined as the angular velocity of any of the three vectors (same for all). The addition of angular velocity vectors for frames is also defined by the usual vector addition (composition of linear movements), and can be useful to decompose the rotation as in a gimbal. Components of the vector can be calculated as derivatives of the parameters defining the moving frames (Euler angles or rotation matrices). As in the general case, addition is commutative: $\omega_1 + \omega_2 = \omega_2 + \omega_1$.

By Euler's rotation theorem, any rotating frame possesses an instantaneous axis of rotation, which is the direction of the angular velocity vector, and the magnitude of the angular velocity is consistent with the two dimensional case.

Components from the Vectors of the Frame

Considering a coordinate vector of the frame as a particle, $\mathbf{r} = \mathbf{e}_i$ with $|\mathbf{e}_i| = 1$, we obtain the angular velocity vector:

$$\omega = \frac{\mathbf{r} \times \mathbf{v}}{r^2} = \mathbf{e}_1 \times \dot{\mathbf{e}}_1 = \mathbf{e}_2 \times \dot{\mathbf{e}}_2 = \mathbf{e}_3 \times \dot{\mathbf{e}}_3 .$$

Here $\mathbf{e}_i$ are the columns of the matrix of the frame, and $\dot{\mathbf{e}}_i$ their time derivatives.

Components from Euler Angles

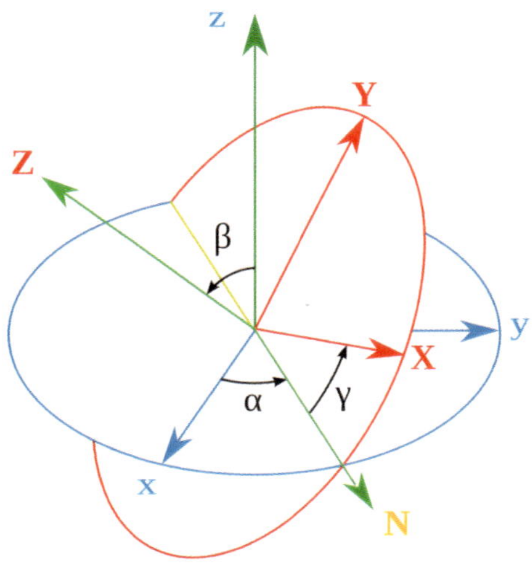

Diagram showing Euler frame in green

The components of the angular velocity pseudovector were first calculated by Leonhard Euler using his Euler angles and the use of an intermediate frame:

- One axis of the reference frame (the precession axis)
- The line of nodes of the moving frame with respect to the reference frame (nutation axis)
- One axis of the moving frame (the intrinsic rotation axis)

Euler proved that the projections of the angular velocity pseudovector on each of these three axes is the derivative of its associated angle (which is equivalent to decomposing the instantaneous rotation into three instantaneous Euler rotations). Therefore:

$$\omega = \dot{\alpha}\mathbf{u}_1 + \dot{\beta}\mathbf{u}_2 + \dot{\gamma}\mathbf{u}_3$$

This basis is not orthonormal and it is difficult to use, but now the velocity vector can be changed to the fixed frame or to the moving frame with just a change of bases. For example, changing to the mobile frame:

$$\omega = (\dot{\alpha}\sin\beta\sin\gamma + \dot{\beta}\cos\gamma)\mathbf{i} + (\dot{\alpha}\sin\beta\cos\gamma - \dot{\beta}\sin\gamma)\mathbf{j} + (\dot{\alpha}\cos\beta + \dot{\gamma})\mathbf{k}$$

where $\mathbf{i}, \mathbf{j}, \mathbf{k}$ are unit vectors for the frame fixed in the moving body. This example has been made using the Z-X-Z convention for Euler angles.

Angular Velocity Tensor

The angular velocity vector $\boldsymbol{\omega} = (\omega_x, \omega_y, \omega_z)$ defined above may be equivalently expressed as an angular velocity tensor, the matrix (or linear mapping) $W = W(t)$ defined by:

$$W = \begin{pmatrix} 0 & -\omega_z & \omega_y \\ \omega_z & 0 & -\omega_x \\ -\omega_y & \omega_x & 0 \end{pmatrix}$$

This is an infinitesimal rotation matrix. The linear mapping W acts as $(\boldsymbol{\omega}\times)$:

$$\boldsymbol{\omega} \times \mathbf{r} = W \cdot \mathbf{r}.$$

Calculation from the Orientation Matrix

A vector undergoing uniform circular motion around an axis satisfies:

$$\frac{d\mathbf{r}}{dt} = \boldsymbol{\omega} \times \mathbf{r} = W \cdot \mathbf{r}$$

Given the orientation matrix $A(t)$ of a frame, whose columns are the moving orthonormal coordinate vectors $\mathbf{e}_1, \mathbf{e}_2, \mathbf{e}_3$, we can obtain its angular velocity tensor $W(t)$ as follows. Angular velocity must be the same for the three vectors $\mathbf{r} = \mathbf{e}_i$, so arranging the three vector equations into columns of a matrix, we have:

$$\frac{dA}{dt} = W \cdot A.$$

(This holds even if $A(t)$ does not rotate uniformly.) Therefore the angular velocity tensor is:

$$W = \frac{dA}{dt} \cdot A^{-1} = \frac{dA}{dt} \cdot A^{\mathrm{T}},$$

since the inverse of the orthogonal matrix A is its transpose A^{T}.

Properties of Angular Velocity Tensors

In general, the angular velocity in an n-dimensional space is the time derivative of the angular displacement tensor, which is a second rank skew-symmetric tensor.

This tensor W will have $n(n-1)/2$ independent components, which is the dimension of the Lie algebra of the Lie group of rotations of an n-dimensional inner product space.

Duality with Respect to the Velocity Vector

In three dimensions, angular velocity can be represented by a pseudovector because second rank tensors are dual to pseudovectors in three dimensions. Since the angular velocity tensor $W = W(t)$ is a skew-symmetric matrix:

$$W = \begin{pmatrix} 0 & -\omega_z & \omega_y \\ \omega_z & 0 & -\omega_x \\ -\omega_y & \omega_x & 0 \end{pmatrix},$$

its Hodge dual is a vector, which is precisely the previous angular velocity vector $\boldsymbol{\omega} = [\omega_x, \omega_y, \omega_z]$.

Exponential of W

If we know an initial frame $A(0)$ and we are given a *constant* angular velocity tensor W, we can obtain $A(t)$ for any given t. Recall the matrix differential equation:

$$\frac{dA}{dt} = W \cdot A.$$

This equation can be integrated to give:

$$A(t) = e^{Wt} A(0),$$

which shows a connection with the Lie group of rotations.

W is Skew-symmetric

We prove that angular velocity tensor is skew symmetric, i.e. $W = \frac{dA(t)}{dt} \cdot A^{\mathrm{T}}$ satisfies $W^{\mathrm{T}} = -W$.

A rotation matrix A is orthogonal, inverse to its transpose, so we have $I = A \cdot A^{\mathrm{T}}$. For $A = A(t)$ a frame matrix, taking the time derivative of the equation gives:

$$0 = \frac{dA}{dt} A^{\mathrm{T}} + A \frac{dA^{\mathrm{T}}}{dt}$$

Applying the formula $(AB)^{\mathrm{T}} = B^{\mathrm{T}} A^{\mathrm{T}}$,

$$0 = \frac{dA}{dt}A^{T} + \left(\frac{dA}{dt}A^{T}\right)^{T} = W + W^{T}$$

Thus, W is the negative of its transpose, which implies it is skew symmetric.

Coordinate-free Description

At any instant t, the angular velocity tensor represents a linear map between the position vector $r(t)$ and the velocity vectors $v(t)$ of a point on a rigid body rotating around the origin:

$$\mathbf{v} = W\mathbf{r}.$$

The relation between this linear map and the angular velocity pseudovector ω is the following.

Because W is the derivative of an orthogonal transformation, the bilinear form

$$B(\mathbf{r},\mathbf{s}) = (W\mathbf{r})\cdot\mathbf{s}$$

is skew-symmetric. Thus we can apply the fact of exterior algebra that there is a unique linear form L on $\Lambda^2 V$ that

$$L(\mathbf{r}\wedge\mathbf{s}) = B(\mathbf{r},\mathbf{s})$$

where $\mathbf{r}\wedge\mathbf{s}\in\Lambda^2 V$ is the exterior product of $\mathbf{r}$ and $\mathbf{s}$.

Taking the sharp $L^{\sharp}$ of L we get

$$(W\mathbf{r})\cdot\mathbf{s} = L^{\sharp}\cdot(\mathbf{r}\wedge\mathbf{s})$$

Introducing $\omega := \star(L^{\sharp})$, as the Hodge dual of $L^{\sharp}$, and applying the definition of the Hodge dual twice supposing that the preferred unit 3-vector is $\star 1$

$$(W\mathbf{r})\cdot\mathbf{s} = \star(\star(L^{\sharp})\wedge\mathbf{r}\wedge\mathbf{s}) = \star(\omega\wedge\mathbf{r}\wedge\mathbf{s}) = \star(\omega\wedge\mathbf{r})\cdot\mathbf{s} = (\omega\times\mathbf{r})\cdot\mathbf{s},$$

where

$$\omega\times\mathbf{r} := \star(\omega\wedge\mathbf{r})$$

by definition.

Because $\mathbf{s}$ is an arbitrary vector, from nondegeneracy of scalar product follows

$$W\mathbf{r} = \omega\times\mathbf{r}$$

Angular Velocity as a Vector Field

For angular velocity tensor maps positions to velocities, it is a vector field. In particular, this vector field is a Killing vector field belonging to an element of the Lie algebra so(3) of the 3-dimensional rotation group SO(3). This element of so(3) can also be regarded as the angular velocity vector.

Rigid Body Considerations

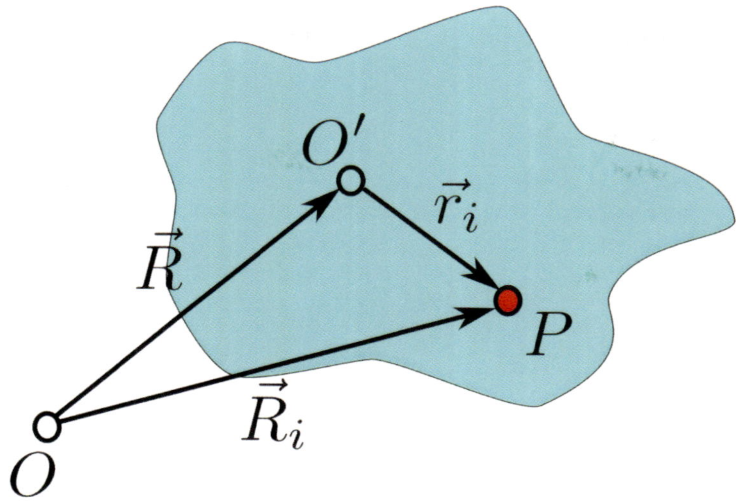

Position of point P located in the rigid body (shown in blue). R_i is the position with respect to the lab frame, centered at O and r_i is the position with respect to the rigid body frame, centered at O'. The origin of the rigid body frame is at vector position R from the lab frame.

The same equations for the angular speed can be obtained reasoning over a rotating rigid body. Here is not assumed that the rigid body rotates around the origin. Instead, it can be supposed rotating around an arbitrary point that is moving with a linear velocity $V(t)$ in each instant.

To obtain the equations, it is convenient to imagine a rigid body attached to the frames and consider a coordinate system that is fixed with respect to the rigid body. Then we will study the coordinate transformations between this coordinate and the fixed "laboratory" system.

As shown in the figure on the right, the lab system's origin is at point O, the rigid body system origin is at O' and the vector from O to O' is R. A particle (i) in the rigid body is located at point P and the vector position of this particle is R_i in the lab frame, and at position r_i in the body frame. It is seen that the position of the particle can be written:

$$\mathbf{R}_i = \mathbf{R} + \mathbf{r}_i$$

The defining characteristic of a rigid body is that the distance between any two points in a rigid body is unchanging in time. This means that the length of the vector $\mathbf{r}_i$ is unchanging. By Euler's rotation theorem, we may replace the vector $\mathbf{r}_i$ with $\mathcal{R}\mathbf{r}_{io}$ where $\mathcal{R}$ is a 3×3 rotation matrix and $\mathbf{r}_{io}$ is the position of the particle at some fixed point in time, say $t = 0$. This replacement is useful, because now it is only the rotation matrix $\mathcal{R}$ that is changing in time and not the reference vector $\mathbf{r}_{io}$, as the rigid body rotates about point O'. Also, since the three columns of the rotation matrix represent the three versors of a reference frame rotating together with the rigid body, any rotation about any axis becomes now visible, while the vector $\mathbf{r}_i$ would not rotate if the rotation axis were parallel to it, and hence it would only describe a rotation about an axis perpendicular to it (i.e., it would not see the component of the angular velocity pseudovector parallel to it, and would only allow the computation of the component perpendicular to it). The position of the particle is now written as:

$$\mathbf{R}_i = \mathbf{R} + \mathcal{R}\mathbf{r}_{io}$$

Taking the time derivative yields the velocity of the particle:

$$\mathbf{V}_i = \mathbf{V} + \frac{d\mathcal{R}}{dt}\mathbf{r}_{io}$$

where V_i is the velocity of the particle (in the lab frame) and V is the velocity of O' (the origin of the rigid body frame). Since $\mathcal{R}$ is a rotation matrix its inverse is its transpose. So we substitute $\mathcal{I} = \mathcal{R}^{\mathrm{T}}\mathcal{R}$:

$$\mathbf{V}_i = \mathbf{V} + \frac{d\mathcal{R}}{dt}\mathcal{I}\mathbf{r}_{io}$$

$$\mathbf{V}_i = \mathbf{V} + \frac{d\mathcal{R}}{dt}\mathcal{R}^{\mathrm{T}}\mathcal{R}\mathbf{r}_{io}$$

$$\mathbf{V}_i = \mathbf{V} + \frac{d\mathcal{R}}{dt}\mathcal{R}^{\mathrm{T}}\mathbf{r}_i$$

or

$$\mathbf{V}_i = \mathbf{V} + W\mathbf{r}_i$$

where $W = \frac{d\mathcal{R}}{dt}\mathcal{R}^{\mathrm{T}}$ is the previous angular velocity tensor.

It can be proved that this is a skew symmetric matrix, so we can take its dual to get a 3 dimensional pseudovector that is precisely the previous angular velocity vector $\vec{\omega}$:

$$\boldsymbol{\omega} = [\omega_x, \omega_y, \omega_z]$$

Substituting ω for W into the above velocity expression, and replacing matrix multiplication by an equivalent cross product:

$$\mathbf{V}_i = \mathbf{V} + \boldsymbol{\omega} \times \mathbf{r}_i$$

It can be seen that the velocity of a point in a rigid body can be divided into two terms – the velocity of a reference point fixed in the rigid body plus the cross product term involving the angular velocity of the particle with respect to the reference point. This angular velocity is the “spin” angular velocity of the rigid body as opposed to the angular velocity of the reference point O' about the origin O.

Consistency

We have supposed that the rigid body rotates around an arbitrary point. We should prove that the angular velocity previously defined is independent from the choice of origin, which means that the angular velocity is an intrinsic property of the spinning rigid body.

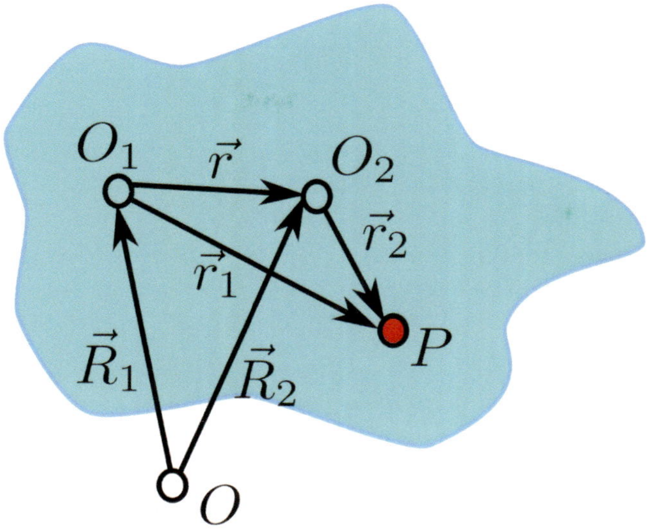

Proving the independence of angular velocity from choice of origin

The origin of lab frame is O, while O_1 and O_2 are two fixed points on the rigid body, whose velocity is $\mathbf{v}_1$ and $\mathbf{v}_2$ respectively. Suppose the angular velocity with respect to O_1 and O_2 is $\boldsymbol{\omega}_1$ and $\boldsymbol{\omega}_2$ respectively. Since point P and O_2 have only one velocity,

$$\mathbf{v}_1 + \boldsymbol{\omega}_1 \times \mathbf{r}_1 = \mathbf{v}_2 + \boldsymbol{\omega}_2 \times \mathbf{r}_2$$
$$\mathbf{v}_2 = \mathbf{v}_1 + \boldsymbol{\omega}_1 \times \mathbf{r} = \mathbf{v}_1 + \boldsymbol{\omega}_1 \times (\mathbf{r}_1 - \mathbf{r}_2)$$

The above two yields that

$$(\boldsymbol{\omega}_1 - \boldsymbol{\omega}_2) \times \mathbf{r}_2 = 0$$

Since the point P (and thus $\mathbf{r}_2$) is arbitrary, it follows that

$$\boldsymbol{\omega}_1 = \boldsymbol{\omega}_2$$

If the reference point is the instantaneous axis of rotation the expression of velocity of a point in the rigid body will have just the angular velocity term. This is because the velocity of instantaneous axis of rotation is zero. An example of instantaneous axis of rotation is the hinge of a door. Another example is the point of contact of a purely rolling spherical (or, more generally, convex) rigid body.

Angular Acceleration

Angular acceleration, also called rotational acceleration, is a quantitative expression of the change in angular velocity that a spinning object undergoes per unit time. It is a vector quantity, consisting of a magnitude component and either of two defined directions or senses.

The magnitude, or length, of the angular acceleration vector is directly proportional to the rate of change of angular velocity, and is measured in radian s per second squared (rad/s ^{2} or rad $\cdot$ s $^{-2}$). Alternatively, the angular acceleration magnitude can be expressed in degrees per second squared (deg/s ^{2} or deg $\cdot$ s $^{-2}$). The direction of the angular acceleration vector is perpendicular to the plane in which the rotation takes place. If the increase in angular velocity appears clockwise with respect to an observer, then the angular acceleration vector points away from the observer. If the increase in angular velocity appears counterclockwise, then the angular acceleration vector points toward the observer.

The angular acceleration vector does not necessarily point in the same direction as the angular velocity vector. Consider a car rolling forward along a highway at increasing speed. The angular acceleration vectors for all four tires point toward the left along the lines containing the wheel axles. If the car stops accelerating and maintains a constant velocity, the angular acceleration vectors disappear. If the car slows down going forward, the vectors reverse their directions, and point toward the right along the lines containing the wheel axles. If the car is put into reverse and increases velocity going backwards, the angular acceleration vectors point toward the right along the lines containing the axles. If the backward velocity is constant, the angular acceleration vectors vanish; if the backward velocity decreases, the angular acceleration vectors point toward the left along the lines containing the wheel axles.

The angular acceleration of a rotating object is the rate at which the angular velocity changes with respect to time. It is the change in the angular velocity, divided by the change in time. The average angular acceleration is the change in the angular velocity, divided by the change in time. The angular acceleration is a vector that points in a direction along the rotation axis. The magnitude of the angular acceleration is given by the formula below. The unit of angular acceleration is radians/s^2.

$$\textit{angular acceleration} = \frac{\textit{change in angular velocity}}{\textit{change in time}}$$
$$= \frac{(\textit{final angular velocity}) - (\textit{initial angular velocity})}{(\textit{final time}) - (\textit{initial time})}$$

$$\alpha = \frac{\Delta\omega}{\Delta t} = \frac{\omega_2 - \omega_1}{t_2 - t_1}$$

α = angular acceleration, (radians/s^2)

$\Delta\omega$ = change in angular velocity (radians/s)

Δt = change in time (s)

ω_1 = initial angular velocity (radians/s)

ω_2= final angular velocity (radians/s)

t_1 = initial time (s)

t_2= final time (s)

Angular Acceleration Formula Problems

Problem 3

A disc in a DVD player starts from rest, and then begins spinning when the user presses “Play”. After 4.00 s, the disc is spinning at 160 radians/s. What was the average angular acceleration of the disc?

Solution

At the initial time (t_1 = 0 s), the angular velocity was ω_1 = 0 radians/s. At the final time (t_2 = 4.00 s), the angular velocity of the disc was ω_1 = 160 radians/s. The average angular acceleration can be found using the formula:

$$\alpha = \frac{\omega_2 - \omega_1}{t_2 - t_1}$$

$$\alpha = \frac{160 \text{ radians} / s - 0 \text{ radians} / s}{4.00s - 0\ s}$$

$$\alpha = \frac{160 \text{ radians} / s}{4.00s}$$

$$\alpha = 40.0 \text{ radians} / s^2$$

Between the initial and final times, the average angular acceleration of the disc was 40.0 radians/s^2.

Problem 4

At certain high-speed roller coaster has a magnetic braking system that is meant so slow the roller coaster cars as they approach the station. The magnetic brakes slow the cars just enough for another braking system to take over. If the wheels of the cars are initially spinning at 400.0 revolutions per second, and the magnetic brakes apply a constant angular acceleration of -440.0 radians/s^2 for 5.00 s, what will be the angular velocity of the roller coaster cars after the magnetic brakes have been applied?

Solution

The angular velocity at the initial time (t_1 = 0 s) is given in terms of revolutions per second. To use this value in the angular acceleration formula, the value must be converted to radians per second. There are 2π radians per revolution, and so the initial angular velocity is:

ω_1 = 400.0 revolutions/s

$$\omega_1 = (400.0 \text{ radians} / s)\left(\frac{2\pi \text{ radians}}{\text{revolution}}\right)$$

$$\omega_1 \cong 2513 \text{ radians/s}$$

The final angular velocity at time t_1 = 5.0 s can be found by rearranging the angular acceleration formula:

$$\alpha = \frac{\omega_2 - \omega_1}{t_2 - t_1} \rightarrow \omega_2 - \omega_1 = \alpha(t_2 - t_1)$$

$$\omega_2 - \omega_1 = \alpha(t_2 - t_1) \rightarrow \omega_2 = \omega_1 + \alpha(t_2 - t_1)$$

$$\omega_2 = \omega_1 + \alpha(t_2 - t_1)$$

$$\omega_2 \cong (2513 \text{ radians/s}) + (-440.0 \text{ radians}/s^2)(5.00s - 0\,s)$$

$$\omega_2 \cong 2513 \text{ radians}/s - 2200 \text{ radians}/s$$

$$\omega_2 \cong 313 \text{ radians}/s$$

The angular velocity after the magnetic brakes have been applied is approximately 313 radians/s. This is equivalent to 49.8 revolutions/s.

Rotating Reference Frame

Suppose that a given object has position vector $\mathbf{r}$ in some *non-rotating inertial* reference frame. Let us observe the motion of this object in a *non-inertial* reference frame which *rotates* with *constant* angular velocity $\mathbf{\Omega}$ about an axis passing through the origin of the inertial frame. Suppose, first of all, that our object appears *stationary* in the rotating reference frame. Hence, in the non-rotating frame, the object's position vector $\mathbf{r}$ will appear to *precess* about the origin with angular velocity $\mathbf{\Omega}$. It follows, from equation, that in the non-rotating reference frame,

$$\frac{\mathrm{dr}}{\mathrm{dt}} = \mathbf{\Omega} \times \mathbf{r}.$$

Suppose, now, that our object appears to move in the rotating reference frame with instantaneous velocity v'. It is fairly obvious that the appropriate generalization of the above equation is simply,

$$\frac{\mathrm{dr}}{\mathrm{dt}} = \mathrm{v}' + \mathbf{\Omega} \times \mathbf{r}.$$

Let and $\mathrm{d/dt}$ and $\mathrm{d/dt'}$ denote apparent time derivatives in the non-rotating and rotating frames of reference, respectively. Since an object which is stationary in the rotating reference frame appears to move in the non-rotating frame, it is clear that $\mathrm{d/dt} \neq \mathrm{d/dt'}$. Writing the apparent velocity, v', of our object in the rotating reference frame as $\mathrm{dr/dt'}$, the above equation takes the form

$$\frac{\mathrm{dr}}{\mathrm{dt}} = \frac{\mathrm{dr}}{\mathrm{dt'}} + \mathbf{\Omega} \times \mathbf{r},$$

or

$$\frac{\mathrm{d}}{\mathrm{dt}} = \frac{\mathrm{dr}}{\mathrm{dt'}} + \Omega \times,$$

since $\mathbf{r}$ is a general position vector. Equation expresses the relationship between apparent time derivatives in the non-rotating and rotating reference frames.

Operating on the general position vector $\mathbf{r}$ with the time derivative, we get

$$\mathbf{v} = \mathbf{v'} + \mathbf{\Omega} \times \mathbf{r}.$$

This equation relates the apparent velocity, $\mathbf{v} = dr/dt$, of an object with position vector $\mathbf{r}$ in the non-rotating reference frame to its apparent velocity, $\mathbf{v'} = dr/dt'$, in the rotating reference frame.

Operating twice on the position vector $\mathbf{r}$ with the time derivative, we obtain:

$$\mathbf{a} = \left(\frac{d}{dt'} + \mathbf{\Omega} \times \right) \left(\mathbf{v'} + \mathbf{\Omega} \times \mathbf{r} \right)$$

or

$$\mathbf{a} = \mathbf{a}' + \Omega \times (\Omega \times \mathbf{r}) + 2\Omega \times \mathbf{v'}.$$

This equation relates the apparent acceleration, $\mathbf{a} = d^2\mathbf{r} / dt^2$, of an object with position vector $\mathbf{r}$ in the non-rotating reference frame to its apparent acceleration, $\mathbf{a} = d^2\mathbf{r} / dt^2$, in the rotating reference frame.

Applying Newton's second law of motion in the inertial (*i.e.*, non-rotating) reference frame, we obtain

$$m\mathbf{a} = \mathbf{f}.$$

Here, m is the mass of our object, and $\mathbf{f}$ is the (non-fictitious) force acting on it. Note that these quantities are the *same* in both reference frames. Making use of equation, the apparent equation of motion of our object in the rotating reference frame takes the form:

$$m\mathbf{a}' = \mathbf{f} - m\Omega \times (\Omega \times \mathbf{r}) - 2m\Omega \times \mathbf{v'}.$$

The last two terms in the above equation are so-called ``fictitious forces". Such forces are always needed to account for motion observed in non-inertial reference frames. Note that fictitious forces can always be distinguished from non-fictitious forces in Newtonian dynamics because the former have no associated reactions. Let us now investigate the two fictitious forces appearing in equation.

Fictitious Forces

All non-inertial reference frames exhibit fictitious forces; rotating reference frames are characterized by three:

- the centrifugal force,
- the Coriolis force,

and, for non-uniformly rotating reference frames,

- the Euler force.

Scientists living in a rotating box can measure the speed and direction of their rotation by measuring these fictitious forces. For example, Léon Foucault was able to show the Coriolis force that results from Earth's rotation using the Foucault pendulum. If Earth were to rotate many times faster, these fictitious forces could be felt by humans, as they are when on a spinning carousel.

Relating Rotating Frames to Stationary Frames

The following is a derivation of the formulas for accelerations as well as fictitious forces in a rotating frame. It begins with the relation between a particle's coordinates in a rotating frame and its coordinates in an inertial (stationary) frame. Then, by taking time derivatives, formulas are derived that relate the velocity of the particle as seen in the two frames, and the acceleration relative to each frame. Using these accelerations, the fictitious forces are identified by comparing Newton's second law as formulated in the two different frames.

Relation between Positions in the Two Frames

To derive these fictitious forces, it's helpful to be able to convert between the coordinates (x',y',z') of the rotating reference frame and the coordinates (x,y,z) of an inertial reference frame with the same origin. If the rotation is about the z axis with a constant angular velocity Ω, or $\theta(t)=\Omega t$, and the two reference frames coincide at time $t=0$, the transformation from rotating coordinates to inertial coordinates can be written

$$x = x'\cos(\theta(t)) - y'\sin(\theta(t))$$
$$y = x'\sin(\theta(t)) + y'\cos(\theta(t))$$

whereas the reverse transformation is

$$x' = x\cos(-\theta(t)) - y\sin(-\theta(t))$$
$$y' = x\sin(-\theta(t)) + y\cos(-\theta(t))$$

This result can be obtained from a rotation matrix.

Introduce the unit vectors $\hat{\imath}, \hat{\jmath}, \hat{k}$ representing standard unit basis vectors in the rotating frame. The time-derivatives of these unit vectors are found next. Suppose the frames are aligned at $t = 0$ and the z-axis is the axis of rotation. Then for a counterclockwise rotation through angle Ωt:

$$\hat{\imath}(t) = (\cos\theta(t), \sin\theta(t))$$

where the (x, y) components are expressed in the stationary frame. Likewise,

$$\hat{\jmath}(t) = (-\sin\theta(t), \cos\theta(t)).$$

Thus the time derivative of these vectors, which rotate without changing magnitude, is

$$\frac{d}{dt}\hat{\imath}(t) = \Omega(-\sin\theta(t), \cos\theta(t)) = \Omega\hat{\jmath};$$
$$\frac{d}{dt}\hat{\jmath}(t) = \Omega(-\cos\theta(t), -\sin\theta(t)) = -\Omega\hat{\imath},$$

where $\Omega \equiv \frac{\mathrm{d}}{\mathrm{d}t}\theta(t)$. This result is the same as found using a vector cross product with the rotation vector Ω pointed along the z-axis of rotation $\mathbf{\Omega} = (0, 0, \Omega)$, namely,

$$\frac{\mathrm{d}}{\mathrm{d}t}\hat{\mathbf{u}} = \mathbf{\Omega} \times \hat{\mathbf{u}},$$

where $\hat{\mathbf{u}}$ is either $\hat{\imath}$ or $\hat{\jmath}$.

Time Derivatives in the two Frames

Introduce the unit vectors $\hat{\imath}, \hat{\jmath}, \hat{\mathbf{k}}$ representing standard unit basis vectors in the rotating frame. As they rotate they will remain normalized. If we let them rotate at the speed of Ω about an axis $\mathbf{\Omega}$ then each unit vector $\hat{\mathbf{u}}$ of the rotating coordinate system abides by the following equation:

$$\frac{\mathrm{d}}{\mathrm{d}t}\hat{\mathbf{u}} = \mathbf{\Omega} \times \hat{\mathbf{u}}.$$

Then if we have a vector function $\boldsymbol{f}$,

$$\mathbf{f}(t) = f_x(t)\hat{\imath} + f_y(t)\hat{\jmath} + f_z(t)\hat{\mathbf{k}},$$

and we want to examine its first derivative we have (using the product rule of differentiation):

$$\begin{aligned}\frac{\mathrm{d}}{\mathrm{d}t}\mathbf{f} &= \frac{\mathrm{d}f_x}{\mathrm{d}t}\hat{\imath} + \frac{\mathrm{d}\hat{y}}{\mathrm{d}t}f_x + \frac{\mathrm{d}f_y}{\mathrm{d}t}\hat{\jmath} + \frac{\mathrm{d}\hat{\jmath}}{\mathrm{d}t}f_y + \frac{\mathrm{d}f_z}{\mathrm{d}t}\hat{\mathbf{k}} + \frac{\mathrm{d}\hat{\mathbf{k}}}{\mathrm{d}t}f_z \\ &= \frac{\mathrm{d}f_x}{\mathrm{d}t}\hat{\imath} + \frac{\mathrm{d}f_y}{\mathrm{d}t}\hat{\jmath} + \frac{\mathrm{d}f_z}{\mathrm{d}t}\hat{\mathbf{k}} + [\mathbf{\Omega} \times (f_x\hat{\imath} + f_y\hat{\jmath} + f_z\hat{\mathbf{k}})] \\ &= \left(\frac{\mathrm{d}\mathbf{f}}{\mathrm{d}t}\right)_r + \mathbf{\Omega} \times f(t),\end{aligned}$$

where $\left(\frac{\mathrm{d}\mathbf{f}}{\mathrm{d}t}\right)_r$ is the rate of change of f as observed in the rotating coordinate system. As a short hand the differentiation is expressed as:

$$\frac{\mathrm{d}}{\mathrm{d}t}f = \left[\left(\frac{\mathrm{d}}{\mathrm{d}t}\right) + \quad \times\right]f.$$

This result is also known as the Transport Theorem in analytical dynamics and is also sometimes referred to as the Basic Kinematic Equation.

Relation between Velocities in the Two Frames

A velocity of an object is the time-derivative of the object's position, or

$$\mathbf{v} \stackrel{\text{def}}{=} \frac{\mathrm{d}\mathbf{r}}{\mathrm{d}t}.$$

The time derivative of a position $\mathbf{r}(t)$ in a rotating reference frame has two components, one from the explicit time dependence due to motion of the particle itself, and another from the frame's own rotation. Applying the result of the previous subsection to the displacement $\mathbf{r}(t)$, the velocities in the two reference frames are related by the equation:

$$\mathbf{v}_{\mathrm{i}} \stackrel{\mathrm{def}}{=} \frac{\mathrm{d}\mathbf{r}}{\mathrm{d}t} = \left(\frac{\mathrm{d}\mathbf{r}}{\mathrm{d}t}\right)_{\mathrm{r}} + \mathbf{\Omega}\times\mathbf{r} = \mathbf{v}_{\mathrm{r}} + \mathbf{\Omega}\times\mathbf{r}\,,$$

where subscript i means the inertial frame of reference, and r means the rotating frame of reference.

Relation between Accelerations in the two Frames

Acceleration is the second time derivative of position, or the first time derivative of velocity

$$\mathbf{a}_{\mathrm{i}} \stackrel{\mathrm{def}}{=} \left(\frac{\mathrm{d}^2\mathbf{r}}{\mathrm{d}t^2}\right)_{\mathrm{i}} = \left(\frac{\mathrm{d}\mathbf{v}}{\mathrm{d}t}\right)_{\mathrm{i}} = \left[\left(\frac{\mathrm{d}}{\mathrm{d}t}\right)_{\mathrm{r}} + \mathbf{\Omega}\times\right]\left[\left(\frac{\mathrm{d}\mathbf{r}}{\mathrm{d}t}\right)_{\mathrm{r}} + \mathbf{\Omega}\times\mathbf{r}\right],$$

where subscript i means the inertial frame of reference. Carrying out the differentiations and re-arranging some terms yields the acceleration in the *rotating* reference frame

$$\mathbf{a}_{\mathrm{r}} = \mathbf{a}_{\mathrm{i}} - 2\mathbf{\Omega}\times\mathbf{v}_{\mathrm{r}} - \mathbf{\Omega}\times(\mathbf{\Omega}\times\mathbf{r}) - \frac{\mathrm{d}\mathbf{\Omega}}{\mathrm{d}t}\times\mathbf{r}$$

where $\mathbf{a}_{\mathrm{r}} \stackrel{\mathrm{def}}{=} \left(\frac{\mathrm{d}^2\mathbf{r}}{\mathrm{d}t^2}\right)_{\mathrm{r}}$ is the apparent acceleration in the rotating reference frame, the term represents centrifugal acceleration, and the term $-\mathbf{\Omega}\times(\mathbf{\Omega}\times\mathbf{r})$ represents centrifugal acceleration, and the term $-2\Omega \times v_r$ is the Coriolis acceleration. The last term ($-\frac{\mathrm{d}\mathbf{\Omega}}{\mathrm{d}t}\times\mathbf{r}$) is the Euler acceleration and is zero in uniformly rotating frames.

Newton's Second Law in the Two Frames

When the expression for acceleration is multiplied by the mass of the particle, the three extra terms on the right-hand side result in fictitious forces in the rotating reference frame, that is, apparent forces that result from being in a non-inertial reference frame, rather than from any physical interaction between bodies.

Using Newton's second law of motion $\mathbf{F} = m\mathbf{a}$, we obtain:

- the Coriolis force

  $$\mathbf{F}_{\mathrm{Coriolis}} = -2m\mathbf{\Omega}\times\mathbf{v}_{\mathrm{r}}$$

- the centrifugal force

  $$\mathbf{F}_{\mathrm{centrifugal}} = -m\mathbf{\Omega}\times(\mathbf{\Omega}\times\mathbf{r})$$

- and the Euler force

$$\mathbf{F}_{\text{Euler}} = -m\frac{d\Omega}{dt}\times\mathbf{r}$$

where m is the mass of the object being acted upon by these fictitious forces. Notice that all three forces vanish when the frame is not rotating, that is, when $\Omega = 0$.

For completeness, the inertial acceleration $\mathbf{a}_i$ due to impressed external forces $\mathbf{F}_{\text{imp}}$ can be determined from the total physical force in the inertial (non-rotating) frame (for example, force from physical interactions such as electromagnetic forces) using Newton's second law in the inertial frame:

$$\mathbf{F}_{\text{imp}} = m\mathbf{a}_i$$

Newton's law in the rotating frame then becomes

$$\mathbf{F}_{\mathbf{r}} = \mathbf{F}_{\text{imp}} + \mathbf{F}_{\text{centrifugal}} + \mathbf{F}_{\text{Coriolis}} + \mathbf{F}_{\text{Euler}} = m\mathbf{a}_{\mathbf{r}}\,.$$

In other words, to handle the laws of motion in a rotating reference frame:

Treat the fictitious forces like real forces, and pretend you are in an inertial frame.

—Louis N. Hand

Obviously, a rotating frame of reference is a case of a non-inertial frame. Thus the particle in addition to the real force is acted upon by a fictitious force...The particle will move according to Newton's second law of motion if the total force acting on it is taken as the sum of the real and fictitious forces.

—HS Hans & SP Pui

This equation has exactly the form of Newton's second law, *except* that in addition to F, the sum of all forces identified in the inertial frame, there is an extra term on the right...This means we can continue to use Newton's second law in the noninertial frame *provided* we agree that in the noninertial frame we must add an extra force-like term, often called the inertial force.

—John R. Taylor

Centrifugal Force

In classical mechanics, centrifugal force is an outward force associated with rotation. Centrifugal force is one of several so-called pseudo-forces (also known as inertial forces), so named because, unlike real forces, they do not originate in interactions with other bodies situated in the environment of the particle upon which they act. Instead, centrifugal force originates in the rotation of the frame of reference within which observations are made.

Coriolis Effect

The mathematical expression for the Coriolis force appeared in an 1835 paper by a French scientist Gaspard-Gustave Coriolis in connection with hydrodynamics, and also in the tidal equations of

Pierre-Simon Laplace in 1778. Early in the 20th century, the term Coriolis force began to be used in connection with meteorology.

Perhaps the most commonly encountered rotating reference frame is the Earth. Moving objects on the surface of the Earth experience a Coriolis force, and appear to veer to the right in the northern hemisphere, and to the left in the southern. Movements of air in the atmosphere and water in the ocean are notable examples of this behavior: rather than flowing directly from areas of high pressure to low pressure, as they would on a non-rotating planet, winds and currents tend to flow to the right of this direction north of the equator, and to the left of this direction south of the equator. This effect is responsible for the rotation of large cyclones.

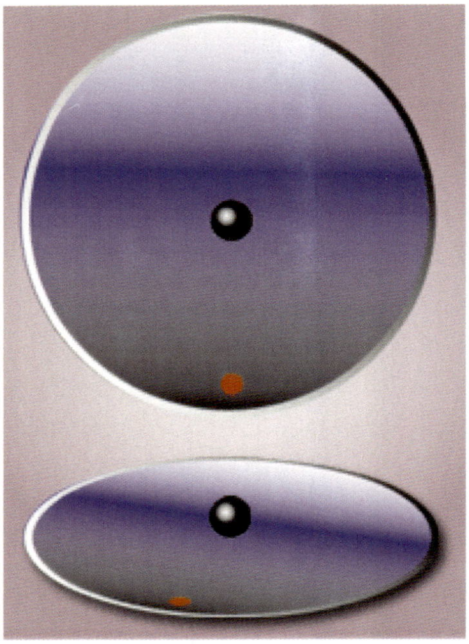

Figure: In the inertial frame of reference (upper part of the picture), the black object moves in a straight line. However, the observer (red dot) who is standing in the rotating frame of reference (lower part of the picture) sees the object as following a curved path.

Euler Force

In classical mechanics, the Euler acceleration (named for Leonhard Euler), also known as azimuthal acceleration or transverse acceleration is an acceleration that appears when a non-uniformly rotating reference frame is used for analysis of motion and there is variation in the angular velocity of the reference frame's axis. This topic is restricted to a frame of reference that rotates about a fixed axis.

The Euler force is a fictitious force on a body that is related to the Euler acceleration by $F = ma$, where a is the Euler acceleration and m is the mass of the body.

Use in Magnetic Resonance

It is convenient to consider magnetic resonance in a frame that rotates at the Larmor frequency of the spins. The rotating wave approximation may also be used.

Centrifugal Force

Centrifugal force, a fictitious force, peculiar to a particle moving on a circular path, that has the same magnitude and dimensions as the force that keeps the particle on its circular path (the centripetal force) but points in the opposite direction.

A stone whirling in a horizontal plane on the end of a string tied to a post on the ground is continuously changing the direction of its velocity and, therefore, has an acceleration toward the post. This acceleration is equal to the square of its velocity divided by the length of the string. According to Newton's second law, an acceleration is caused by a force, which in this case is the tension in the string. If the stone is moving at a constant speed and gravity is neglected, the inward-pointing string tension is the only force acting on the stone. If the string breaks, the stone, because of inertia, will keep on going in a straight line tangent to its previous circular path; it does not move in the outward direction as it would if the centrifugal force were real.

Although it is not a real force according to Newton's laws, the centrifugal-force concept is a useful one. For example, when analyzing the behaviour of the fluid in a cream separator or a centrifuge, it is convenient to study the fluid's behaviour relative to the rotating container rather than relative to the Earth; and, in order that Newton's laws be applicable in such a rotating frame of reference, an inertial force, or a fictitious force (the centrifugal force), equal and opposite to the centripetal force, must be included in the equations of motion. In a frame of reference attached to the whirling stone, the stone is at rest; to obtain a balanced force system, the outward-acting centrifugal force must be included.

Centrifugal force can be increased by increasing either the speed of rotation or the mass of the body or by decreasing the radius, which is the distance of the body from the centre of the curve. Increasing the mass or decreasing the radius increases the centrifugal force in direct or inverse proportion, respectively, but increasing the speed of rotation increases it in proportion to the square of the speed; that is, an increase in speed of 10 times, say from 10 to 100 revolutions per minute, increases the centrifugal force by a factor of 100. Centrifugal force is expressed as a multiple of *g*, the symbol for normal gravitational force (strictly speaking, the acceleration due to gravity). Centrifugal fields of more than 1,000,000,000*g* have been produced in the laboratory by devices called centrifuges.

It is not a real force (according to my definition of real above). It does, however, have a useful role in some physics. Let me start with an example. Suppose a boy (or girl, it really could be a girl too) was sitting on a stationary merry-go-round with a ball. He then rolls the ball across to a girl on the other side. No problem, right? Speaking of merry-go-round, they just don't put these things in parks anymore do they? Anyway, here is a diagram.

In this case, with the merry-go-round not rotating, the ball goes straight across to the girl. No problem. Now suppose the merry-go-round is spinning. If the boy tries to roll the ball straight across to the girl, the ball will indeed still go in a straight line (assume small frictional forces) as seen from someone not on the merry-go-round. The ball will not make it to the girl because she will rotate out of the way.

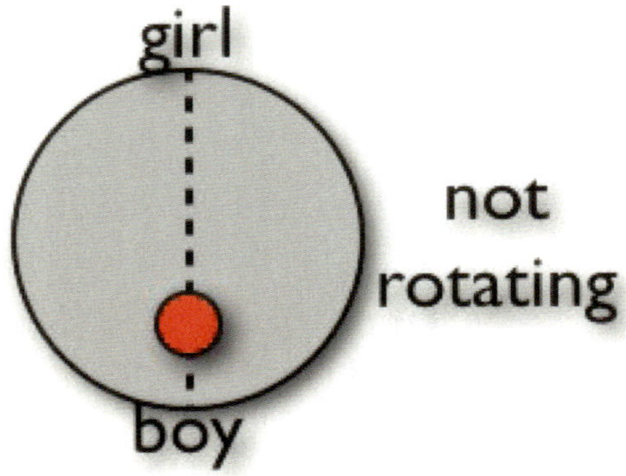

The boy and girl sitting on the spinning merry-go-round will see the ball not moving in a straight line (in their reference frame of the merry-go-round). Here are two diagrams that might help.

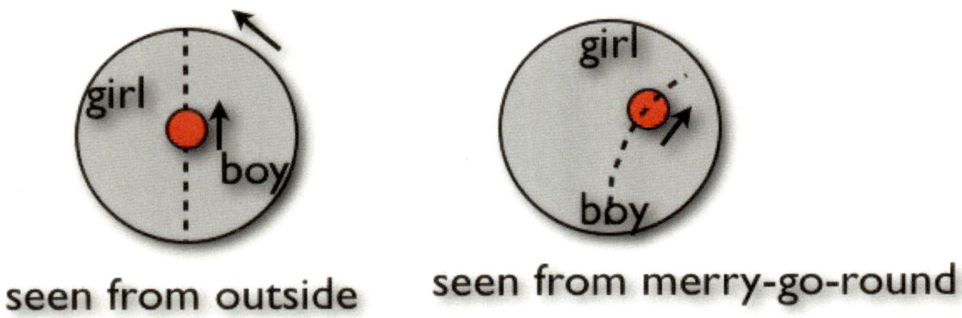

The problem is with the boy and the girl on the merry-go-round. They see the ball NOT going in a straight line. Since they understand force and motion, they are saying "hey - if it is not moving in a straight line, there must be a force on it". They make a good point. However, they are not in an inertial frame of reference. Their frame is actually accelerating (because it is moving in a circle). In order for things to work out in a non-intertial frame, a force needs to be added (a non-real force). This force is the centrifugal force.

The centrifugal force is the force (non-real force) that is needed to make things work as you would think in a reference frame that is accelerating. The best example is when you are in a car that is turning. If the car is turning left, it feels like there is a force pushing you to the right (in your frame of reference inside the car). This would correctly be called the centrifugal force. The only real problem is that if you call the centrifugal force a real (as in fundamental interaction) force.

Vehicle Driving Round a Curve

A common experience that gives rise to the idea of a centrifugal force is encountered by passengers riding in a vehicle, such as a car, that is changing direction. If a car is traveling at a constant speed along a straight road, then a passenger inside is not accelerating and, according to Newton's second law of motion, the net force acting on him is therefore zero (all forces acting on him cancel each other out). If the car enters a curve that bends to the left, the passenger experiences an apparent force that seems to be pulling him towards the right. This is the fictitious centrifugal force. It is needed within the passenger's local frame of reference to explain his sudden tendency to start accelerating to the right relative to the car—a tendency which he must resist by applying a rightward force to the car (for instance, a frictional force against the seat) in order to remain in a fixed position inside. Since he pushes the seat toward the right, Newton's third law says that the seat pushes him toward the left. The centrifugal force must be included in the passenger's reference frame (in which the passenger remains at rest): it counteracts the leftward force applied to the passenger by the seat, and explains why this otherwise unbalanced force does not cause him to accelerate. However, it would be apparent to a stationary observer watching from an overpass above that the frictional force exerted on the passenger by the seat is not being balanced; it constitutes a net force to the left, causing the passenger to accelerate toward the inside of the curve, as he must in order to keep moving with the car rather than proceeding in a straight line as he otherwise would. Thus the "centrifugal force" he feels is the result of a "centrifugal tendency" caused by inertia. Similar effects are encountered in aeroplanes and roller coasters where the magnitude of the apparent force is often reported in "G's".

Stone on a String

If a stone is whirled round on a string, in a horizontal plane, the only real force acting on the stone in the horizontal plane is applied by the string (gravity acts vertically). There is a net force on the stone in the horizontal plane which acts toward the center.

In an inertial frame of reference, were it not for this net force acting on the stone, the stone would travel in a straight line, according to Newton's first law of motion. In order to keep the stone moving in a circular path, a centripetal force, in this case provided by the string, must be continuously applied to the stone. As soon as it is removed (for example if the string breaks) the stone moves in a straight line. In this inertial frame, the concept of centrifugal force is not required as all motion can be properly described using only real forces and Newton's laws of motion.

In a frame of reference rotating with the stone around the same axis as the stone, the stone is stationary. However, the force applied by the string is still acting on the stone. If one were to apply Newton's laws in their usual (inertial frame) form, one would conclude that the stone should accelerate in the direction of the net applied force—towards the axis of rotation—which it does not do. The centrifugal force and other fictitious forces must be included along with the real forces in order to apply Newton's laws of motion in the rotating frame.

Earth

The Earth constitutes a rotating reference frame because it rotates once a day on its axis. Because the rotation is slow, the fictitious forces it produces are small, and in everyday situations can generally be neglected. Even in calculations requiring high precision, the centrifugal force is generally not explicitly included, but rather lumped in with the gravitational force: the strength and direction of the local "gravity" at any point on the Earth's surface is actually a combination of gravitational and centrifugal forces.

Weight of an Object at the Poles and on the Equator

If an object is weighed with a simple spring balance at one of the Earth's poles, there are two forces acting on the object: the Earth's gravity, which acts in a downward direction, and the equal and opposite restoring force in the spring, acting upward. Since the object is stationary and not accelerating, there is no net force acting on the object and the force from the spring is equal in magnitude to the force of gravity on the object. In this case, the balance shows the value of the force of gravity on the object.

When the same object is weighed on the equator, the same two real forces act upon the object. However, the object is moving in a circular path as the Earth rotates and therefore experiencing a centripetal acceleration. When considered in an inertial frame (that is to say, one that is not rotating with the Earth), the non-zero acceleration means that force of gravity will not balance with the force from the spring. In order to have a net centripetal force, the magnitude of the restoring force of the spring must be less than the magnitude of force of gravity. Less restoring force in the spring is reflected on the scale as less weight — about 0.3% less at the equator than at the poles. In the Earth reference frame (in which the object being weighed is at rest), the object does not appear to be accelerating, however the two real forces, gravity and the force from the spring, are the same magnitude and do not balance. The centrifugal force must be included to make the sum of the forces be zero to match the apparent lack of acceleration.

In fact, the observed weight difference is more — about 0.53%. Earth's gravity is a bit stronger at the poles than at the equator, because the Earth is not a perfect sphere, so an object at the poles is slightly closer to the center of the Earth than one at the equator; this effect combines with the centrifugal force to produce the observed weight difference.

Equatorial Railway

This thought experiment is more complicated than the previous examples in that it requires the use of the Coriolis force as well as the centrifugal force.

If there were a railway line running round the Earth's equator, a train moving westward along it fast enough would remain stationary in a frame moving (but not rotating) with the Earth; it would stand still as the Earth spun beneath it. In this inertial frame the situation is easy to analyze. The only forces acting on the train (assuming no wind resistance or other horizontal forces) are its gravity (downward) and the equal and opposite (upward) force from the track. There is no net force on the train and it therefore remains stationary.

In a frame rotating with the Earth the train moves in a circular orbit as it travels round the Earth. In this frame, the upward reaction force from the track and the force of gravity on the train remain the same, as they are real forces. However, in the Earth's (rotating) frame, the train is traveling in a circular path and therefore requires a centripetal (downward) force to keep it on this path. Because this uses a rotating frame, the (fictitious) centrifugal force must be applied to the train. This is equal in value to the required centripetal force but acts in an upward direction — the opposite direction to that required. It would seem that there is a net upward force on the train and it should therefore accelerate upward.

The resolution to this paradox lies in the fact that the train is in motion with respect to the rotating frame and is subject to (in addition to the centrifugal force) the Coriolis force, which, in this example, acts downward direction and is twice as strong as centrifugal force.

Derivation

For the following formalism, the rotating frame of reference is regarded as a special case of a non-inertial reference frame that is rotating relative to an inertial reference frame denoted the stationary frame.

Time Derivatives in a Rotating Frame

In a rotating frame of reference, the time derivatives of any vector function *P* of time—such as the velocity and acceleration vectors of an object—will differ from its time derivatives in the stationary frame. If P_1 P_2, P_3 are the components of *P* with respect to unit vectors *i*, *j*, *k* directed along the axes of the rotating frame, then the first time derivative [d*P*/d*t*] of *P* with respect to the rotating frame is, by definition, $dP_1/dt\, i + dP_2/dt\, j + dP_3/dt\, k$. If the absolute angular velocity of the rotating frame is *ω* then the derivative d*P*/d*t* of *P* with respect to the stationary frame is related to [d*P*/d*t*] by the equation:

$$\frac{d\mathbf{P}}{dt} = \left[\frac{d\mathbf{P}}{dt}\right] + \boldsymbol{\omega} \times \mathbf{P},$$

where × denotes the vector cross product. In other words, the rate of change of *P* in the stationary frame is the sum of its apparent rate of change in the rotating frame and a rate of rotation $\boldsymbol{\omega} \times \mathbf{P}$ attributable to the motion of the rotating frame. The vector *ω* has magnitude *ω* equal to the rate of rotation and is directed along the axis of rotation according to the right-hand rule.

Acceleration

Newton's law of motion for a particle of mass *m* written in vector form is:

$$\mathbf{F} = m\mathbf{a},$$

where *F* is the vector sum of the physical forces applied to the particle and *a* is the absolute acceleration (that is, acceleration in an inertial frame) of the particle, given by:

$$a = \frac{d^2 r}{dt^2},$$

where *r* is the position vector of the particle.

By applying the transformation above from the stationary to the rotating frame three times, the absolute acceleration of the particle can be written as:

$$\mathbf{a} = \frac{d^2\mathbf{r}}{dt^2} = \frac{d}{dt}\frac{d\mathbf{r}}{dt} = \frac{d}{dt}\left(\left[\frac{d\mathbf{r}}{dt}\right] + \boldsymbol{\omega} \times \mathbf{r}\right)$$

$$= \left[\frac{d^2\mathbf{r}}{dt^2}\right] + \boldsymbol{\omega} \times \left[\frac{d\mathbf{r}}{dt}\right] + \frac{d\boldsymbol{\omega}}{dt} \times \mathbf{r} + \boldsymbol{\omega} \times \frac{d\mathbf{r}}{dt}$$

$$= \left[\frac{d^2\mathbf{r}}{dt^2}\right] + \boldsymbol{\omega}\times\left[\frac{d\mathbf{r}}{dt}\right] + \frac{d\boldsymbol{\omega}}{dt}\times\mathbf{r} + \boldsymbol{\omega}\times\left(\left[\frac{d\mathbf{r}}{dt}\right] + \boldsymbol{\omega}\times\mathbf{r}\right)$$

$$= \left[\frac{d^2\mathbf{r}}{dt^2}\right] + \frac{d\boldsymbol{\omega}}{dt}\times\mathbf{r} + 2\boldsymbol{\omega}\times\left[\frac{d\mathbf{r}}{dt}\right] + \boldsymbol{\omega}\times(\boldsymbol{\omega}\times\mathbf{r}).$$

Force

The apparent acceleration in the rotating frame is $[d^2r/dt^2]$. An observer unaware of the rotation would expect this to be zero in the absence of outside forces. However, Newton's laws of motion apply only in the inertial frame and describe dynamics in terms of the absolute acceleration d^2r/dt^2. Therefore, the observer perceives the extra terms as contributions due to fictitious forces. These terms in the apparent acceleration are independent of mass; so it appears that each of these fictitious forces, like gravity, pulls on an object in proportion to its mass. When these forces are added, the equation of motion has the form:

$$\mathbf{F} - m\frac{d\boldsymbol{\omega}}{dt}\times r - 2m\boldsymbol{\omega}\times\left[\frac{d\mathbf{r}}{dt}\right] - m\boldsymbol{\omega}\times(\boldsymbol{\omega}\times\mathbf{r}) = m\left[\frac{d^2\mathbf{r}}{dt^2}\right].$$

From the perspective of the rotating frame, the additional force terms are experienced just like the real external forces and contribute to the apparent acceleration. The additional terms on the force side of the equation can be recognized as, reading from left to right, the Euler force $-m\,d\boldsymbol{\omega}/dt\times\mathbf{r}$, the Coriolis force $-2m\boldsymbol{\omega}\times[d\mathbf{r}/dt]$, and the centrifugal force $-m\boldsymbol{\omega}\times(\boldsymbol{\omega}\times\mathbf{r})$, respectively. Unlike the other two fictitious forces, the centrifugal force always points radially outward from the axis of rotation of the rotating frame, with magnitude $m\omega^2 r$, and unlike the Coriolis force in particular, it is independent of the motion of the particle in the rotating frame. As expected, for a non-rotating inertial frame of reference $(\boldsymbol{\omega} = 0)$ the centrifugal force and all other fictitious forces disappear. Similarly, as the centrifugal force is proportional to the distance from object to the axis of rotation of the frame, the centrifugal force vanishes for objects that lie upon the axis.

Absolute Rotation

The interface of two immiscible liquids rotating around a vertical axis is an upward-opening circular paraboloid.

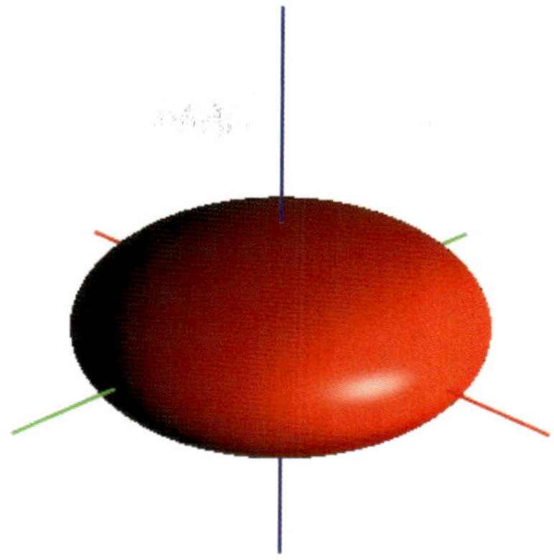

When analysed in a rotating reference frame of the planet, centrifugal force causes rotating planets to assume the shape of an oblate spheroid

Three scenarios were suggested by Newton to answer the question of whether the absolute rotation of a local frame can be detected; that is, if an observer can decide whether an observed object is rotating or if the observer is rotating.

- The shape of the surface of water rotating in a bucket. The shape of the surface becomes concave to balance the centrifugal force against the other forces upon the liquid.
- The tension in a string joining two spheres rotating about their center of mass. The tension in the string will be proportional to the centrifugal force on each sphere as it rotates around the common center of mass.

In these scenarios, the effects attributed to centrifugal force are only observed in the local frame (the frame in which the object is stationary) if the object is undergoing absolute rotation relative to an inertial frame. By contrast, in an inertial frame, the observed effects arise as a consequence of the inertia and the known forces without the need to introduce a centrifugal force. Based on this argument, the privileged frame, wherein the laws of physics take on the simplest form, is a stationary frame in which no fictitious forces need to be invoked.

Within this view of physics, any other phenomenon that is usually attributed to centrifugal force can be used to identify absolute rotation. For example, the oblateness of a sphere of freely flowing material is often explained in terms of centrifugal force. The oblate spheroid shape reflects, following Clairaut's theorem, the balance between containment by gravitational attraction and dispersal by centrifugal force. That the Earth is itself an oblate spheroid, bulging at the equator where the radial distance and hence the centrifugal force is larger, is taken as one of the evidences for its absolute rotation.

Applications

The operations of numerous common rotating mechanical systems are most easily conceptualized in terms of centrifugal force. For example:

- A centrifugal governor regulates the speed of an engine by using spinning masses that move radially, adjusting the throttle, as the engine changes speed. In the reference frame of the spinning masses, centrifugal force causes the radial movement.

- A centrifugal clutch is used in small engine-powered devices such as chain saws, go-karts and model helicopters. It allows the engine to start and idle without driving the device but automatically and smoothly engages the drive as the engine speed rises. Inertial drum brake ascenders used in rock climbing and the inertia reels used in many automobile seat belts operate on the same principle.
- Centrifugal forces can be used to generate artificial gravity, as in proposed designs for rotating space stations. The Mars Gravity Biosatellite would have studied the effects of Mars-level gravity on mice with gravity simulated in this way.
- Spin casting and centrifugal casting are production methods that uses centrifugal force to disperse liquid metal or plastic throughout the negative space of a mold.
- Centrifuges are used in science and industry to separate substances. In the reference frame spinning with the centrifuge, the centrifugal force induces a hydrostatic pressure gradient in fluid-filled tubes oriented perpendicular to the axis of rotation, giving rise to large buoyant forces which push low-density particles inward. Elements or particles denser than the fluid move outward under the influence of the centrifugal force. This is effectively Archimedes' principle as generated by centrifugal force as opposed to being generated by gravity.
- Some amusement rides make use of centrifugal forces. For instance, a Gravitron's spin forces riders against a wall and allows riders to be elevated above the machine's floor in defiance of Earth's gravity.

Nevertheless, all of these systems can also be described without requiring the concept of centrifugal force, in terms of motions and forces in a stationary frame, at the cost of taking somewhat more care in the consideration of forces and motions within the system.

Centripetal Force

Centripetal force is the force needed to make something move in a circle. Well, maybe it would be better to say that centripetal force is the force needed to make something change direction. A constant magnitude centripetal force that is always perpendicular to the direction of motion will make the object move in a circle.

It is important to note that centripetal force is still some force. For instance, suppose you swing a rock around your head with a string. There is a force acting on the rock to make it move in a circle. This force is the tension in the string (which is a real, fundamental force based on the electrostatic interactions between atoms). In this case, the tension would be the centripetal force. Another example - a car moving in a circular path. In this case there is a centripetal force, it would be the frictional force between the tires and the road. In both of these examples, if you take away the centripetal force (break the string or drive on ice), the object would move in a straight line.

Centripetal acceleration is the acceleration of an object moving in a circle:

$$a_{\text{centripetal}} = \frac{v^2}{r}$$

This form does not include the vector nature of the acceleration. The direction of this acceleration should always be pointing towards the center of the circle.

Tethered Ball

One apparatus that clearly illustrates the centripetal force consists of a tethered mass (m_1) swung in a horizontal circle by a lightweight string which passes through a vertical tube to a counter-weight (m_2) as shown in figure.

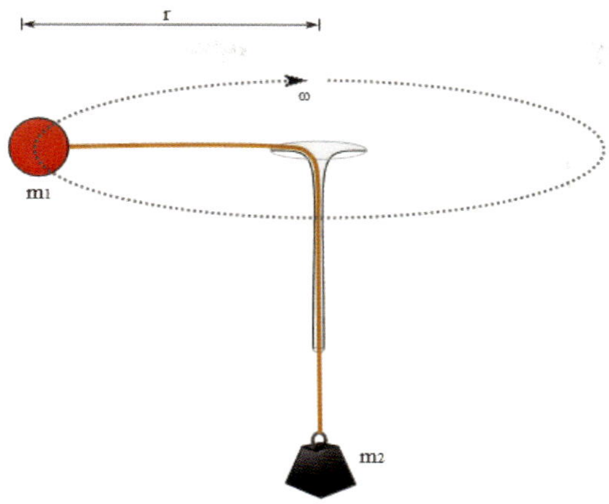

Figure: Demonstration of a centripetal force provided by a mass m2 holding a spinning tethered ball.

Calculation of Centripetal Force

The mathematical representation of centripetal force was derived by Dutch physicist Christiaan Huygens in 1659. For a body following a circular path at constant speed, the radius of the circle (r) equals the mass of the body (m) times the square of the velocity (v) divided by the centripetal force (F):

$$r = mv^2/F$$

The equation may be rearranged to solve for centripetal force:

$$F = mv^2/r$$

An important point you should note from the equation is that centripetal force is proportional to the square of velocity. This means doubling the speed of an object needs four times the centripetal force to keep the object moving in a circle. A practical example of this is seen when taking a sharp curve with an automobile. Here, friction is the only force keeping the vehicle's tires on the road. Increasing speed greatly increases force, so a skid becomes more likely.

Centripetal Acceleration Formula

Another common calculation is centripetal acceleration, which is the change in velocity divided by the change in time. Acceleration is the square of velocity divided by the radius of the circle:

$$\Delta v/\Delta t = a = v^2/r$$

Practical Applications of Centripetal Force

- The classic example of centripetal force is the case of an object being swung on a rope. Here, the tension on the rope supplies the centripetal "pull" force.
- Centripetal force is the "push" force in the case of a Wall of Death motorcycle rider.
- Centripetal force is used for laboratory centrifuges. Here, particles that are suspended in a liquid are separated from the liquid by accelerating tubes oriented so the heavier particles (i.e., objects of higher mass) are pulled toward the bottom of the tubes. While centrifuges commonly separate solids from liquids, they may also fractionate liquids, as in blood samples, or separate components of gases. Gas centrifuges are used to separate the heavier isotope uranium-238 from the lighter isotope uranium-235. The heavier isotope is drawn toward the outside of a spinning cylinder. The heavy fraction is tapped and sent to another centrifuge. The process is repeated until the gas is sufficiently "enriched".
- A liquid mirror telescope (LMT) may be made by rotating a reflective liquid metal, such as mercury. The mirror surface assumes a paraboloid shape because the centripetal force depends on the square of the velocity. Because of this, the height of the spinning liquid metal is proportional to the square of its distance from the center. The interesting shape assumed by spinning liquids may be observed by spinning a bucket of water at a constant rate.

Difference between Centripetal and Centrifugal Force

While centripetal force acts to draw a body toward the center of the point of rotation, the centrifugal force (center-fleeing force) pushes away from the center. According to Newton's First Law, "a body at rest will remain at rest, while a body in motion will remain in motion unless acted upon by an external force". The centripetal force allows a body to follow a circular path without flying off at a tangent by continuously acting at a right angle to the path.

The *centripetal force requirement* is a consequence of Newton's Second Law, which says an object being accelerated undergoes a net force, with the direction of the net force the same as the direction of the acceleration. For an object moving in a circle, the centripetal force must be present to counter the centrifugal force.

From the standpoint of a stationary object on the rotating frame of reference (e.g., a seat on a swing), the centripetal and centrifugal are equal in magnitude, but opposite in direction. The centripetal force acts on the body in motion, while the centrifugal force does not. For this reason, centrifugal force is sometimes called a "virtual" force.

Circular Motion

Two-dimensional motion that is motion of a particle on a circular path. This type of motion is called circular motion.The motion of a body is said to be circular if it moves in such a way that its distance from a certain fixed point always remains the same.

Direction of Motion of Body at any Instant

If the string breaks suddenly, the stone shall fly tangentially to the path of motion. So, instantaneous direction of motion of the body is always along the tangent to the curve at that point.

Consider a particle P is moving on circle of radius r on X-Y plane with origin O as centre.

The position of the particle at a given instant may be described by angle θ, called angular position of the particle, measured in radian. As the particle moves on the path, its angular position θ changes. The rate of change of angular position is called angular velocity, ω, measured in radian per second.

$$w = \lim_{\Delta t \to 0} \frac{\Delta\theta}{\Delta t}$$

$= d\theta/dt = ds/rdt = v/r$

The rate of change of angular velocity is called angular acceleration, measured in rad/s^2. Thus, the angular acceleration is

$\alpha = d\omega/dt = d^2\theta/dt^2$

Relation between these Parameters

$\omega = \omega_0 + \alpha t$

$? = ?_0 + \omega_0 t + \alpha t^2/2$

$\omega^2 = \omega_0 2 + 2\alpha\,(? - ?_0)$

$? = ?_0 + (\omega_0 + \omega)/(2t)$

Here, $?_0$ is the initial angle and ω_0 is the initial angular speed.

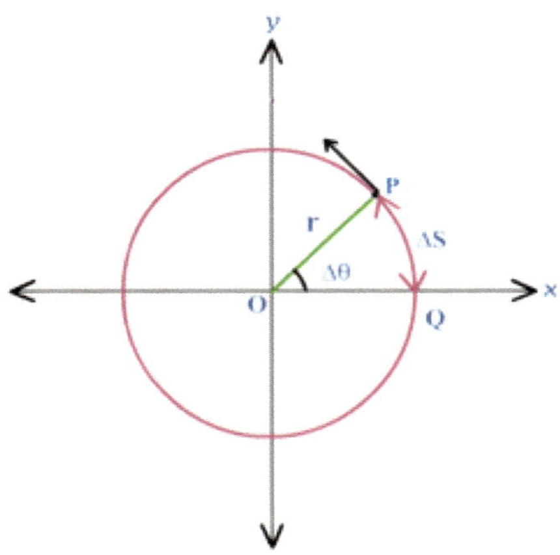

It is easy to derive the equations of rotational kinematics for the case of constant angular acceleration with fixed axis of rotation. These equations are of the same form as those for on-dimensional transitional motion.

Problem 5

(a) What is the angular velocity of the minute and hour hands of a clock?

(b) Suppose the clock starts malfunctioning at 7 AM which decelerates the minute hand at the rate of 4Π radians/day. How much time would the clock loose by 7 AM next day?

Solution

(a) Angular speed of,

minute hand : $\omega_{mh} = 2\pi$ rad/hr = 48π rad/day = (Π/1800) rad/sec

hour hand : $\omega_{hh} = (\pi/6)$ rad/hr = 4π rad/day = (Π/21600) rad/sec

(b) Assume at t = 0, $?_0 = 0$, when the clock begins to malfunction.

Use equation (ii) to get the angle covered by the minute hand in one day.

So, ? = ω_0(1 day) 1/2α(1 day)^2 = 46π rad

Hence the minute hand complete 23 revolutions, so the clock losses 1 hour.

Problem 6

A particle is rotating in a circular path having initial angular velocity 5 rad/sec and the angular acceleration $\alpha = 0.5\ \omega$, where ω is angular velocity at that instant. Find the angular velocity, after it moved an angle π?

Solution

Here angular acceleration is

$\alpha = 0.5\ \omega$

=> $d\omega/dt = 0.5\omega$

=> $(d\omega/d\theta)\ (d\theta/dt) = 0.5\omega$

=> $\omega\ d\omega/d\theta = 0.5\omega$

Or,

$$\int_5^w dw = 0.5 \int_0^\pi d\theta$$

=> $\omega - 5 = 0.5 \times \pi$

=> $\omega = 5 + 0.5 \times \pi = 6.57$ rad/sec.

Hence, when acceleration is not constant, use the method of calculus as shown in above illustration

- Conservation of energy.

Conservation means that these three quantities, mass, momentum and energy, cannot change over time, unless an external agent intervenes. Other properties such as density, pressure, temperature, and velocity are assumed to be well-defined at small "material points", and to vary continuously from point to point. The material point is the geometric center of a small elementary cube large enough to smooth out inter granular "empty" discontinuities but sufficiently small to discount effects of macroscopic changes.

Explanation

Modeling an object as a continuum assumes that the substance of the object completely fills the space it occupies. Modeling objects in this way ignores the fact that matter is made of atoms, and so is not continuous; however, on length scales much greater than that of inter-atomic distances, such models are highly accurate. Fundamental physical laws such as the conservation of mass, the conservation of momentum, and the conservation of energy may be applied to such models to derive differential equations describing the behavior of such objects, and some information about the particular material studied is added through constitutive relations.

Continuum mechanics deals with physical properties of solids and fluids which are independent of any particular coordinate system in which they are observed. These physical properties are then represented by tensors, which are mathematical objects that have the required property of being independent of coordinate system. These tensors can be expressed in coordinate systems for computational convenience.

Concept of a Continuum

Materials, such as solids, liquids and gases, are composed of molecules separated by space. On a microscopic scale, materials have cracks and discontinuities. However, certain physical phenomena can be modeled assuming the materials exist as a *continuum, meaning the matter in the body is continuously distributed and fills the entire region of space it occupies*. A continuum is a body that can be continually sub-divided into infinitesimal elements with properties being those of the bulk material.

The validity of the continuum assumption may be verified by a theoretical analysis, in which either some clear periodicity is identified or statistical homogeneity and ergodicity of the microstructure exists. More specifically, the continuum hypothesis/assumption hinges on the concepts of a representative elementary volume and separation of scales based on the Hill–Mandel condition. This condition provides a link between an experimentalist's and a theoretician's viewpoint on constitutive equations (linear and nonlinear elastic/inelastic or coupled fields) as well as a way of spatial and statistical averaging of the microstructure.

When the separation of scales does not hold, or when one wants to establish a continuum of a finer resolution than that of the representative volume element (RVE) size, one employs a *statistical volume element* (SVE), which, in turn, leads to random continuum fields. The latter then provide a micromechanics basis for stochastic finite elements (SFE). The levels of SVE and RVE link continuum mechanics to statistical mechanics. The RVE may be assessed only in a limited way via experimental testing: when the constitutive response becomes spatially homogeneous.

Specifically for fluids, the Knudsen number is used to assess to what extent the approximation of continuity can be made.

Car Traffic as an Introductory Example

Consider car traffic on a highway, with just one lane for simplicity. Somewhat surprisingly, and in a tribute to its effectiveness, continuum mechanics effectively models the movement of cars via a partial differential equation (PDE) for the density of cars. The familiarity of this situation empowers us to understand a little of the continuum-discrete dichotomy underlying continuum modelling in general.

To start modelling define that: x measure distance (in km) along the highway; t is time (in minutes); $\rho(x,t)$ is the density of cars on the highway (in cars/km in the lane); and $u(x,t)$ is the flow velocity (average velocity) of those cars 'at' position x.

Conservation Derives a PDE (Partial Differential Equation)

Cars do not appear and disappear. Consider any group of cars: from the particular car at the back of the group located at $x = a(t)$ to the particular car at the front located at $x = b(t)$. The total number of cars in this group $N = \int_{a(t)}^{b(t)} \rho(x,t)dx$. Since cars are conserved (if there is overtaking, then the car at the front \ back' may become a different car) $dN/dt = 0$. But via the Leibniz integral rule

$$
\begin{aligned}
\frac{dN}{dt} &= \frac{d}{dt}\int_{a(t)}^{b(t)} \rho(x,t)dx \\
&= \int_a^b \frac{\partial\rho}{\partial t}dx + \rho(b,t)\frac{db}{dt} - \rho(a,t)\frac{da}{dt} \\
&= \int_a^b \frac{\partial\rho}{\partial t}dx + \rho(b,t)u(b,t) - \rho(a,t)u(a,t) \\
&= \int_a^b \left[\frac{\partial\rho}{\partial t} + \frac{\partial}{\partial x}(\rho u)\right]dx.
\end{aligned}
$$

This integral being zero holds for all groups, that is, for all intervals $[a,b]$. The only way an integral can be zero for all intervals is if the integrand is zero for all x. Consequently, conservation derives the first order nonlinear conservation PDE:

$$\frac{\partial\rho}{\partial t} + \frac{\partial}{\partial x}(\rho u) = 0$$

for all positions on the highway.

This conservation PDE applies not only to car traffic but also to fluids, solids, crowds, animals, plants, bushfires, financial traders, and so on.

Observation Closes the Problem

The previous PDE is one equation with two unknowns, so another equation is needed to form a well posed problem. Such an extra equation is typically needed in continuum mechanics and

Simulation for Circular Motion's Component

Radius vector, velocity vector, centripetal force vector and centripetal acceleration vector. It then animates them traveling in a circle.

Motion of a Particle in a Circular Path

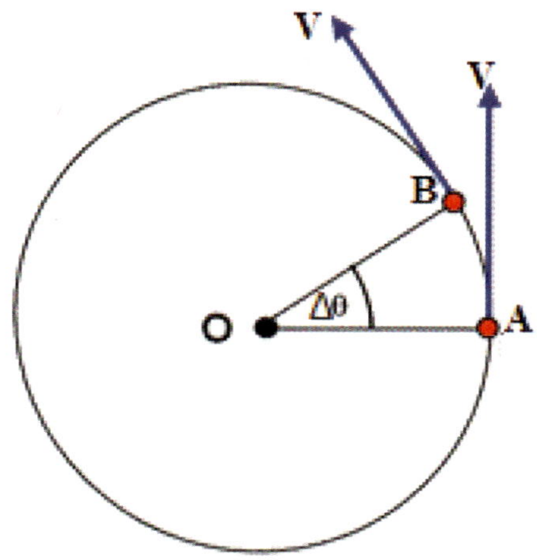

It is a special kind of two-dimensional motion in which the particle's position vector always lies on the circumference of a circle. In order to calculate the acceleration parameter it is helpful to first consider circular motion with constant speed, called uniform circular motion. Let there be a particle moving along a circle of radius r with a velocity $\vec{v}'$, as shown in figure, such that $|\vec{v}'|$ = v= constant. For this particle, it is our aim to calculate the magnitude and direction of its acceleration. We know that,

$$\vec{a} = \lim_{\Delta t \to o} \frac{\Delta \vec{v}}{\Delta t}$$

Now, we have to find an expression for $\Delta \vec{v}$ in terms of known quantities. For this, consider the particle velocity vector at two points A and B. Displacing $\vec{v}\,B$, parallel to itself and placing it back to back with $\vec{v}\,A$, as shown in figure given below. We have

$$\Delta \vec{v} = \vec{v}\,B - \vec{v}\,A$$

Consider ΔAOB, angle between OA and OB is same as angle between $\vec{v}\,A$ and $\vec{v}\,B$ because $\vec{v}\,A$ is perpendicular to vector OA and $\vec{v}\,B$ is also perpendicular to vector OB.

OB = OA = r and

$$|\vec{v}\,A| = |\vec{v}\,B| = v$$

So, ΔAOB is similar to the triangle formed by $\vec{v}\,A, \vec{v}\,B$ and $\Delta \vec{v}$

Thus, From geometry we have, Δv/v = AB/r. Now AB is approximately equal to vΔt.

$$\frac{\Delta v}{v} \cong \frac{v\Delta t}{r}$$

$$\Rightarrow \frac{v^2}{r} \cong \frac{\Delta v}{\Delta t}$$

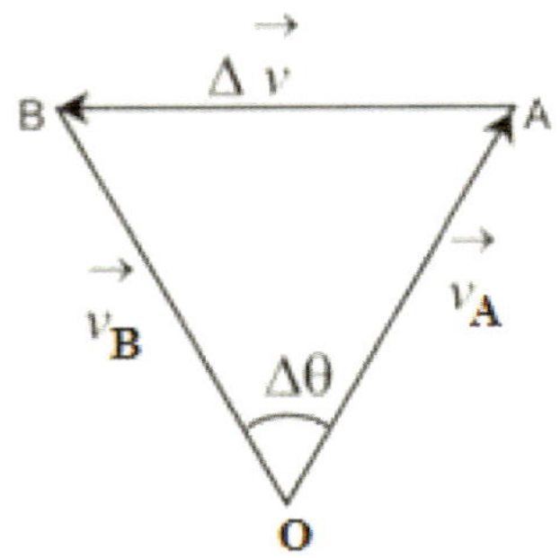

In the limit Δt → 0 the above relation becomes exact, we have

$$|\vec{a}| = \lim_{\Delta t \to 0} \frac{\Delta \vec{v}}{\Delta t} = \frac{v^2}{r}$$

$$a = \frac{v^2}{r}$$

This is the magnitude of the acceleration. The direction $\vec{A}$ is instantaneously along a radius inward towards the centre of the circle, because of this $\vec{A}$ is called radial or centripetal acceleration. The acceleration vector of a particle in uniform circular motion averaged over one cycle is a null vector.

Problem 7

The moon revolved about the earth making a complete revolution in 2.36 mega second. Assume that the orbit is circular and has a radius of 385 mega meter. What is the magnitude of the acceleration of the moon towards the earth?

Solution

Here first of all we calculate speed v of the moon which is given by $v = 2\pi R/T$

where R = 385 mega meter = 385×10^6 m

and T = 2.36 mega second = 2.36×10^6 sec.

Hence v = 1020 m/sec.

The magnitude of centripetal acceleration is

$a = v^2/R = 0.00273$ m/sec^2.

From the above observation we conclude that, the magnitude of the acceleration of the moon towards the earth would be 0.00273 m/sec^2.

In the previous enquiry we have discussed the uniform circular motion in which the particle has constant speed. If the particle's speed varies with time then the motion will be no more uniform but a non-uniform circular motion. Let us discuss about this motion using the concept of vectors.

Uniform Circular Motion

Uniform circular motion can be described as the motion of an object in a circle at a constant speed. As an object moves in a circle, it is constantly changing its direction. At all instances, the object is moving tangent to the circle. Since the direction of the velocity vector is the same as the direction of the object's motion, the velocity vector is directed tangent to the circle as well.

An object moving in a circle is accelerating. Accelerating objects are objects which are changing their velocity - either the speed (i.e., magnitude of the velocity vector) or the direction. An object undergoing uniform circular motion is moving with a constant speed. Nonetheless, it is accelerating due to its change in direction. The direction of the acceleration is inwards.

The final motion characteristic for an object undergoing uniform circular motion is the net force. The net force acting upon such an object is directed towards the center of the circle. The net force is said to be an inward or *centripetal* force. Without such an inward force, an object would continue in a straight line, never deviating from its direction. Yet, with the inward net force directed perpendicular to the velocity vector, the object is always changing its direction and undergoing an inward acceleration.

Formulas

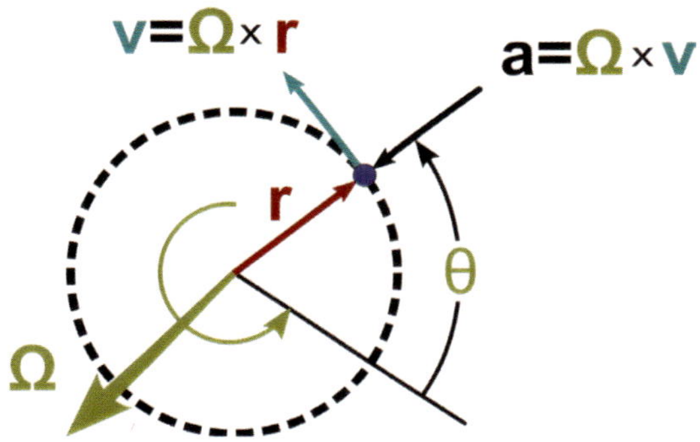

Figure: Vector relationships for uniform circular motion; vector ω representing the rotation is normal to the plane of the orbit

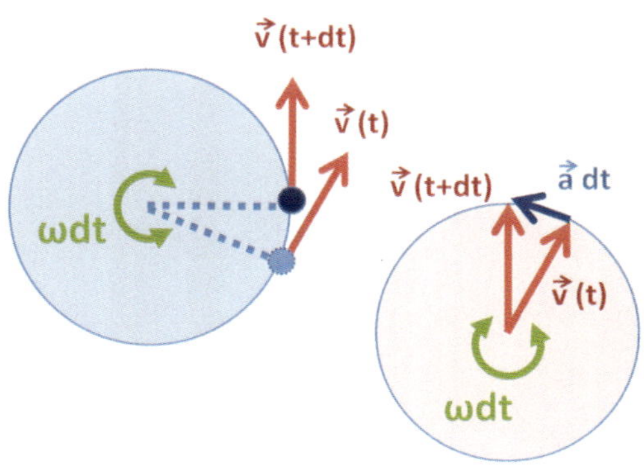

In the figure above, the velocity vectors at time t and time $t + dt$ are moved from the orbit on the left to new positions where their tails coincide, on the right. Because the velocity is fixed in magnitude at $v = r\,\omega$, the velocity vectors also sweep out a circular path at angular rate ω. As $dt \to 0$, the acceleration vector a becomes perpendicular to v, which means it points toward the center of the orbit in the circle on the left. Angle $\omega\, dt$ is the very small angle between the two velocities and tends to zero as $dt \to 0$.

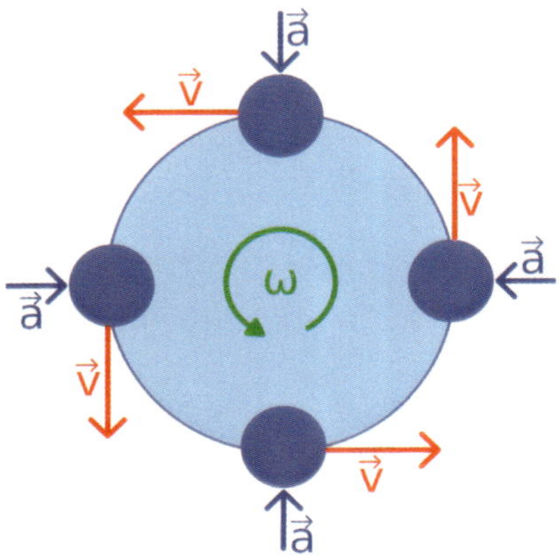

Figure: Velocity v and acceleration a in uniform circular motion at angular rate ω; the speed is constant, but the velocity is always tangent to the orbit; the acceleration has constant magnitude, but always points toward the center of rotation.

For motion in a circle of radius r, the circumference of the circle is $C = 2\pi\, r$. If the period for one rotation is T, the angular rate of rotation, also known as angular velocity, ω is:

$$\omega = \frac{2\pi}{T} = 2\pi f = \frac{d\theta}{dt} \quad \text{and the units are radians/second}$$

The speed of the object travelling the circle is:

$$v = \frac{2\pi r}{T} = \omega r$$

The angle θ swept out in a time t is:

$$\theta = 2\pi \frac{t}{T} = \omega t$$

The angular acceleration, α, of the particle is:

$$\alpha = \frac{d\omega}{dt}$$

In the case of uniform circular motion, α will be zero.

The acceleration due to change in the direction is:

$$a = \frac{v^2}{r} = \omega^2 r$$

The centripetal and centrifugal force can also be found out using acceleration:

$$F_c = \dot{p} \overset{\dot{m}=0}{=} ma = \frac{mv^2}{r}$$

The vector relationships are shown in figure. The axis of rotation is shown as a vector ω perpendicular to the plane of the orbit and with a magnitude $\omega = d\theta / dt$. The direction of ω is chosen using the right-hand rule. With this convention for depicting rotation, the velocity is given by a vector cross product as

$$\mathbf{v} = \boldsymbol{\omega} \times \mathbf{r},$$

which is a vector perpendicular to both ω and r(t), tangential to the orbit, and of magnitude ω r. Likewise, the acceleration is given by

$$\mathbf{a} = \boldsymbol{\omega} \times \mathbf{v} = \boldsymbol{\omega} \times (\boldsymbol{\omega} \times \mathbf{r}),$$

which is a vector perpendicular to both ω and v(t) of magnitude $\omega\,|v| = \omega^2 r$ and directed exactly opposite to r(t).

In the simplest case the speed, mass and radius are constant.

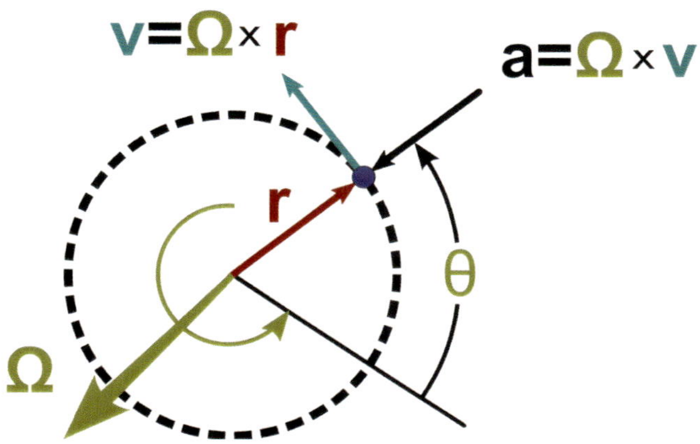

Figure: Vector relationships for uniform circular motion; vector ω representing the rotation is normal to the plane of the orbit.

Consider a body of one kilogram, moving in a circle of radius one metre, with an angular velocity of one radian per second.

- The speed is one metre per second.
- The inward acceleration is one metre per square second, v^2/r.
- It is subject to a centripetal force of one kilogram metre per square second, which is one newton.
- The momentum of the body is one $kg \cdot m \cdot s^{-1}$.
- The moment of inertia is one $kg \cdot m^2$.
- The angular momentum is one $kg \cdot m^2 \cdot s^{-1}$.

- The kinetic energy is 1/2 joule.
- The circumference of the orbit is 2π (~6.283) metres.
- The period of the motion is 2π seconds per turn.
- The frequency is $(2\pi)^{-1}$ hertz.

In Polar Coordinates

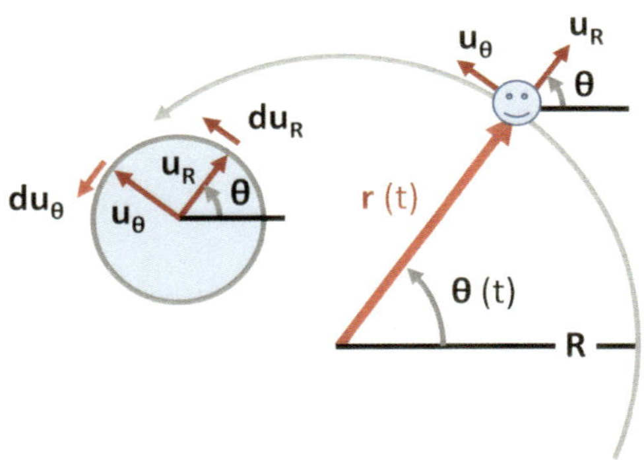

Figure: Polar coordinates for circular trajectory. On the left is a unit circle showing the changes $\mathbf{d\hat{u}_R}$ and $\mathbf{d\hat{u}_è}$ in the unit vectors $\mathbf{\hat{u}_R}$ and $\mathbf{\hat{u}_è}$ for a small increment $d\theta$ in angle θ.

During circular motion the body moves on a curve that can be described in polar coordinate system as a fixed distance *R* from the center of the orbit taken as origin, oriented at an angle $\theta(t)$ from some reference direction in figure. The displacement *vector* $\vec{r}$ is the radial vector from the origin to the particle location:

$$\vec{r}(t) = R\hat{u}_R(t),$$

where $\hat{u}_R(t)$ is the unit vector parallel to the radius vector at time *t* and pointing away from the origin. It is convenient to introduce the unit vector orthogonal to $\hat{u}_R(t)$ as well, namely $\hat{u}_\theta(t)$. It is customary to orient $\hat{u}_\theta(t)$ to point in the direction of travel along the orbit.

The velocity is the time derivative of the displacement:

$$\vec{v}(t) = \frac{d}{dt}\vec{r}(t) = \frac{dR}{dt}\hat{u}_R(t) + R\frac{d\hat{u}_R}{dt}.$$

Because the radius of the circle is constant, the radial component of the velocity is zero. The unit vector $\hat{u}_R(t)$ has a time-invariant magnitude of unity, so as time varies its tip always lies on a circle of unit radius, with an angle θ the same as the angle of $\vec{r}(t)$. If the particle displacement rotates through an angle $d\theta$ in time *dt*, so does $\hat{u}_R(t)$, describing an arc on the unit circle of magnitude $d\theta$.

$$\frac{d\hat{u}_R}{dt} = \frac{d\theta}{dt}\hat{u}_\theta(t),$$

where the direction of the change must be perpendicular to $\hat{u}_R(t)$ (or, in other words, along $\hat{u}_\theta(t)$ because any change $d\hat{u}_R(t)$ in the direction of $\hat{u}_R(t)$ would change the size of $\hat{u}_R(t)$. The sign is positive, because an increase in $d\theta$ implies the object and $\hat{u}_R(t)$ have moved in the direction of $\hat{u}_\theta(t)$. Hence the velocity becomes:

$$\vec{v}(t)=\frac{d}{dt}\vec{r}(t)=R\frac{d\hat{u}_R}{dt}=R\frac{d\theta}{dt}\hat{u}_\theta(t)=R\omega\hat{u}_\theta(t).$$

The acceleration of the body can also be broken into radial and tangential components. The acceleration is the time derivative of the velocity:

$$\begin{aligned}\vec{a}(t)&=\frac{d}{dt}\vec{v}(t)=\frac{d}{dt}\left(R\omega\hat{u}_\theta(t)\right)\\&=R\left(\frac{d\omega}{dt}\hat{u}_\theta(t)+\omega\frac{d\hat{u}_\theta}{dt}\right).\end{aligned}$$

The time derivative of $\hat{u}_\theta(t)$ is found the same way as for $\hat{u}_R(t)$. Again, $\hat{u}_\theta(t)$ is a unit vector and its tip traces a unit circle with an angle that is $\pi/2+\theta$. Hence, an increase in angle $d\theta$ by $\vec{r}(t)$ implies $\hat{u}_\theta(t)$ traces an arc of magnitude $d\theta$, and as $\hat{u}_\theta(t)$ is orthogonal to $\hat{u}_R(t)$, we have:

$$\frac{d\hat{u}_\theta}{dt}=-\frac{d\theta}{dt}\hat{u}_R(t)=-\omega\hat{u}_R(t),$$

where a negative sign is necessary to keep $\hat{u}_\theta(t)$ orthogonal to $\hat{u}_R(t)$. (Otherwise, the angle between $\hat{u}_\theta(t)$ and $\hat{u}_R(t)$ would *decrease* with increase in $d\theta$.) Consequently, the acceleration is:

$$\begin{aligned}\vec{a}(t)&=R\left(\frac{d\omega}{dt}\hat{u}_\theta(t)+\omega\frac{d\hat{u}_\theta}{dt}\right)\\&=R\frac{d\omega}{dt}\hat{u}_\theta(t)-\omega^2R\hat{u}_R(t).\end{aligned}$$

The centripetal acceleration is the radial component, which is directed radially inward:

$$\vec{a}_R(t)=-\omega^2R\hat{u}_R(t),$$

while the tangential component changes the magnitude of the velocity:

$$\vec{a}_\theta(t)=R\frac{d\omega}{dt}\hat{u}_\theta(t)=\frac{dR\omega}{dt}\hat{u}_\theta(t)=\frac{d|\vec{v}(t)|}{dt}\hat{u}_\theta(t).$$

Using Complex Numbers

Circular motion can be described using complex numbers. Let the x axis be the real axis and the y axis be the imaginary axis. The position of the body can then be given as z, a complex "vector":

$$z=x+iy=R(cos[\theta(t)]+i\,sin[\theta(t)])=Re^{i\theta(t)},$$

where i is the imaginary unit, and $\theta(t)$ is the argument of the complex number as a function of time, t.

Since the radius is constant:

$$\dot{R} = \ddot{R} = 0\,,$$

where a *dot* indicates differentiation in respect of time.

With this notation the velocity becomes:

$$v = \dot{z} = \frac{d\left(Re^{i\theta[t]}\right)}{dt} = R\frac{d\left(e^{i\theta[t]}\right)}{dt} = Re^{i\theta(t)}\frac{d(i\theta[t])}{dt} = iR\dot{\theta}(t)e^{i\theta(t)} = i\omega Re^{i\theta(t)} = i\omega z$$

and the acceleration becomes:

$$\begin{aligned} a = \dot{v} = i\dot{\omega}z + i\omega\dot{z} &= \left(i\dot{\omega} - \omega^2\right)z \\ &= \left(i\dot{\omega} - \omega^2\right)Re^{i\theta(t)} \\ &= -\omega^2 Re^{i\theta(t)} + \dot{\omega}e^{i\frac{\pi}{2}}Re^{i\theta(t)}\,. \end{aligned}$$

The first term is opposite in direction to the displacement vector and the second is perpendicular to it, just like the earlier results shown before.

Velocity

Velocity and acceleration vectors for uniform motion at four different points in the orbit. Because the velocity v is tangent to the circular path, no two velocities point in the same direction. Although the object has a constant *speed*, its *direction* is always changing. This change in velocity is caused by an acceleration a, whose magnitude is (like that of the velocity) held constant, but whose direction also is always changing. The acceleration points radially inwards (centripetally) and is perpendicular to the velocity. This acceleration is known as centripetal acceleration.

For a path of radius r, when an angle θ is swept out, the distance travelled on the periphery of the orbit is $s = r\theta$. Therefore, the speed of travel around the orbit is

$$v = r\frac{d\theta}{dt} = r\omega\,,$$

where the angular rate of rotation is ω. (By rearrangement, $\omega = v/r$.) Thus, v is a constant, and the velocity vector v also rotates with constant magnitude v, at the same angular rate ω.

Relativistic Circular Motion

In this case the three-acceleration vector is perpendicular to the three-velocity vector,

$$\vec{u}\cdot\vec{a} = 0.$$

and the square of proper acceleration, expressed as a scalar invariant, the same in all reference frames,

$$\alpha^2 = \gamma^4 a^2 + \gamma^6 (\vec{u}\cdot\vec{a})^2,$$

becomes the expression for circular motion,

$$\alpha^2 = \gamma^4 a^2.$$

or, taking the positive square root and using the three-acceleration, we arrive at the proper acceleration for circular motion:

$$\alpha = \gamma^2 \frac{v^2}{r}.$$

Acceleration

The left-hand circle in figure is the orbit showing the velocity vectors at two adjacent times. On the right, these two velocities are moved so their tails coincide. Because speed is constant, the velocity vectors on the right sweep out a circle as time advances. For a swept angle $d\theta = \omega\, dt$ the change in v is a vector at right angles to v and of magnitude $v\, d\theta$, which in turn means that the magnitude of the acceleration is given by

$$a = v\frac{d\theta}{dt} = v\omega = \frac{v^2}{r}$$

Centripetal acceleration for some values of radius and magnitude of velocity								
	$\lvert \boldsymbol{v} \rvert$	1 m/s 3.6 km/h 2.2 mph	2 m/s 7.2 km/h 4.5 mph	5 m/s 18 km/h 11 mph	10 m/s 36 km/h 22 mph	20 m/s 72 km/h 45 mph	50 m/s 180 km/h 110 mph	100 m/s 360 km/h 220 mph
$\boldsymbol{r}$		Slow walk		Bicycle		City car		Aerobatics
10 cm 3.9 in	Laboratory centrifuge	10 m/s² 1.0 *g*	40 m/s² 4.1 *g*	250 m/s² 25 *g*	1.0 km/s² 100 *g*	4.0 km/s² 410 *g*	25 km/s² 2500 *g*	100 km/s² 10000 *g*
20 cm 7.9 in		5.0 m/s² 0.51 *g*	20 m/s² 2.0 *g*	130 m/s² 13 *g*	500 m/s² 51 *g*	2.0 km/ s² 200 *g*	13 km/s² 1300 *g*	50 km/s² 5100 *g*
50 cm 1.6 ft		2.0 m/s² 0.20 *g*	8.0 m/s² 0.82 *g*	50 m/s² 5.1 *g*	200 m/s² 20 *g*	800 m/s² 82 *g*	5.0 km/s² 510 *g*	20 km/s² 2000 *g*
1 m 3.3 ft	Playground carousel	1.0 m/s² 0.10 *g*	4.0 m/s² 0.41 *g*	25 m/s² 2.5 *g*	100 m/s² 10 *g*	400 m/s² 41 *g*	2.5 km/s² 250 *g*	10 km/s² 1000 *g*
2 m 6.6 ft		500 mm/s² 0.051 *g*	2.0 m/s² 0.20 *g*	13 m/s² 1.3 *g*	50 m/s² 5.1 *g*	200 m/s² 20 *g*	1.3 km/s² 130 *g*	5.0 km/s² 510 *g*
5 m 16 ft		200 mm/s² 0.020 *g*	800 mm/ s² 0.082 *g*	5.0 m/s² 0.51 *g*	20 m/s² 2.0 *g*	80 m/s² 8.2 *g*	500 m/s² 51 *g*	2.0 km/s² 200 *g*
10 m 33 ft	Roll- er-coaster vertical loop	100 mm/s² 0.010 *g*	400 mm/ s² 0.041 *g*	2.5 m/s² 0.25 *g*	10 m/s² 1.0 *g*	40 m/s² 4.1 *g*	250 m/s² 25 *g*	1.0 km/s² 100 *g*
20 m 66 ft		50 mm/s² 0.0051 *g*	200 mm/ s² 0.020 *g*	1.3 m/s² 0.13 *g*	5.0 m/s² 0.51 *g*	20 m/s² 2 *g*	130 m/s² 13 *g*	500 m/s² 51 *g*

50 m 160 ft		20 mm/s2 0.0020 *g*	80 mm/s2 0.0082 *g*	500 mm/s2 0.051 *g*	2.0 m/s2 0.20 *g*	8.0 m/s2 0.82 *g*	50 m/s2 5.1 *g*	200 m/s2 20 *g*
100 m 330 ft	Freeway on-ramp	10 mm/s2 0.0010 *g*	40 mm/s2 0.0041 *g*	250 mm/s2 0.025 *g*	1.0 m/s2 0.10 *g*	4.0 m/s2 0.41 *g*	25 m/s2 2.5 *g*	100 m/s2 10 *g*
200 m 660 ft		5.0 mm/s2 0.00051 *g*	20 mm/s2 0.0020 *g*	130 m/s2 0.013 *g*	500 mm/s2 0.051 *g*	2.0 m/s2 0.20 *g*	13 m/s2 1.3 *g*	50 m/s2 5.1 *g*
500 m 1600 ft		2.0 mm/s2 0.00020 *g*	8.0 mm/s2 0.00082 *g*	50 mm/s2 0.0051 *g*	200 mm/s2 0.020 *g*	800 mm/s2 0.082 *g*	5.0 m/s2 0.51 *g*	20 m/s2 2.0 *g*
1 km 3300 ft	High-speed railway	1.0 mm/s2 0.00010 *g*	4.0 mm/s2 0.00041 *g*	25 mm/s2 0.0025 *g*	100 mm/s2 0.010 *g*	400 mm/s2 0.041 *g*	2.5 m/s2 0.25 *g*	10 m/s2 1.0 *g*

Non Uniform Circular Motion

Non-uniform circular motion denotes a change in the speed of a particle moving along a circular path.

What do we mean by non-uniform circular motion? The answer lies in the definition of uniform circular motion, which is a circular motion with constant speed. It follows then that non-uniform circular motion denotes a change in the speed of the particle moving along the circular path. Note especially the change in the velocity vector sizes, denoting change in the magnitude of velocity.

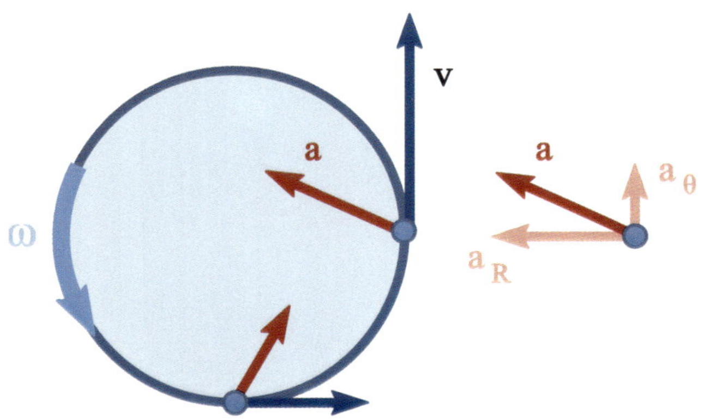

Diagram of non-uniform circular motion: In non-uniform circular motion, the magnitude of the angular velocity changes over time.

The change in direction is accounted by radial acceleration (centripetal acceleration), which is given by following relation: $a_r = \frac{v^2}{r}$. The change in speed has implications for radial (centripetal) acceleration. There are two possibilities:

1: The radius of circle is constant (like in the motion along a circular rail or motor track). A change in v will change the magnitude of radial acceleration. This means that the centripetal acceleration is not constant, as is the case with uniform circular motion. The greater the speed, the greater the radial acceleration. A particle moving at higher speed will need a greater radial force to change direction and vice-versa when the radius of the circular path is constant.

2: The radial (centripetal) force is constant (like a satellite rotating about the earth under the influence of a constant force of gravity). The circular motion adjusts its radius in response to changes in speed. This means that the radius of the circular path is variable, unlike the case of uniform circular motion. In any eventuality, the equation of centripetal acceleration in terms of "speed" and "radius" must be satisfied. The important thing to note here is that, although change in speed of the particle affects radial acceleration, the change in speed is not affected by radial or centripetal force. We need a tangential force to affect the change in the magnitude of a tangential velocity. The corresponding acceleration is called tangential acceleration.

In either case, the angular velocity in non-uniform circular motion is not constant as $w = \frac{v}{r}$ and v is varying.

Let us use the vector method to discuss non-uniform circular motion.

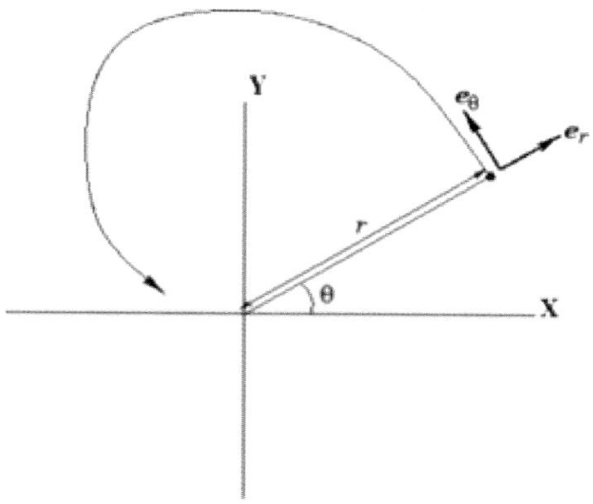

In the figure, $\hat{e}_r$ and $\hat{e}_\theta$ are unit vectors along radius and tangent vector respectively. In terms of er and eθ the motion of a particle moving counter clockwise in a circle about the origin in figure can be described be the vector equation.

$$\vec{v} = \hat{e}_\theta v$$

In this case, not only $\hat{e}_\theta$ but v also varies with time. We can obtain instantaneous acceleration as,

$$\vec{a} = \frac{d\vec{v}}{dt} = \hat{e}_\theta \frac{d\theta}{dt} + v\left(\frac{d\hat{e}_\theta}{dt}\right)$$

$$\vec{a} = (d\,\vec{v})/dt = \hat{e}\,\theta\, d\theta/dt + v(d\,\hat{e}\,\theta)/dt$$

Again,

$$\frac{d\hat{e}_\theta}{dt} = \frac{\hat{e}_r v^2}{r}$$

$$\vec{a} = \hat{e}_\theta a_T - \hat{e}_r \frac{v^2}{r} = a_T \hat{e}_\theta - a_R \hat{e}_r$$

Here, $a_T = dv/dt$ and $a_R = v^2/r$

The first term, $\hat{e}_\theta a_T$ is the vector component of $\vec{A}$ that is tangential to the path of the particle and arises from a change in the magnitude of the velocity in circular motion, called tangentiWal acceleration whereas a_R centripetal acceleration. The magnitude of $\vec{A}$ is

$$|\vec{a}| = \sqrt{\left(a_T{}^2 + a_N{}^2\right)}$$

Problem 8

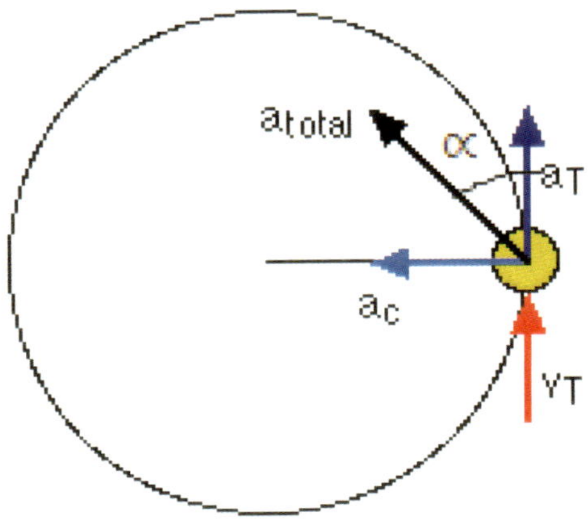

Point A travels along an arc of a circle of radius r as shown in figure. Its velocity depends on the arc coordinates l as $V=A\sqrt{1}$ where A is a constant. Let us calculate the angle α between the vectors of the total acceleration and of the velocity of the point as a function of the coordinate l.

Solution

It is seen from figure shown above that the angle α can be found by means of the formula tan α = a_R/a_T. Let us find a_R and a_T.

$a_R = v^2/r = (A^2\, l)/r$; $a_T = dv/dt = dv/dl = A/(2\sqrt{l}) \times A\sqrt{l} = A^2/2$

Hence tan α = 2l/r.

Problem 9

Position vector of a particle moving in x-y palne at time t is

$$\vec{r} = a(1 - \cos wt)\hat{i} + a \sin wt\hat{j}.$$

What will be the path of the particle?

Solution

x = a (1-cos ωt)

y = a sinωt

From first equation, we get

$a\cos \omega t = a - X$

From second equations, we get,

$a \sin \omega t = y$

Squaring and adding Equations, we get

$(X-a)^2 + y^2 = a^2$

This is an equation of a circle of radius a and ceter at (a,0).

References

- Mobberley, Martin (2009-03-01). Cataclysmic Cosmic Events and How to Observe Them. Springer Science & Business Media. ISBN 9780387799469
- David P. Stern (2006). "Frames of Reference: The Basics". From Stargazers to Starships. Goddard Space Flight Center Space Physics Data Facility. Retrieved 20 April 2017
- Kobayashi, Yukio (2008). "Remarks on viewing situation in a rotating frame". European Journal of Physics. 29 (3): 599–606. Bibcode:2008EJPh...29..599K. doi:10.1088/0143-0807/29/3/019
- Gomez, R W; Hernandez-Gomez, J J; Marquina, V (25 July 2012). "A jumping cylinder on an inclined plane". Eur. J. Phys. IOP. 33 (5): 1359–1365. arXiv:1204.0600. Bibcode:2012EJPh...33.1359G. doi:10.1088/0143-0807/33/5/1359. Retrieved 25 April 2016
- Hibbeler, Russell C. (2009). Engineering Mechanics. Upper Saddle River, New Jersey: Pearson Prentice Hall. pp. 314, 153. ISBN 978-0-13-607791-6.(EM1)
- Boynton, Richard (2001). "Precise Measurement of Mass"(PDF). Sawe Paper No. 3147. Arlington, Texas: S.A.W.E., Inc. Retrieved 2007-01-21

3

Translational and Oscillatory Motion

A translation is a displacement in the position of an object. It can be a linear or an angular displacement. Translational motion can be rectilinear or curvilinear. The repetitive variation of some measure about a point of equilibrium is termed as an oscillation. Simple harmonic motion is a type of oscillatory motion. This chapter has been carefully written to provide an easy understanding of translational and oscillatory motion.

Translational Motion

Translational motion is the motion by which a body shifts from one point in space to another. One example of translational motion is the the motion of a bullet fired from a gun.

An object has a rectilinear motion when it moves along a straight line. At any time, t, the object occupies a position along the line as shown in the following figure. The distance x, with appropriate sign, define the position of the object. When the position of the object at particular time is known, the motion of the particle will be known, and generally is expressed in a form of an equation which relates distance x, to time t, for example $x = 6t - 4$, or a graph.

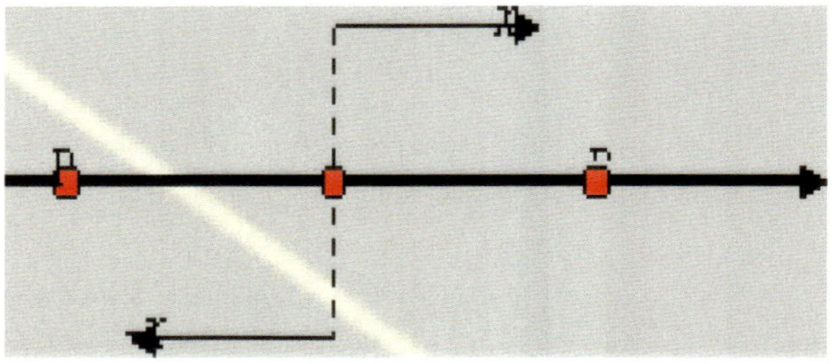

Motion in two or three dimensions is more complicated. In two dimensions, we need to specify two coordinates in order to fix the position of any object. The following figure shows a simple example of projectile motion: a ball rolling off a table. Let us define the horizontal direction as the x-axis and the vertical direction as the y -axis. Consider a ball initially rolling on off a flat table with an initial velocity of 10 m/s.

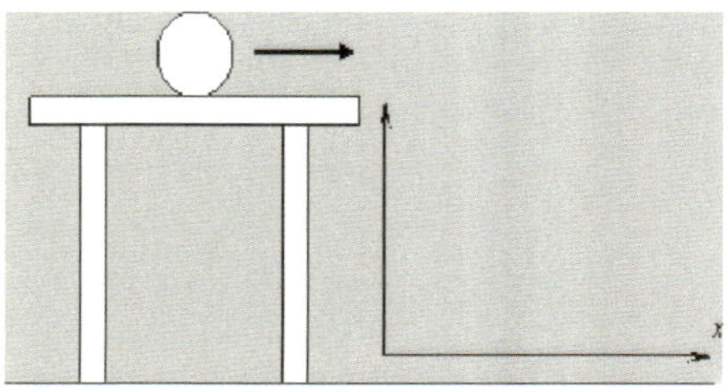

While the ball is on the table we observe that the initial x-component of velocity (v_{0x}) is 10 m/s (constant), the initial y-component of velocity is 0 m/s, the x-component of acceleration is 0 m/s^2 and the y-component of acceleration is 0 m/s^2. The components of acceleration and velocity are those parts of the velocity or acceleration that points in the x or y direction. Let us observe what happens the instant the ball leaves the table. The initial velocity in the y-direction is still zero and the initial velocity in the x-direction remains 10 m/s. However, the ball is no longer in contact with the table and it falls freely. The gravitational acceleration of the ball is down. In this case, the motions in the horizontal and vertical directions should be analyzed independently. Horizontally, there is no acceleration in the horizontal direction, therefore, the x-component of velocity is constant,

$$a_x = 0$$

$$v_x = \text{constant}$$

$$x = v_{0x} t$$

In the vertical direction there is an acceleration equal to the acceleration of gravity. Therefore, the velocity in the vertical direction changes as below

$$a_y = g$$

$$v_y = v_{0y} + gt$$

Translational Dynamics

It is the study of translational motion which investigates the force effect on the motion which uses the basic equations and analyze what is movement of objects and how they get affected by various kind forces. The tools that are used to study movement are many among which Newton's laws of motion is one. The First law states that any body either in rest or motion will not change its state unless we apply any external force upon it, while the second law gives the relationship between mass and acceleration where the force is product of mass and acceleration where its difficult to change translational motion of larger mass object than the smaller one.

Translational Movement

Movement itself is nothing but the motion. Translational movement of any body can occur along a curved path or along the straight lines. Since the body undergoes this changes in its motion when a force acts onit, the movement takes place depending on the mass, the body having larger mass would have less translational motion whereas the body having lighter mass has much movement. This study of translational movement is known as translational dynamics.

Rectilinear Motion

In Motion, we went through many types of motion like Uniform Motion, non uniform Motion and also velocity, acceleration etc. Now, we will study the Rectilinear Motion where a body moves in a straight line path.

Let us Consider a body moving in a straight line, if we can see every particle (which the body is made of) is traveling the same path, we can call it as Rectilinear Motion.

or

A body is said to experience rectilinear motion, if the two particles in that body travel the same distance along two parallel straight lines.

or

The Motion attained by a body so that every particle of a body follows a straight line path is defined as the motion which occurs in a straight line.

The parameters involved in a rectilinear motion are displacement, velocity, acceleration and time.

Rectilinear Motion

Rectilinear Motion Equations

The equations for rectilinear motion can be derived from the first principle. Let us assume that an object starts with an initial velocity **U** and with a constant acceleration **a**. A simple calculation tells us that after a time **t**, the increase in velocity during that time t is acceleration×time, that is just at. Since the object already had an initial velocity u at the start, the quantity is also added to the increase in the velocity to determine the actual final velocity **v**. In other words, the first equation for final velocity is framed as,

$$v = u + at$$

Now, let us try to know what could be the final velocity of an object after certain displacement, instead of, after certain time. Let us assume that the object has traveled a distance of S in a time t. Since we are discussing the linear motion, the distance traveled is given by,

S = average velocity during time t^2.

$$= \frac{1}{2}(u + v) \times t$$

From the first equation we derived, we can solve for t as,

$$t = \frac{(v-u)}{a}$$

Plugging this for t in the equation,

$$s = \left(\frac{1}{2}\right)(\mathrm{u+v})\frac{(u-v)}{a}$$ which becomes as,

$$\mathbf{2aS} = \left(\mathbf{v}^2 - \mathbf{u}^2\right)\text{or,}$$

$$v^2 - u^2 = 2aS.$$

This is another equation for final velocity in terms of initial velocity and distance.

Now, let us derive an equation for the displacement in terms of initial velocity, time and acceleration.

As mentioned earlier,

S = Average velocity during time t^2

$$= \frac{1}{2}(\mathrm{u+v})\times \mathrm{t}$$

Since v = u + at,

$$S = \frac{1}{2}[\mathrm{u+(u+at)}]\times \mathrm{t}$$

$$= \left(\frac{1}{2}\right)(2\mathrm{u+at})(\mathrm{t})$$

or,

$$\mathrm{S = ut} + \left(\frac{1}{2}\right)\mathrm{at}^2$$

Rectilinear Motion Problems

Example 1

A man is rotating a sling with a stone in a 2 m long at the rate of 18 revolutions every 9 seconds. Find the rectilinear speed of the stone.

Solution

Now the distance traveled by the stone in 9 s is d $= (18)\times 2\pi \mathrm{r}$

$$= 18\times 2\pi \times 2\mathrm{m}$$

$$= 288\pi \mathrm{m}.$$

So the Rectilinear speed of the stone is $v = \frac{d}{t}$

$$= \frac{288\pi m}{9s}$$

$$= 32\pi m / s.$$

Example 2

A Clown riding a uni-cycle has the wheel base diameter as 60 cm. if the wheel rotates at the rate of 120 revolutions per minute, then find the Rectilinear speed at which the clown is traveling.

Solution

Step 1:

A clown riding a uni-cycle has the wheel base diameter as 60 cm. if the wheel rotates at the rate of 120 revolutions per minute, then find the Rectilinear speed at which the clown is traveling.

Step 2:

The Angular speed (ω) of the wheel is $(2\pi)\times 120{=}240\pi\times\text{min}$

The wheels have a radius of 30 cm $[\frac{60\text{m}}{2}{=}30\text{cm}]$

The Rectilinear speed = $v = r\times$ (angular speed)

$$v = r\omega$$

Rectilinear speed $v = r\omega = 30\times 240\pi/\text{min} = 7200\text{cm/min}.$

Example 3

Below is given some practical problem for Rectilinear motion under gravity. Determine Velocity and Height h.

Solution

The most important practical problem of rectilinear motion is the motion under gravity.

For motions under gravity, the equations can be simplified, as the acceleration in such cases is the acceleration due to gravity, which is known to us. In customary units, the equations can be rewritten (taking the approximate value of acceleration due to gravity as 32ft/sec²) as,

$$\text{v} \quad \text{u+32t and}$$

$$\text{v}^2 - \text{u}^2 = 64\text{h or}$$

$$\text{v} = \sqrt{(\text{u}^2 + 64\text{h}}$$

Considering vertical motion under gravity, we replace the distance s by height h in feet.

As a special case, when an object is freely dropped vertical (which means the initial velocity is 0) from a height h in feet, the final velocity of the object v in ft/sec can be simply given as,

$$v = 8\sqrt{(h)}$$

On the other hand, we can estimate the maximum height that an object can reach in feet, when thrown up vertically as,

$$0^2 - u^2 = -64h$$

It may be noted that the final velocity in this case is 0 and we used the negative sign on the right because the gravity effect is a negative acceleration in this case.

Therefore, the maximum height is given by,

$$h = \frac{(u^2)}{(64)}.$$

It may be difficult to know the initial velocity of an upward throw. It is not an easily measurable quantity where as the time of travel of the object can easily be measured.

The Same height can also be figured out without knowing the initial velocity in the following way,

$$s = ut + \frac{1}{2}at^2 \text{ or}$$

$h = 0+16t^2$ (replacing s by h and plugging in the approximate value of acceleration due to gravity n ft/sec²)

$$= 16t^2.$$

Curvilinear Motion

General Curvilinear Motion

Curvilinear motion is defined as motion that occurs when a particle travels along a curved path. The curved path can be in two dimensions (in a plane), or in three dimensions. This type of motion is more complex than rectilinear (straight-line) motion.

Three-dimensional curvilinear motion describes the most general case of motion for a particle.

To find the velocity and acceleration of a particle experiencing curvilinear motion one only needs to know the position of the particle as a function of time.

Let's say we are given the position of a particle *P* in three-dimensional Cartesian (*x*,*y*,*z*) coordinates, with respect to time, where

$$x_p = x_p(t)$$

$$y_p = y_p(t)$$

$$z_p = z_p(t)$$

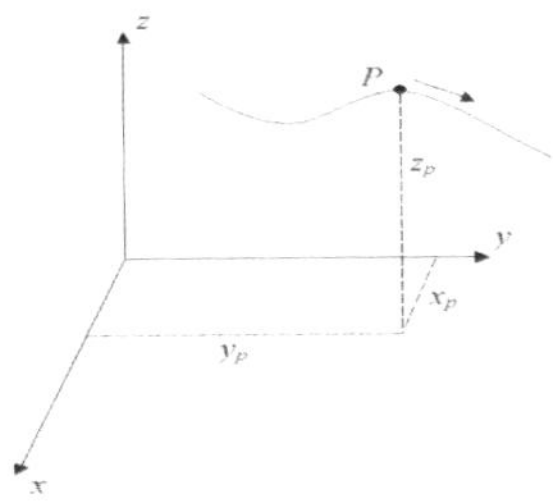

The velocity of the particle *P* is given by

$$v_x = \frac{dx_p}{dt}$$

$$v_y = \frac{dy_p}{dt}$$

$$v_z = \frac{dz_p}{dt}$$

The acceleration of the particle *P* is given by

$$a_x = \frac{d^2x_p}{dt^2}$$

$$a_y = \frac{d^2y_p}{dt^2}$$

$$a_z = \frac{d^2z_p}{dt^2}$$

As you can see, if we know the position of a particle as a function of time, it is a fairly simple exercise to find the velocity and acceleration. You simply take the first derivative to find the velocity and the second derivative to find the acceleration.

The magnitude of the velocity of particle *P* is given by

$$v_p = \sqrt{v_x^2 + v_y^2 + v_z^2}$$

The magnitude of the acceleration of particle *P* is given by

$$a_p = \sqrt{a_x^2 + a_y^2 + a_z^2}$$

Note that the direction of velocity of the particle *P* is always tangent to the curve. But the direction of acceleration is generally *not* tangent to the curve.

However, the acceleration component tangent to the curve is equal to the time derivative of the magnitude of velocity of the particle P (along the curve). In other words, if v_t is the magnitude of the particle velocity (tangent to the curve), the acceleration component of the particle tangent to the curve (a_t) is simply

$$a_t = \frac{dv_t}{dt}.$$

In addition, the acceleration component normal to the curve (a_n) is given by

$$a_n = \frac{v_t^2}{R}.$$

where R is the radius of curvature of the curve at a given point on the curve (x_p, y_p, z_p).

The figure below illustrates the acceleration components a_t and a_n at a given point on the curve (x_p, y_p, z_p).

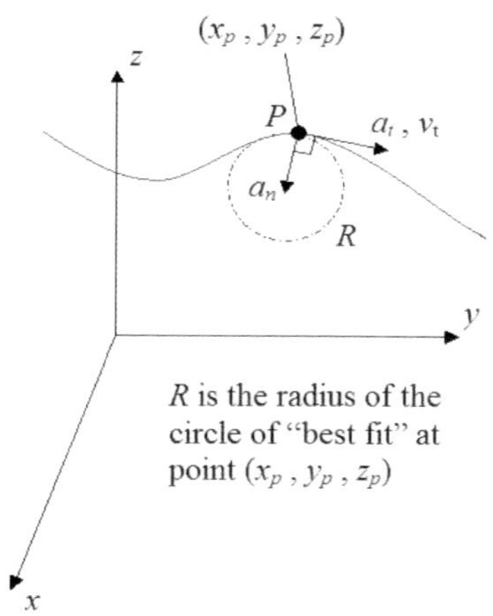

For the specific case where the path of the blue curve is given by $y = f(x)$ (two-dimensional motion), the radius of curvature R is given by:

$$R = \left| \frac{\left(1 + \left(\frac{dy}{dx}\right)^2\right)^{3/2}}{\frac{d^2y}{dx^2}} \right|.$$

where $|x|$ means the "absolute value" of x. For example, $|-2.5| = 2.5$, and $|3.1| = 3.1$.

However, it is usually not necessary to know the radius of curvature R along a curve. But nonetheless, it is informative to understand it on the basis of its relationship to the normal acceleration (a_n).

Curvilinear Motion in Polar Coordinates

It is sometimes convenient to express the planar (two-dimensional) motion of a particle in terms of polar coordinates (R, θ), so that we can explicitly determine the velocity and acceleration of the particle in the radial (R-direction) and circumferential (θ -direction). For this type of motion, a particle is only allowed to move along the radial R-direction for a given angle θ.

For a particle P defined in polar coordinates (as shown below), we can derive a general equation for its radial velocity (v_r), radial acceleration (a_r), circumferential velocity (v_c), and circumferential acceleration (a_c).

Note that the circumferential direction is perpendicular to the radial direction.

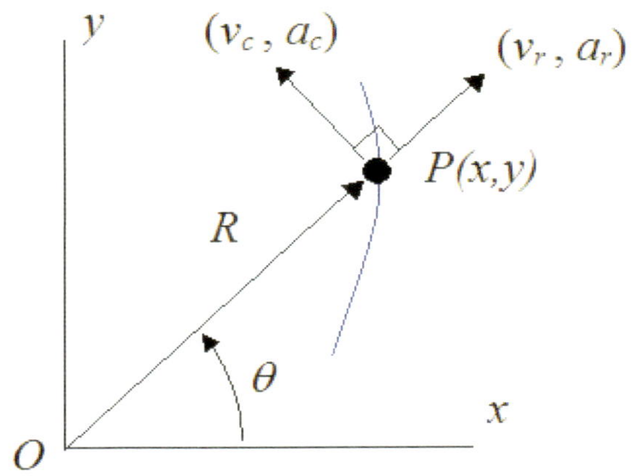

The position of the particle P is given with respect to time, where

$$x(t) = R\cos\theta$$

$$y(t) = R\sin\theta$$

To find the velocity, take the first derivative of *x(t)* and *y(t)* with respect to time:

$$\frac{dx}{dt} = \frac{dR}{dt}\cos\theta - R\sin\theta.\frac{d\theta}{dt}$$

$$\frac{dy}{dt} = \frac{dR}{dt}\sin\theta - R\cos\theta.\frac{d\theta}{dt}$$

To find the acceleration, take the second derivative of *x(t)* and *y(t)* with respect to time:

$$\frac{d^2x}{dt^2} = \frac{d^2R}{dt^2}\cos\theta - 2\frac{dR}{dt}\sin\theta.\frac{d\theta}{dt}$$
$$- R\cos\theta.\left(\frac{d\theta}{dt}\right)^2 - R\sin\theta.\frac{d^2\theta}{dt^2}$$

$$\frac{d^2y}{dt^2} = \frac{d^2R}{dt^2}\sin\theta + 2\frac{dR}{dt}\cos\theta.\frac{d\theta}{dt}$$

$$-R\sin\theta.\left(\frac{d\theta}{dt}\right)^2 - R\cos\theta.\frac{d^2\theta}{dt^2}$$

Without loss of generality we can evaluate the velocities and accelerations at angle $\theta = 0$, knowing that (at this angle) radial velocity and radial acceleration is in the x-direction, and circumferential velocity and circumferential acceleration is in the y-direction.

Setting $\theta = 0$ we have:

$$v_r = \frac{dx}{dt}$$

$$v_r = \frac{dR}{dt}$$

$$a_r = \frac{d^2x}{dt^2}$$

$$a_r = \frac{d^2R}{dt^2} - R\left(\frac{d\theta}{dt}\right)^2$$

$$v_c = \frac{dy}{dt}$$

$$v_c = R\frac{d\theta}{dt}$$

$$a_c = \frac{d^2y}{dt^2}$$

$$a_c = 2\frac{dR}{dt}.\frac{d\theta}{dt} + R\frac{d^2\theta}{dt^2}$$

Equations describe the curvilinear motion of a particle P in polar coordinates.

The term $d\theta/dt$ is called angular velocity. It has units of rad/s. One rad (radian) = 57.296 degrees.

The term $d^2\theta/dt^2$ is called angular acceleration. It has units of rad/s^2.

Since v_r and v_c are perpendicular to each other, the magnitude of the velocity of particle P is given by

$$v_p = \sqrt{v_r^2 + v_c^2}\,.$$

Since a_r and a_c are perpendicular to each other, the magnitude of the acceleration of particle P is given by

$$a_p = \sqrt{a_r^2 + a_c^2}\cdot$$

Example Problem for Curvilinear Motion

A slotted link is rotating about fixed pivot O with a counterclockwise angular velocity of 3 rad/s, and a clockwise angular acceleration of 2 rad/s². The movement of the link is causing a rod to slide along the curved channel, as shown. The radius of the channel as a function of θ is given by, $R = 0.7\theta$ (with R in meters and θ in radians). Determine the velocity and acceleration components of the rod at $\theta = 45°$.

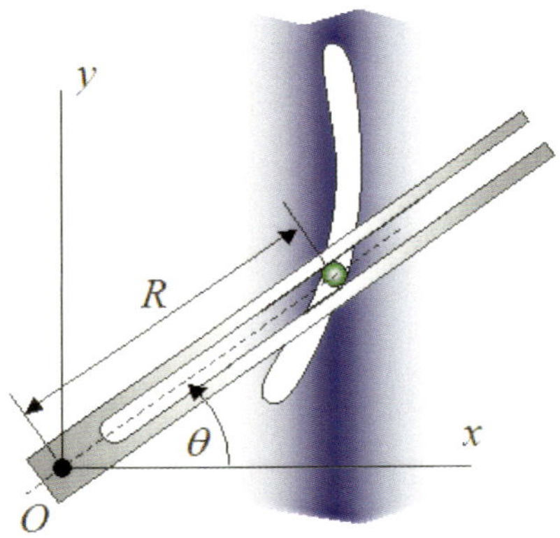

The angle $\theta = 45°$ is equal to $\pi/4$ radians. In the equations, counterclockwise angular velocity is positive, and clockwise angular acceleration is negative.

The radial velocity of the rod is given by equation:

$$v_r = \frac{dR}{dt} = 0.7\frac{d\theta}{dt} = 0.7(3) = 2.1\,\text{m/s}$$

(The radial velocity is in the direction of increasing R).

The circumferential velocity of the rod is given by equation:

$$v_c = R\frac{d\theta}{dt} = 0.7\left(\frac{\pi}{4}\right).(3) = 1.65\,\text{m/s}$$

(The circumferential velocity is in the direction of increasing θ).

The radial acceleration of the rod is given by equation:

$$a_r = \frac{d^2R}{dt^2} - R\left(\frac{d\theta}{dt}\right)^2$$

$$= 0.7\frac{d^2\theta}{dt^2} - 0.7\left(\frac{\pi}{4}\right).(3)^2$$

$$= 0.7(-2) - 0.7\left(\frac{\pi}{4}\right).(3)^2 = -6.35\,\text{m/s}^2$$

(The radial acceleration is in the direction of decreasing R).

The circumferential acceleration of the rod is given by equation:

$$a_c = 2\frac{dR}{dt}.\frac{d\theta}{dt} + R\frac{d^2\theta}{dt^2}$$

$$= 2.\left(0.7\frac{d\theta}{dt}\right).\frac{d\theta}{dt} - 0.7\left(\frac{\pi}{4}\right).(-2)$$

$$= 2.(0.7(3)).(3) - 0.7\left(\frac{\pi}{4}\right).(-2)$$

$$= 13.7\,\text{m/s}^2$$

(The circumferential acceleration is in the direction of increasing θ).

Oscillatory Motion

An oscillating motion in a mechanical system is swinging side to side. It can be translated into a rotary motion (turning around in a circle) by a peg-and-slot. Likewise, rotary motion can be changed to oscillating motion by the same method.

Every body at rest is in its equilibrium position. At this position, no external force is acting on it. Therefore, the net force acting on the body is zero. Now, if this body is displaced a little from its equilibrium position, a force acts on the body which tries to bring back the body to its equilibrium position. This force is the restoring force and it gives rise to oscillations or vibrations.

For example, consider a ball that is placed in a bowl. It will be in its equilibrium position. If displaced a little from this position, it will perform oscillations in the bowl. Therefore, every oscillatory motion is periodic but all periodic motions are not oscillatory. Circular motion is a periodic motion but not oscillatory motion.

There is no significant difference between oscillations and vibrations. When the frequency is low, we call it oscillatory motion and when the frequency is high, we call it vibrations. Simple harmonic motion is the simplest form of oscillatory motion. This motion takes place when the restoring force acting on the body is directly proportional to its displacement from its equilibrium position.

In practice, Oscillatory motion eventually comes to rest due to damping or frictional forces. However, we can force them by means of some external forces. A number of oscillatory motions together form waves like electromagnetic waves.

Variables of Oscillation

- Amplitude is the maximum displacement from the equilibrium point. If a pendulum swings one centimeter from the equilibrium point before beginning its return journey, the amplitude of oscillation is one centimeter.
- Period is the time it takes for a complete round trip by the object, returning to its initial position. If a pendulum starts on the right and takes one second to travel all the way to the left and another second to return to right, its period is two seconds. Period is usually measured in seconds.
- Frequency is the number of cycles per unit of time. Frequency equals one divided by the period. Frequency is measured in Hertz, or cycles per second.

Displacement in Oscillatory Motion

Displacement of a particle is a change in its position vector. In oscillatory motion, it simply displacement means a change in any physical property with time.

Consider a block attaches to a spring, which in turn is fixed to a rigid wall. We measure the displacement of the block from its equilibrium position. In oscillatory motion, we can represent the displacement by a mathematical function of time. One of the simplest periodic functions is given by,

$$f(t) = A\cos\omega t$$

If the argument, ωt, is increased by an integral multiple of 2π radians, the value of the function remains the same. Therefore, it is periodic in nature and its period T is given by,

$$T = 2\pi/\omega$$

Thus, the function f(t) is periodic with period T. ∴ f(t) = f(t + T). Now, if we consider a sine function, the result will be same. Further, taking a linear combination of sine and cosine functions is also a periodic function with period T.

$$f(t) = A\sin\omega t + B\cos\omega t$$

Taking A = Dcos∅ and B = Dsin∅ equation V becomes, f(t) = Dsin(ωt + ∅). In this equation D and ∅ are constant and they are given by,

$$D = \sqrt{A^2 + B^2} \text{ and } \varnothing = \tan^{-2}(B/A)$$

Therefore, we can express any periodic function as a superposition of sine and cosine functions of different time periods with suitable coefficients. The period of the function is 2π/ω.

Simple Harmonic Motion

Simple harmonic motion, in physics, repetitive movement back and forth through an equilibrium, or central, position, so that the maximum displacement on one side of this position is equal to the maximum displacement on the other side. The time interval of each complete vibration is the same, and the force responsible for the motion is always directed toward the equilibrium position and is directly proportional to the distance from it.

Many physical systems exhibit simple harmonic motion (assuming no energy loss): an oscillating pendulum, the electrons in a wire carrying alternating current, the vibrating particles of the medium in a sound wave, and other assemblages involving relatively small oscillations about a position of stable equilibrium.

A specific example of a simple harmonic oscillator is the vibration of a mass attached to a vertical spring, the other end of which is fixed in a ceiling. At the maximum displacement $-x$, the spring is under its greatest tension, which forces the mass upward. At the maximum displacement $+x$, the spring reaches its greatest compression, which forces the mass back downward again. At either position of maximum displacement, the force is greatest and is directed toward the equilibriumposition, the velocity (v) of the mass is zero, its acceleration is at a maximum, and the mass changes direction. At the equilibrium position, the velocity is at its maximum and the acceleration (a) has fallen to zero. Simple harmonic motion is characterized by this changing acceleration that always is directed toward the equilibrium position and is proportional to the displacement from the equilibrium position. Furthermore, the interval of time for each complete vibration is constant and does not depend on the size of the maximum displacement. In some form, therefore, simple harmonic motion is at the heart of timekeeping.

The motion is called harmonic because musical instruments make such vibrations that in turn cause corresponding sound waves in air. Musical sounds are actually a combination of many simple harmonic waves corresponding to the many ways in which the vibrating parts of a musical instrument oscillate in sets of superimposed simple harmonic motions, the frequencies of which are multiples of a lowest fundamental frequency. In fact, any regularly repetitive motion and any wave, no matter how complicated its form, can be treated as the sum of a series of simple harmonic motions or waves.

Linear Simple Harmonic Motion

Linear simple harmonic motion is defined as the linear periodic motion of a body in which the restoring force is always directed towards the equilibrium position or mean position and its magnitude is directly proportional to the displacement from the equilibrium position. All simple harmonic motions are periodic in nature but all periodic motions are not simple harmonic motions.

Now, take the previous example of the string. Let its mass be m. The acceleration of the body is given by,

$$a = F / m = -kx / m = -\omega^2 x$$

Here, $k / m = \omega^2$ (ω is the angular frequency of the body).

Concepts of Simple Harmonic Motion (S.H.M)

- Amplitude: The maximum displacement of a particle from its equilibrium position or mean position is its amplitude. Its S.I. unit is the metre. The dimensions are $[L^1M^0T^0]$. Its direction is always away from the mean or equilibrium position.
- Period: The time taken by a particle to complete one oscillation is its period. Therefore,

period if S.H.M. is the least time after which the motion will repeat itself. Thus, the motion will repeat itself after nT. where n is an integer.

- Frequency: Frequency of S.H.M. is the number of oscillations that a particle performs per unit time. S.I. unit of frequency is hertz or r.p.s(rotations per second). Its dimensions are $[L^oM^oT^{-1}]$.
- Phase: Phase of S.H.M. is its state of oscillation. Magnitude and direction of displacement of particle represent the phase. The phase at the beginning of the motion is known as Epoch(α).

Simple Harmonic Motion Equations

The motion equation for simple harmonic motion contains a complete description of the motion, and other parameters of the motion can be calculated from it.

$$y = A \sin \omega t = A \sin \sqrt{\frac{k}{m}} t$$

The velocity and acceleration are given by

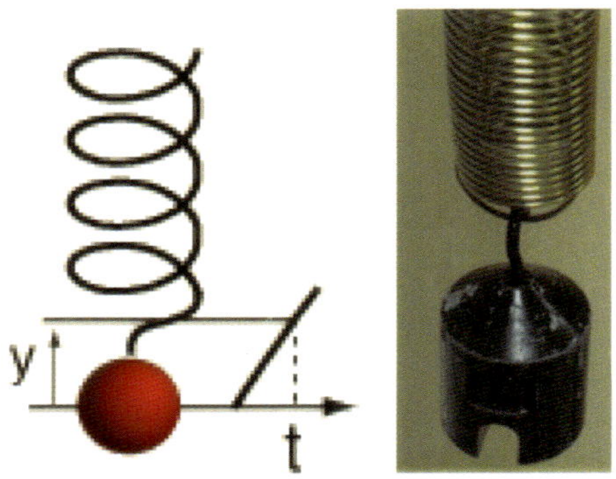

$$v = \omega A \cos \omega t$$

$$a = -\omega A \sin \omega t = -\omega^2 y$$

The total energy for an undamped oscillator is the sum of its kinetic energy and potential energy, which is constant at

$$E = KE + PE = \frac{1}{2kA^2}.$$

Simple Harmonic Motion Calculation

The motion equations for simple harmonic motion provide for calculating any parameter of the motion if the others are known.

$$y = A \sin \omega t = A \sin \sqrt{\frac{k}{m}} t$$

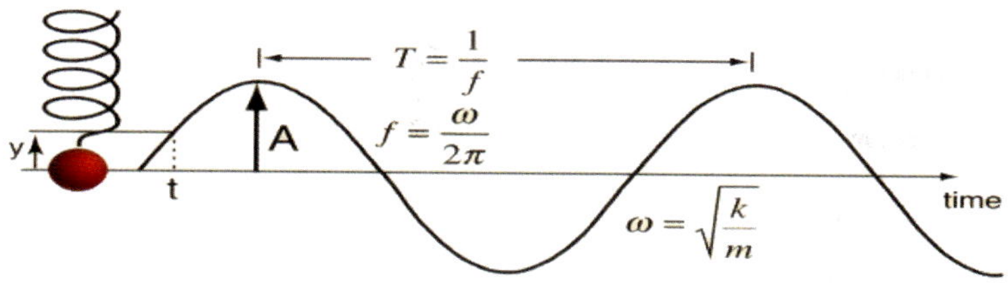

Displacement, Velocity and Acceleration Vectors of a Particle Undergoing Simple Harmonic Motion

We set this out mathematically, using a differential equation as in equation (4). We specify the equation in terms of the forces acting on the object. The acceleration is the second derivative of the position with respect to time and this is proportional to the position with respect to time. The minus sign indicates that the position is in the opposite direction to the acceleration.

$$m\, y''(t) = -k\, y(t)$$

The derivation of the solution can be found here

For which the general solution is a wave like solution.

$$y(t) = c_1 \cos(\omega t) + c_2 \sin(\omega t)$$

Where, ω is the angular frequency. ($\omega = 2\pi f$) The values of c_1 and c_2 are determined by the initial conditions. Specifically, $c_1 = y_0$ and $c_2 = v_0/\omega$ These two initial conditions specify the starting position and the initial velocity.

The general solution can also be written more compactly as

$$y(t) = A\cos(\omega t - \varphi)$$

Where $\varphi = \tan^{-1}(\omega y_0 / v_0), A = (y_0^{\,2} + (v_0/\omega)^2)^{1/2}$

Differentiating once with respect to time, we obtain the velocity. (The derivative of cos x = - sin x)

$$v = y'(t) = -\,\omega A \sin(\omega t - \varphi)$$

Finally, the acceleration is the derivative of the velocity with respect to time. (The derivitive of -sin x = - cos x)

$$a = y''(t) = -\,\omega^2 A\cos(\omega t - \varphi)$$

Substituting equations (5) and (7) into equation (4) we verify that this does indeed satisfy the equation for simple harmonic motion. With the constant of proportionality $k = \omega^2$.

Thus,

$$a(t) = -\,\omega^2 y(t)\cdot$$

The time for the maximum velocity and acceleration can be determined from these equations. From equation the maximum magnitude of the velocity occurs when sin(ωt - φ) is 1 or -1. Therefore the maximum velocity is $\pm\omega A$. Intuitively, we can imagine that this velocity occurs when the oscillating system has reached the equilibrium position and is about to overshoot. The minus sign indicates the direction of travel is in the opposite direction.

The maximum acceleration occurs where the argument of cosine in equation is also -1 or 1. Thus the maximum acceleration is $\pm\omega^2 A$ which occurs at the ends of the oscillations, as this is where the direction changes.

Examples of SHM

Spring Mass System

Consider a spring of spring constant *k* conected to a mass *m*. If the mass is displaced from its equilibrium position by a distance *x* a force *F* will act in the opposite direction to the displacement. From Hooke's Law the magnitude of the force is given by

$$F = -k\,x$$

When the mass is released it the force will act on the mass to bring it back to its equilibrium position. However, if there is no friction the inertia of the mass will cause it to overshoot the equilibrium position and the force will act in the opposite direction slowing it down and pulling it back.

The action of this force on the mass keeps it oscillating backwards and forwards.

The spring mass system consists of a spring with a spring constant of *k* attached to a mass, *m*.The mass is displaced a distance *x* from its equilibrium position work is done and potential energy is stored in the spring. If the mass is displaced by a small distance *dx*, the work done in stretching the spring is given by $dW = F\,dx$.

The force on the spring is assumed to obey Hooke's law, therefore, the restoring force is proportional to the extension. The work done is then $dW = -k\,x\,dx$

$$\text{Making the total work } W = -\int k\,x\,dx = -1/2k\,x^2 + C$$

At the time of release, the energy of the system will consist totally of potential energy. PE= $-1/2\,kx^2$ and the potential energy as a function of time is

$$PE(t) = 1/2k\,A\cos^2(\omega t - \varphi)$$

As the spring pulls the mass toward the equilibrium position, the potential energy is transformed into kinetic energy until at the equilibrium position the kinetic energy will be maximised. The *KE* is $1/2\,mv^2$ and the maximum velocity occurs at x=0. From equation (6), the velocity will be ω A

Therefore the kinetic energy is $1/2\,m\omega^2 A^2 \sin^2(\omega t - \varphi)$

From the conservation of mechanical energy, the sum of energy between kinetic energy and potential energy will always be the same.

$$U = KE(t) + PE(t) = 1/2m\omega^2 A^2 \sin^2(\omega t - \varphi) + 1/2k\ A\cos^2(\omega t - \varphi)$$
$$1/2m\omega^2 x^2 = 1/2k\ x^2$$

is transferred into kinetic energy by moving the mass. For a particle undergoing simple harmonic motion, the displacement x is given by equation 1. The potential energy is gradually transferred to kinetic energy. Kinetic energy is given by 1/2 mv^2. The velocity is given by equation (6). Summing the kinetic energy and potential energy we obtain,

$$U = (1/2)A^2(k+m)$$
$$= 1/2\ kA^2$$

Since $k = m\omega^2$ and $\cos^2 x + \sin^2 x = 1$

Simple Pendulum

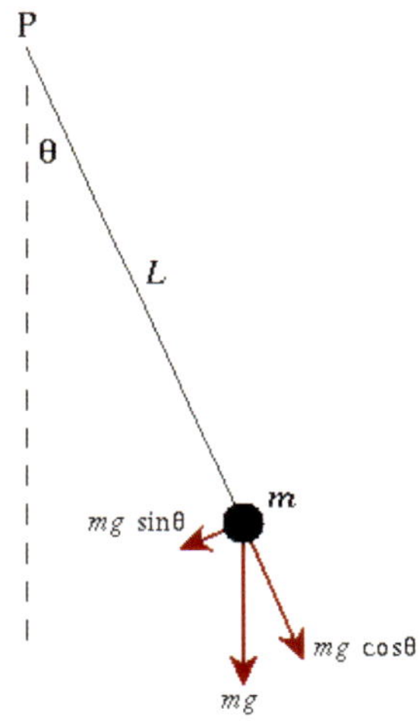

Another common example used to illustrate simple harmonic motion is the simple pendulum. This idealised system has a one end massless string suspended a mass *m* and the other end fixed to a stationary point. If the mass is displaced by a small distance, the angle moved is small.

The torque on the fixed point P is $\tau = I\alpha$

$$- mg \sin\theta(t)L = mL^2\,\theta''(t)$$

$$\theta''(t) + g/L\,\theta(t) = 0.$$

This has the same form as simple harmonic motion equation, $x''(t) - \omega^2 x(t)$, and so the solution is $\theta(t) = \theta_0 \cos(\omega t$ - &phi).

The angular frequency is $\omega = (g/L)^{1/2}$.

It is interesting to note that the mass does not appear in this equation. This means that the frequency of the period only depends on the length of the string and the force of gravity. Pendulums with shorter strings will oscillate faster than pendulums with longer strings. And the same pendulum on the moon, where the force of gravity is 1/6th that of the gravity on the Earth, will also take longer to oscillate.

We have glossed over one important aspect, in that this analysis is true only for small angles of theta. We had to make the approximation that sin θ is aproximately the same as θ which is true only for small angles. The real differential equation

$$\theta''(t) + g/L \sin\theta(t) = 0$$

is non-linear and cannot be solved analytically.

Torsional Pendulum

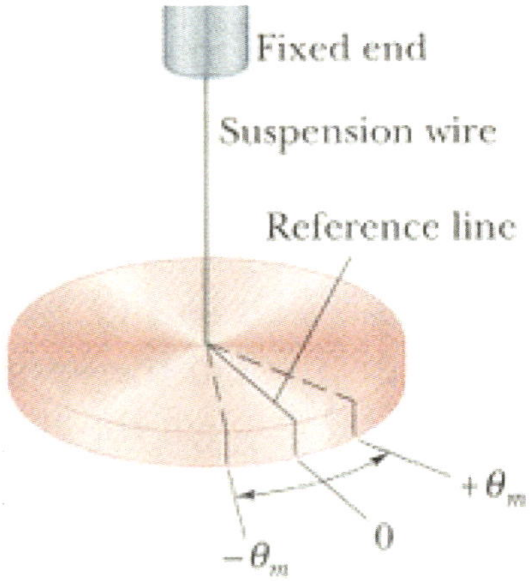

Figure. A torsional pendulum.

A mass suspended to a fixed support by a thin wire can be made to twist about its axis. This is known as a torsional pendulum. The mass attached to the wire rotates in the horizontal plane. In this case θ is the angle of rotation. When the wire is untwisted and in equilibrium, the angle θ is 0 degrees. It is the twisting of the wire that creates a restoring torque due to the resistance of the wire to the deformation. For small angles of θ the magnitude of the torque is proportional to the angle θ.

$$\tau = -k\theta$$

Where k is the torque constant of the wire. As with the simple pendulum the equation of motion is

$$\tau = I\alpha$$

Where I is the moment of inertia

$$k\theta(t) = -I\theta''(t)$$

Once again we have formed the equation for simple harmonic motion and can write the solution as $\theta(t) = A\cos(\omega t - \varphi)$

The angular frequency, ω is given by

$$\omega = (k/I)^{1/2}$$

4

Formulations of Classical Mechanics

In order to develop a complete understanding of classical mechanics, it is necessary to understand the formulations and concepts central to it, such as constraints, Lagrangian and Hamiltonian mechanics, Euler-Lagrange equation, canonical transformation, Koopman-von Neumann classical mechanics, analytical mechanics, etc. All these topics have been presented in detail in this chapter.

Constraint

The constraints are introduced as relations between the co- ordinates. Since there are a variety of different types of constraints, we will examine some of the possibilities and how to incorporate them into the Lagrangian or Hamiltonian. The simplest constraints are those that can be put into the following form

$$g_j(q_i) = 0$$

where g_j can be a function of all the generalized coordinates. Constraints of this form are called holonomic. Holonomic constraints can be either time independent (scleronomic) or time dependent (rheonomic). A few examples of Holonomic constraints are

Block sliding on a wedge $y - ax - b = 0$

Block sliding on an incline plane angle varies $y - \tan(\omega t) = 0$

Mass attached to a rod (pendulum) $r - \ell = 0$

Bead on a rigid wire $f(q_i) = 0$

All constriants that are not given by $g_j(q_i) = 0$ are called non-holonomic. One example is a particle confined to a box, the constraint is given by $f(\vec{r}) > 0$. Another far more useful example is one that is given in terms of derivatives

$$\sum_i a_{ji}(q_i)dq_i + a_{jt}dt = 0$$

where the sum is over the total number of generalized coordinates. There are two cases that we have to consider for this type of constraint. The first is if the sum can be written as a total differential

$$\sum_i a_i(q_i)dq_i \; where \; a_i = \frac{\partial g}{\partial q_i}$$

where g is the constraint. This equation can be integrated and therefore put into holonomic form. Note that the condition of integrability is given by

$$\frac{\partial a_j}{\partial q_i} = \frac{\partial^2 g}{\partial q_i \partial q_j} = \frac{\partial a_i}{\partial q_j}$$

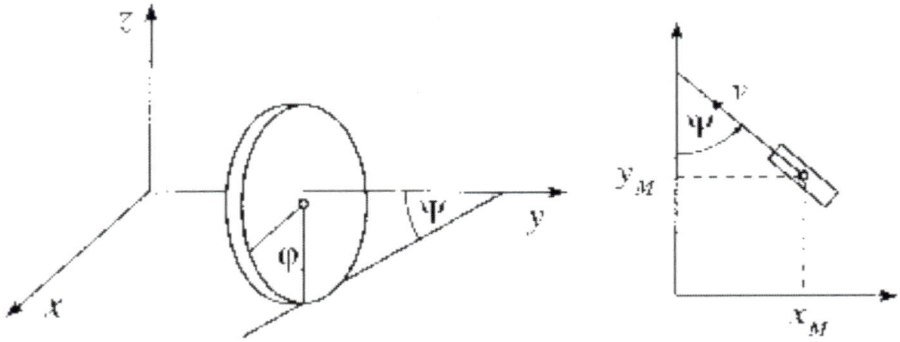

Wheel constrained to roll on a plane surface.

if this condition is not satisfied, then the constraint is non-holonomic. As an example, consider a wheel rolling on a plane. The condition for rolling is given by

$$v = a\dot{\phi}$$

where a is the radius of the wheel. The x and y components of the center of mass velocity are given by

$$\dot{x}_m = -v\sin\psi \quad \dot{y}_m = v\cos\psi$$

Combining these equations together, we arrive at

$$dx_m + a\sin(\psi)d\phi = 0$$

$$dy_m + a\cos(\psi)d\phi = 0$$

if ψ is a constant, then this equation is integrable, if it is only known after we solve the equations of motion, then this is a non-holonomic constraint.

Constrained Systems

In many cases in mechanics, the motion of bodies is constrained in some way: for example, a massive bead may be constrained to move along a bent wire of certain shape; a massive cylinder may be rolling along a surface (but not sliding or flying around); or two masses may be connected by a rigid stick of fixed length.

In each of these cases, there are forces acting on the constrained bodies. In the above examples, the wire produces a force on the bead, the plane acts by the force of friction on the cylinder, and the stick pulls or pushes on the two masses. These forces may vary in time and we do not know the magnitude of these forces in advance. We know, however, that these forces are at every time exactly such as to guarantee that the constraints hold. The bead would fly away if there were no forces acting on it, but the wire provides a force that keeps the bead in place. The two masses connected by a rigid stick experience a force from the stick that is exactly necessary to keep them at a constant distance from each other. (This is what it means that the stick is "rigid").

In the Newtonian approach to mechanics, these systems are treated by introducing variables $F_1, F_2, \ldots$ representing the unknown forces, and solving the system of equations for the unknown forces and accelerations. This procedure might be complicated; moreover, we are not always interested in the magnitudes of these unknown forces.

In the Lagrangian approach, there are two straightforward ways to treat constrained systems:

- The method of solving the constraints. In this method, we introduce the generalized coordinates in such a way that the constraints are automatically satisfied. For example, suppose a point mass is constrained to move along a circle of radius R. We might describe this situation by saying that the Cartesian coordinates x, y are constrained to satisfy the equation $x^2 + y^2 = R^2$. Now we can introduce the angle ϕ as the "generalized coordinate" and express the Cartesian coordinates of the point mass as $x = R\cos\phi, y = R\sin\phi$. These coordinates solve the constraint for all ϕ. The power of the Lagrangian approach is that any generalized coordinates are good enough; so we can now directly write the Lagrangian in terms of the function $\phi(t)$ and forget about the fact that the system is constrained. We shall automatically obtain the correct equations of motion.

- The method of Lagrange multipliers. In this method, we do not try to introduce new generalized coordinates that solve the constraints. (This may be difficult; not all algebraic equations can be solved!) Instead, we formulate the variational problem *in the presence of constraints*: the correct trajectory $q_i(t)$ is such that the action functional has an extremum while the constraint equations are satisfied. For example, if a point mass is constrained to move along a circle of radius R, but is otherwise unforced, then the Lagrangian is $L = \frac{m}{2}(\dot{x}^2 + \dot{y}^2)$ and the constraint is $x^2 + y^2 = R^2$. The correct trajectory $x(t), y(t)$ will be such that the integral $\int L dt$ has the minimum value while the constraint holds at every t. Thus we need to solve the problem of conditional minimization.

Types of Constraint

- First class constraints and second class constraints.
- Primary constraints, secondary constraints, tertiary constraints, quaternary constraints.
- Holonomic constraints, also called integrable constraints, (depending on time and the coordinates but not on the momenta) and Nonholonomic system.
- Pfaffian constraints.
- Scleronomous constraints (not depending on time) and rheonomous constraints (depending on time).

- Ideal constraints: those for which the work done by the constraint forces under a virtual displacement vanishes.

Koopman–von Neumann Classical Mechanics

Koopman and von Neumann developed a mathematical formulation of classical mechanics using concepts usually associated with quantum theory. In this formulation we have wave functions, Hilbert spaces, operators. The Koopman-von Neumann (KvN) formulation has proved useful as a technical device in several applications, notably in ergodic theory. It deserves to be better known among physicists, because it gives a new perspective on the conceptual foundations of quantum theory, and it may suggest new kinds of approximations and even new kinds of theories.

Let us recall the phase space formulation of distributions in classical mechanics, as is often used in statistical mechanics and kinetic theory. For simplicity in notation I'll write equations for a system with one degree of freedom, but their generalization is straightforward, and when it becomes necessary we'll add in complications without much ado.

The quantity of interest is the density ρ(x, p, t). It is a non-negative real quantity, to be interpreted as the probability that we'll find a particle with momentum p at point x at time t. using the measure R dx dp. Liouville's theorem in mechanics tells us that the flow of this fluid is incompressible. So we have

$$\text{or} \quad 0 = \frac{d\rho}{dt} = \frac{\partial \rho}{\partial t} + \dot{x}\frac{\partial \rho}{\partial x} + \dot{p}\frac{\partial \rho}{\partial \rho}$$

$$\begin{aligned} \frac{\partial \rho}{\partial t} &= -\dot{x}\frac{\partial \rho}{\partial x} - \dot{p}\frac{\partial \rho}{\partial \rho} \\ &= -\frac{\partial H}{\partial t}\frac{\partial \rho}{\partial x} + \frac{\partial H}{\partial t}\frac{\partial \rho}{\partial x} \end{aligned}$$

which serves as a dynamical equation for ρ. If we multiply both sides by i, we get an equation having a family resemblance to the Schrödinger equation.

Derivation Starting from the Liouville Equation

In the approach of Koopman and von Neumann (KvN), dynamics in phase space is described by a (classical) probability density, recovered from an underlying wavefunction – the Koopman–von Neumann wavefunction – as the square of its absolute value (more precisely, as the amplitude multiplied with its own complex conjugate). This stands in analogy to the Born rule in quantum mechanics. In the KvN framework, observables are represented by commuting self-adjoint operators acting on the Hilbert space of KvN wavefunctions. The commutativity physically implies that all observables are simultaneously measurable. Contrast this with quantum mechanics, where

$$\frac{d}{dt}\langle\Psi(t)|\hat{p}|\Psi(t)\rangle \quad =\langle\Psi(t)|-U'(\hat{x})|\Psi(t)\rangle.$$

Notice a close resemblance with Ehrenfest theorems in quantum mechanics. Applications of the product rule leads to

$$\begin{aligned}\langle d\Psi/dt|\hat{x}|\Psi\rangle+\langle\Psi|\hat{x}|d\Psi/dt\rangle &=\langle\Psi|\hat{p}/m|\Psi\rangle,\\ \langle d\Psi/dt|\hat{p}|\Psi\rangle+\langle\Psi|\hat{p}|d\Psi/dt\rangle &=\langle\Psi|-U'(\hat{x})|\Psi\rangle,\end{aligned}$$

into which we substitute a consequence of Stone's theorem $i|d\Psi(t)/dt\rangle=\hat{L}|\Psi(t)\rangle$ and obtain

$$\begin{aligned} im\langle\Psi(t)|[\hat{L},\hat{x}]|\Psi(t)\rangle &=\langle\Psi(t)|\hat{p}|\Psi(t)\rangle,\\ i\langle\Psi(t)|[\hat{L},\hat{p}]|\Psi(t)\rangle &=-\langle\Psi(t)|U'(\hat{x})|\Psi(t)\rangle.\end{aligned}$$

Since these identities must be valid for any initial state, the averaging can be dropped and the system of commutator equations for the unknown $\hat{L}$ is derived

$$im[\hat{L},\hat{x}]=\hat{p}, \qquad i[\hat{L},\hat{p}]=-U'(\hat{x}). \text{ (commutator eqs for L)}$$

Assume that the coordinate and momentum commute $[\hat{x},\hat{p}]=0$.

This assumption physically means that the classical particle's coordinate and momentum can be measured simultaneously, implying absence of the uncertainty principle.

The solution $\hat{L}$ cannot be simply of the form $\hat{L}=L(\hat{x},\hat{p})$ because it would imply the contractions $im[L(\hat{x},\hat{p}),\hat{x}]=0=\hat{p}$ and $i[L(\hat{x},\hat{p}),\hat{p}]=0=-U'(\hat{x})$. Therefore, we must utilize additional operators $\hat{\lambda}_x$ and $\hat{\lambda}_p$ obeying

$$[\hat{x},\hat{\lambda}_x]=[\hat{p},\hat{\lambda}_p]=i, \quad [\hat{x},\hat{p}]=[\hat{x},\hat{\lambda}_p]=[\hat{p},\hat{\lambda}_x]=[\hat{\lambda}_x,\hat{\lambda}_p]=0. \text{ (*KvN algebra*)}$$

The need to employ these auxiliary operators arises because all classical observables commute. Now we seek $\hat{L}$ in the form $\hat{L}=L(\hat{x},\hat{\lambda}_x,\hat{p},\hat{\lambda}_p)$. Utilizing KvN algebra, the commutator eqs for L can be converted into the following differential equations

$$mL'_{\lambda_x}(x,\lambda_x,p,\lambda_p)=p, \qquad L'_{\lambda_p}(x,\lambda_x,p,\lambda_p)=-U'(x).$$

Whence, we conclude that the classical KvN wave function $|\Psi(t)\rangle$ evolves according to the Schrödinger-like equation of motion

$$i\frac{d}{dt}|\Psi(t)\rangle=\hat{L}|\Psi(t)\rangle, \qquad \hat{L}=\frac{\hat{p}}{m}\hat{\lambda}_x-U'(\hat{x})\hat{\lambda}_p. \text{ (*KvN dynamical eq*)}$$

Let us explicitly show that KvN dynamical eq is equivalent to the classical Liouville mechanics.

Since $\hat{x}$ and $\hat{p}$ commute, they share the common eigenvectors

$$\hat{x}\,|\,x,p\rangle = x\,|\,x,p\rangle,\quad \hat{p}\,|\,x,p\rangle = p\,|\,x,p\rangle,\quad A(\hat{x},\hat{p})\,|\,x,p\rangle = A(x,p)\,|\,x,p\rangle, \text{ (xp eigenvec)}$$

with the resolution of the identity $1 = \int dxdp|\,x,p\rangle\langle x,p\,|$. Then, one obtains from equation (KvN algebra)

$$\langle x,p\,|\,\hat{\lambda}_x\,|\,\Psi\rangle = -i\frac{\partial}{\partial x}\langle x,p\,|\,\Psi\rangle,\qquad \langle x,p\,|\,\hat{\lambda}_p\,|\,\Psi\rangle = -i\frac{\partial}{\partial p}\langle x,p\,|\,\Psi\rangle.$$

Projecting equation (KvN dynamical eq) onto $\langle x,p\,|$, we get the equation of motion for the KvN wave function in the xp-representation

$$\left[\frac{\partial}{\partial t}+\frac{p}{m}\frac{\partial}{\partial x}-U'(x)\frac{\partial}{\partial p}\right]\langle x,p\,|\,\Psi(t)\rangle = 0.$$

(KvN dynamical eq in xp)

The quantity $\langle x,p\,|\,\Psi(t)\rangle$ is the probability amplitude for a classical particle to be at point with momentum x at time p. According to the axioms above, the probability density is given by $\rho(x,p;t) = \left|\langle x,p\,|\,\Psi(t)\rangle\right|^2$. Utilizing the identity

$$\frac{\partial}{\partial t}\rho(x,p;t) = \langle\Psi(t)\,|\,x,p\rangle\frac{\partial}{\partial t}\langle x,p\,|\,\Psi(t)\rangle + \langle x,p\,|\,\Psi(t)\rangle\left(\frac{\partial}{\partial t}\langle x,p\,|\,\Psi(t)\rangle\right)^*$$

as well as (KvN dynamical eq in xp), we recover the classical Liouville equation

$$\left[\frac{\partial}{\partial t}+\frac{p}{m}\frac{\partial}{\partial x}-U'(x)\frac{\partial}{\partial p}\right]\rho(x,p;t) = 0.$$

(Liouville eq)

Moreover, according to the operator axioms and (xp eigenvec),

$$\begin{aligned}\langle A\rangle &= \langle\Psi(t)\,|\,A(\hat{x},\hat{p})\,|\,\Psi(t)\rangle = \int dxdp\langle\Psi(t)\,|\,x,p\rangle A(x,p)\langle x,p\,|\,\Psi(t)\rangle\\ &= \int dxdp\,A(x,p)\langle\Psi(t)\,|\,x,p\rangle\langle x,p\,|\,\Psi(t)\rangle = \int dxdp\,A(x,p)\rho(x,p;t).\end{aligned}$$

Therefore, the rule for calculating averages of observable $A(x,p)$ in classical statistical mechanics has been recovered from the operator axioms with the additional assumption $[\hat{x},\hat{p}]=0$. As a result, the phase of a classical wave function does not contribute to observable averages. Contrary to quantum mechanics, the phase of a KvN wave function is physically irrelevant. Hence, nonexistence of the double-slit experiment as well as Aharonov–Bohm effect is established in the KvN mechanics.

Projecting KvN dynamical eq onto the common eigenvector of the operators $\hat{x}$ and $\hat{\lambda}_p$ (i.e., $x\lambda_p$ -representation), one obtains classical mechanics in the doubled configuration space, whose generalization leads to the phase space formulation of quantum mechanics.

Quantum Mechanics from the Operator Axioms (Derivation)

As in the derivation of classical mechanics, we begin from the following equations for averages of coordinate x and momentum p

$$m\frac{d}{dt}\langle x\rangle = \langle p\rangle, \qquad \frac{d}{dt}\langle p\rangle = \langle -U'(x)\rangle.$$

With the help of the operator axioms, they can be rewritten as

$$\begin{aligned} m\frac{d}{dt}\langle \Psi(t)|\,\hat{x}\,|\Psi(t)\rangle &= \langle \Psi(t)|\,\hat{p}\,|\Psi(t)\rangle, \\ \frac{d}{dt}\langle \Psi(t)|\,\hat{p}\,|\Psi(t)\rangle &= \langle \Psi(t)|-U'(\hat{x})|\Psi(t)\rangle. \end{aligned}$$

These are the Ehrenfest theorems in quantum mechanics. Applications of the product rule leads to

$$\begin{aligned} \langle d\Psi/dt\,|\,\hat{x}\,|\,\Psi\rangle + \langle \Psi\,|\,\hat{x}\,|\,d\Psi/dt\rangle &= \langle \Psi\,|\,\hat{p}/m\,|\,\Psi\rangle, \\ \langle d\Psi/dt\,|\,\hat{p}\,|\,\Psi\rangle + \langle \Psi\,|\,\hat{p}\,|\,d\Psi/dt\rangle &= \langle \Psi\,|-U'(\hat{x})|\,\Psi\rangle, \end{aligned}$$

into which we substitute a consequence of Stone's theorem

$$i\hbar\,|\,d\Psi(t)/dt\rangle = \hat{H}\,|\,\Psi(t)\rangle,$$

where $\hbar$ was introduced as a normalization constant to balance dimensionality. Since these identities must be valid for any initial state, the averaging can be dropped and the system of commutator equations for the unknown quantum generator of motion $\hat{H}$ are derived

$$im[\hat{H},\hat{x}] = \hbar\hat{p}, \qquad i[\hat{H},\hat{p}] = -\hbar U'(\hat{x}).$$

Contrary to the case of classical mechanics, we assume that observables of the coordinate and momentum obey the canonical commutation relation $[\hat{x},\hat{p}] = i\hbar$. Setting $\hat{H} = H(\hat{x},\hat{p})$, the commutator equations can be converted into the differential equations

$$mH'_p(x,p) = p, \qquad H'_x(x,p) = U'(x),$$

whose solution is the familiar quantum Hamiltonian

$$\hat{H} = \frac{\hat{p}^2}{2m} + U(\hat{x}).$$

Whence, the Schrödinger equation was derived from the Ehrenfest theorems by assuming the canonical commutation relation between the coordinate and momentum. This derivation as well as the derivation of classical KvN mechanics shows that the difference between quantum and classical mechanics essentially boils down to the value of the commutator $[\hat{x},\hat{p}]$.

Measurements

In the Hilbert space and operator formulation of classical mechanics, the Koopman von Neumann–wavefunction takes the form of a superposition of eigenstates, and measurement collapses the

KvN wavefunction to the eigenstate which is associated the measurement result, in analogy to the wave function collapse of quantum mechanics.

However, it can be shown that for Koopman–von Neumann classical mechanics *non-selective measurements* leave the KvN wavefunction unchanged.

KvN vs Liouville Mechanics

The KvN dynamical equation (KvN dynamical eq in xp) and Liouville equation (Liouville eq) are first-order linear partial differential equations. One recovers Newton's laws of motion by applying the method of characteristics to either of these equations. Hence, the key difference between KvN and Liouville mechanics lies in weighting individual trajectories: Arbitrary weights, underlying the classical wave function, can be utilized in the KvN mechanics, while only positive weights, representing the probability density, are permitted in the Liouville mechanics.

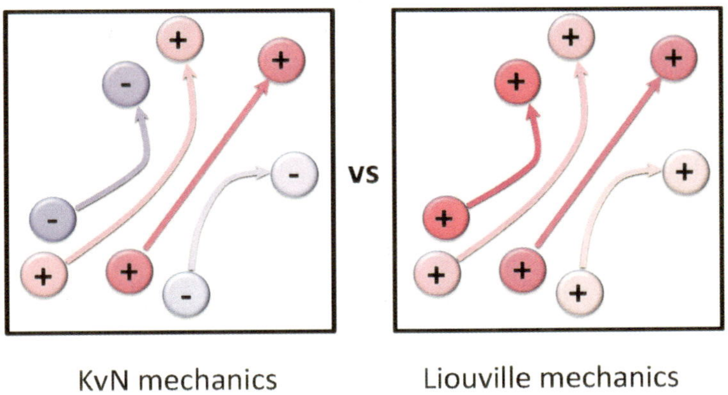

The essential distinction between KvN and Liouville mechanics lies in weighting (coloring) individual trajectories: Any weights can be utilized in KvN mechanics, while only positive weights are allowed in Liouville mechanics. Particles move along Newtonian trajectories in both cases.

Quantum Analogy

Being explicitly based on the Hilbert space language, the KvN classical mechanics adopts many techniques from quantum mechanics, for example, perturbation and diagram techniques as well as functional integral methods . The KvN approach is very general, and it has been extended to dissipative systems, relativistic mechanics, and classical field theories.

The KvN approach is fruitful in studies on the quantum-classical correspondence as it reveals that the Hilbert space formulation is not exclusively quantum mechanical. Even Dirac spinors are not exceptionally quantum as they are utilized in the relativistic generalization of the KvN mechanics. Similarly as the more well-known phase space formulation of quantum mechanics, the KvN approach can be understood as an attempt to bring classical and quantum mechanics into a common mathematical framework. In fact, the time evolution of the Wigner function approaches, in the classical limit, the time evolution of the KvN wavefunction of a classical particle. However, a mathematical resemblance to quantum mechanics does not imply the presence of hallmark quantum effects. In particular, impossibility of double-slit experiment and Aharonov–Bohm effect are explicitly demonstrated in the KvN framework.

Analytical Mechanics

Analytical Mechanics form the 'other' branch of classical mechanics and build upon Newtonian mechanics. It can be divided into three major steps: *Lagrangian Mechanics, Hamiltonian Mechanics* and mechanics based on the Hamilton-Jacobi equation.

The *analytical mechanics* of Euler, Lagrange, Hamilton and Jacobi was the outstanding theoretical invention of the 18th and early 19th centuries. It is an invention, a product of the mind, not a discovery or 'uncovering' of something that existed previously. The *analytical* means that it is an application of the infinitesimal calculus invented by Newton and Leibniz at the end of the 17th century. Its seed was the inchoate concept of energy and work in Leibniz, extending the Newtonian ideas of momentum and force. The concept of energy was not complete until the recognition of internal energy later in the 19th century, when the general principle of the conservation of energy was established at last. In the present essay, energy is simply a constant of the motion, depending on certain conditions, but is still very significant and useful. Analytical mechanics is essentially a mathematical, not a physical, development, though it is completely equivalent to Newton's theory of motion. Beyond this, questions of its 'validity' do not arise. It is, however, a model to which physical theories are referred. The most significant of these theories are the modern Quantum Mechanics of Heisenberg and Schrödinger, and the Theory of Relativity of Einstein, as well as the classical theory of elasticity. Quantum Mechanics is presented in terms of analytical mechanics in its most advanced form, depending on the Hamiltonian function, and Quantum Field Theory uses the Lagrangian density to achieve covariance. An appreciation of analytical mechanics is necessary to the understanding of any of these subjects.

As already mentioned, analytical mechanics uses the calculus, with differentiation and integration at every turn, so facility with the calculus is a pre-requisite. It is characterized by the employment of certain functions of a set of variables that describes the state of system in the necessary detail, but not beyond that, and by the use of variational principles as the fundamental statement of the the theory. A typical variational principle might say that a certain integral takes an extremal or stationary value for the actual motion, or that a certain algebraic expression vanishes for an arbitrary variation in the state of the system. The variational principles are used to derive the algebraic and differential equations that must then be solved to determine the motion of the system. The beauty of the theory is in its generality and symmetry, which was of inestimable value in the development of quantum mechanics and relativity, as outstanding examples.

In analytical mechanics, one can use any set of parameters that fully describe the state of the system, and call them the *generalized coordinates* q. For simplicity, we shall suppress a subscript running from 1 to N, where N is the number of coordinates in the set. N is called the number of *degrees of freedom* of the system. When these coordinates change by infinitesimal amounts dq, the work done on the system is $dW = \Sigma\, Fdq$. The coefficient F of each dq is the *generalized force* associated with q. In most cases treated by analytical mechanics, this expression is an exact differential (the cross partial derivatives of the F's are equal), and can be integrated to give a function W(q), called the *work function* by Lanczos,w with the property that $F = dW/dq$ (of course, this is a partial derivative for the coordinate concerned). By comparing these expressions with those for the work done by the Newtonian forces, they can be expressed in explicit form. For example, the generalized force associated with a rotation is the *torque* about the axis of rotation. W(q) encapsulates all the force information for any configuration in a single function of the coordinates.

Intrinsic Motion

Generalized Coordinates and Constraints

In Newtonian mechanics, one customarily uses all three Cartesian coordinates, or other 3D coordinate system, to refer to a body's position during its motion. In physical systems, however, some structure or other system usually constrains the body's motion from taking certain directions and pathways. So a full set of Cartesian coordinates is often unneeded, as the constraints determine the evolving relations among the coordinates, which relations can be modeled by equations corresponding to the constraints. In the Lagrangian and Hamiltonian formalisms, the constraints are incorporated into the motion's geometry, reducing the number of coordinates to the minimum needed to model the motion. These are known as *generalized coordinates*, denoted q_i ($i = 1, 2, 3...$).

Difference between Curvillinear and Generalized Coordinates

Generalized coordinates incorporate constraints on the system. There is one generalized coordinate q_i for each degree of freedom (for convenience labelled by an index $i = 1, 2...N$), i.e. each way the system can change its configuration; as curvilinear lengths or angles of rotation. Generalized coordinates are not the same as curvilinear coordinates. The number of *curvilinear* coordinates equals the dimension of the position space in question (usually 3 for 3d space), while the number of *generalized* coordinates is not necessarily equal to this dimension; constraints can reduce the number of degrees of freedom (hence the number of generalized coordinates required to define the configuration of the system), following the general rule:

[dimension of position space (usually 3)] × *[number of constituents of system ("particles")]* – *(number of constraints)*

= *(number of degrees of freedom)* = *(number of generalized coordinates)*

For a system with N degrees of freedom, the generalized coordinates can be collected into an N-tuple:

$$\mathrm{q} = (q_1, q_2, \cdots q_N)$$

and the time derivative (here denoted by an overdot) of this tuple give the *generalized velocities*:

$$\frac{d\mathrm{q}}{dt} = \left(\frac{dq_1}{dt}, \frac{dq_2}{dt}, \cdots \frac{dq_N}{dt}\right) \equiv \dot{\mathrm{q}} = (\dot{q}_1, \dot{q}_2, \cdots \dot{q}_N).$$

D'Alembert's Principle

The foundation which the subject is built on is *D'Alembert's principle.*

This principle states that infinitesimal virtual work done by a force across reversible displacements is zero, which is the work done by a force consistent with ideal constraints of the system. The idea of a constraint is useful - since this limits what the system can do, and can provide steps to solving for the motion of the system. The equation for D'Alembert's principle is:

$$\delta W = \mathcal{Q} \cdot \delta \mathrm{q} = 0,$$

where

$$\mathcal{Q} = (\mathcal{Q}_1, \mathcal{Q}_2, \cdots \mathcal{Q}_N)$$

are the generalized forces (script Q instead of ordinary Q is used here to prevent conflict with canonical transformations below) and q are the generalized coordinates. This leads to the generalized form of Newton's laws in the language of analytical mechanics:

$$\mathcal{Q} = \frac{\mathrm{d}}{\mathrm{d}t}\left(\frac{\partial T}{\partial \dot{\mathrm{q}}}\right) - \frac{\partial T}{\partial \mathrm{q}},$$

where T is the total kinetic energy of the system, and the notation

$$\frac{\partial}{\partial \mathrm{q}} = \left(\frac{\partial}{\partial q_1}, \frac{\partial}{\partial q_2}, \cdots \frac{\partial}{\partial q_N}\right)$$

is a useful shorthand.

Holonomic Constraints

If the curvilinear coordinate system is defined by the standard position vector r, and if the position vector can be written in terms of the generalized coordinates q and time t in the form:

$$\mathrm{r} \quad \mathrm{r}(\mathrm{q}(t), t)$$

and this relation holds for all times t, then q are called *Holonomic constraints*. Vector r is explicitly dependent on t in cases when the constraints vary with time, not just because of q(t). For time-independent situations, the constraints are also called scleronomic, for time-dependent cases they are called rheonomic.

Lagrangian Mechanics

Lagrangian and Euler–Lagrange equations

The introduction of generalized coordinates and the fundamental Lagrangian function:

$$L(\mathrm{q}, \dot{\mathrm{q}}, t) = T(\mathrm{q}, \dot{\mathrm{q}}, t) - V(\mathrm{q}, \dot{\mathrm{q}}, t)$$

where T is the total kinetic energy and V is the total potential energy of the entire system, then either following the calculus of variations or using the above formula - lead to the Euler–Lagrange equations;

$$\frac{d}{dt}\left(\frac{\partial L}{\partial \dot{\mathrm{q}}}\right)=\frac{\partial L}{\partial \mathrm{q}},$$

which are a set of N second-order ordinary differential equations, one for each $q_i(t)$.

This formulation identifies the actual path followed by the motion as a selection of the path over which the time integral of kinetic energy is least, assuming the total energy to be fixed, and imposing no conditions on the time of transit.

Configuration Space

The Lagrangian formulation uses the configuration space of the system, the set of all possible generalized coordinates:

$$\mathcal{C}=\{\mathrm{q}\in\mathbb{R}^N\},$$

where $\mathbb{R}^N$ is N-dimensional real space. The particular solution to the Euler–Lagrange equations is called a *(configuration) path or trajectory*, i.e. one particular q(t) subject to the required initial conditions. The general solutions form a set of possible configurations as functions of time:

$$\{\mathrm{q}(t)\in\mathbb{R}^N : t\geq 0, t\in\mathbb{R}\}\subseteq\mathcal{C},$$

The configuration space can be defined more generally, and indeed more deeply, in terms of topological manifolds and the tangent bundle.

Hamiltonian Mechanics

Hamiltonian and Hamilton's Equations

The Legendre transformation of the Lagrangian replaces the generalized coordinates and velocities (q, $\dot{\mathbf{q}}$) with (q, p); the generalized coordinates and the generalized momenta conjugate to the generalized coordinates:

$$\mathrm{p}=\frac{\partial L}{\partial \dot{\mathrm{q}}}=\left(\frac{\partial L}{\partial \dot{q}_1},\frac{\partial L}{\partial \dot{q}_2},\cdots\frac{\partial L}{\partial \dot{q}_N}\right)=(p_1,p_2\cdots p_N),$$

and introduces the Hamiltonian (which is in terms of generalized coordinates and momenta):

$$H(\mathrm{q},\mathrm{p},t)=\mathrm{p}\cdot\dot{\mathrm{q}}-L(\mathrm{q},\dot{\mathrm{q}},t)$$

where • denotes the dot product, also leading to Hamilton's equations:

$$\dot{\mathrm{p}}=-\frac{\partial H}{\partial \mathrm{q}},\quad \dot{\mathrm{q}}=+\frac{\partial H}{\partial \mathrm{p}},$$

which are now a set of $2N$ first-order ordinary differential equations, one for each $q_i(t)$ and $p_i(t)$. Another result from the Legendre transformation relates the time derivatives of the Lagrangian and Hamiltonian:

$$\frac{dH}{dt} = -\frac{\partial L}{\partial t},$$

which is often considered one of Hamilton's equations of motion additionally to the others. The generalized momenta can be written in terms of the generalized forces in the same way as Newton's second law:

$$\dot{\mathrm{p}} = \mathcal{Q}.$$

Generalized Momentum Space

Analogous to the configuration space, the set of all momenta is the *momentum space* (technically in this context; *generalized momentum space*):

$$\mathcal{M} = \{\mathrm{p} \in \mathbb{R}^N\}.$$

"Momentum space" also refers to "k-space"; the set of all wave vectors (given by De Broglie relations) as used in quantum mechanics and theory of waves: this is not referred to in this context.

Phase Space

The set of all positions and momenta form the *phase space*;

$$\mathcal{P} = \mathcal{C} \times \mathcal{M} = \{(\mathrm{q}, \mathrm{p}) \in \mathbb{R}^{2N}\},$$

that is, the Cartesian product × of the configuration space and generalized momentum space.

A particular solution to Hamilton's equations is called a phase path, a particular curve (q(t),p(t)) subject to the required initial conditions. The set of all phase paths, the general solution to the differential equations, is the phase portrait:

$$\{(\mathrm{q}(t), \mathrm{p}(t)) \in \mathbb{R}^{2N} : t \geq 0, t \in \mathbb{R}\} \subseteq \mathcal{P},$$

The Poisson Bracket

All dynamical variables can be derived from position r, momentum p, and time t, and written as a function of these: A = A(q, p, t). If A(q, p, t) and B(q, p, t) are two scalar valued dynamical variables, the Poisson bracket is defined by the generalized coordinates and momenta:

$$\{A, B\} \equiv \{A, B\}_{\mathbf{q},\mathbf{p}} = \frac{\partial A}{\partial \mathrm{q}} \cdot \frac{\partial B}{\partial \mathrm{p}} - \frac{\partial A}{\partial \mathrm{p}} \cdot \frac{\partial B}{\partial \mathrm{q}}$$

$$\equiv \sum_k \frac{\partial A}{\partial q_k}\frac{\partial B}{\partial p_k} - \frac{\partial A}{\partial p_k}\frac{\partial B}{\partial q_k},$$

Calculating the total derivative of one of these, say A, and substituting Hamilton's equations into the result leads to the time evolution of A:

$$\frac{dA}{dt} = \{A, H\} + \frac{\partial A}{\partial t}.$$

This equation in A is closely related to the equation of motion in the Heisenberg picture of quantum mechanics, in which classical dynamical variables become quantum operators (indicated by hats (^)), and the Poisson bracket is replaced by the commutator of operators via Dirac's canonical quantization:

$$\{A, B\} \to \frac{1}{i\hbar}[\hat{A}, \hat{B}].$$

Properties of the Lagrangian and Hamiltonian Functions

Following are overlapping properties between the Lagrangian and Hamiltonian functions.

- All the individual generalized coordinates $q_i(t)$, velocities $\dot{q}_i(t)$ and momenta $p_i(t)$ for every degree of freedom are mutually independent. Explicit time-dependence of a function means the function actually includes time t as a variable in addition to the q(t), p(t), not simply as a parameter through q(t) and p(t), which would mean explicit time-independence.
- The Lagrangian is invariant under addition of the total time derivative of any function of q and t, that is:

  $$L' = L + \frac{d}{dt}F(\mathrm{q}, t),$$

  so each Lagrangian L and L' describe *exactly the same motion*. In other words, the Lagrangian of a system is not unique.
- Analogously, the Hamiltonian is invariant under addition of the partial time derivative of any function of q, p and t, that is:

  $$K = H + \frac{\partial}{\partial t}G(\mathrm{q}, \mathrm{p}, t),$$

  (K is a frequently used letter in this case). This property is used in canonical transformations.
- If the Lagrangian is independent of some generalized coordinates, then the generalized momenta conjugate to those coordinates are constants of the motion, i.e. are conserved, this immediately follows from Lagrange's equations:

$$\frac{\partial L}{\partial q_j} = 0 \rightarrow \frac{dp_j}{dt} = \frac{d}{dt}\frac{\partial L}{\partial \dot{q}_j} = 0$$

Such coordinates are "cyclic" or "ignorable". It can be shown that the Hamiltonian is also cyclic in exactly the same generalized coordinates.

- If the Lagrangian is time-independent the Hamiltonian is also time-independent (i.e. both are constant in time).
- If the kinetic energy is a homogeneous function of degree 2 of the generalized velocities, *and* the Lagrangian is explicitly time-independent, then:

$$T((\lambda \dot{q}_i)^2, (\lambda \dot{q}_j \lambda \dot{q}_k), \mathrm{q}) = \lambda^2 T((\dot{q}_i)^2, \dot{q}_j \dot{q}_k, \mathrm{q}), \quad L(\mathrm{q}, \dot{\mathrm{q}}),$$

where λ is a constant, then the Hamiltonian will be the *total conserved energy*, equal to the total kinetic and potential energies of the system:

$$H = T + V = E.$$

This is the basis for the Schrödinger equation, inserting quantum operators directly obtains it.

Principle of Least Action

Action is another quantity in analytical mechanics defined as a functional of the Lagrangian:

$$\mathcal{S} = \int_{t_1}^{t_2} L(\mathrm{q}, \dot{\mathrm{q}}, t)\, dt.$$

A general way to find the equations of motion from the action is the principle of least action:

$$\delta \mathcal{S} = \delta \int_{t_1}^{t_2} L(\mathrm{q}, \dot{\mathrm{q}}, t)\, dt = 0,$$

where the departure t_1 and arrival t_2 times are fixed. The term "path" or "trajectory" refers to the time evolution of the system as a path through configuration space $\mathcal{C}$, in other words q(t) tracing out a path in $\mathcal{C}$. The path for which action is least is the path taken by the system.

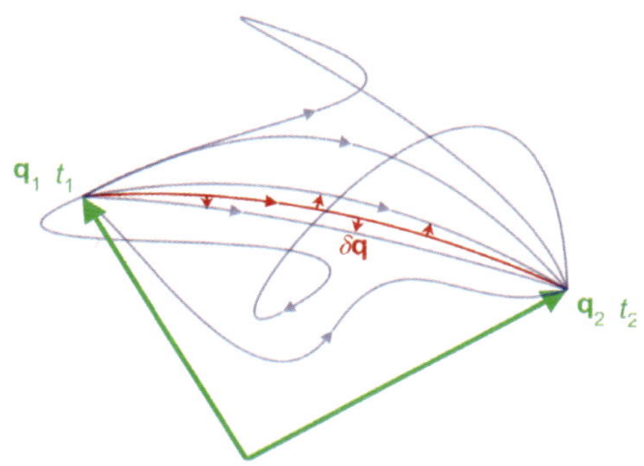

As the system evolves, q traces a path through configuration space. The path taken by the system (red) has a stationary action ($\delta S = 0$) under small changes in the configuration of the system (δq).

From this principle, *all* equations of motion in classical mechanics can be derived. This approach can be extended to fields rather than a system of particles, and underlies the path integral formulation of quantum mechanics, and is used for calculating geodesic motion in general relativity.

Hamiltonian-jacobi Mechanics

Canonical Transformations

The invariance of the Hamiltonian (under addition of the partial time derivative of an arbitrary function of p, q, and t) allows the Hamiltonian in one set of coordinates q and momenta p to be transformed into a new set Q = Q(q, p, t) and P = P(q, p, t), in four possible ways:

$$K(\mathrm{Q},\mathrm{P},t) = H(\mathrm{q},\mathrm{p},t) + \frac{\partial}{\partial t} G_1(\mathrm{q},\mathrm{Q},t)$$

$$K(\mathrm{Q},\mathrm{P},t) = H(\mathrm{q},\mathrm{p},t) + \frac{\partial}{\partial t} G_2(\mathrm{q},\mathrm{P},t)$$

$$K(\mathrm{Q},\mathrm{P},t) = H(\mathrm{q},\mathrm{p},t) + \frac{\partial}{\partial t} G_3(\mathrm{p},\mathrm{Q},t)$$

$$K(\mathrm{Q},\mathrm{P},t) = H(\mathrm{q},\mathrm{p},t) + \frac{\partial}{\partial t} G_4(\mathrm{p},\mathrm{P},t)$$

With the restriction on P and Q such that the transformed Hamiltonian system is:

$$\dot{\mathrm{P}} = -\frac{\partial K}{\partial \mathrm{Q}}, \quad \dot{\mathrm{Q}} = +\frac{\partial K}{\partial \mathrm{P}},$$

the above transformations are called *canonical transformations*, each function G_n is called a generating function of the "nth kind" or "type-n". The transformation of coordinates and momenta can allow simplification for solving Hamilton's equations for a given problem.

The choice of Q and P is completely arbitrary, but not every choice leads to a canonical transformation. One simple criterion for a transformation q → Q and p → P to be canonical is the Poisson bracket be unity,

$$\{Q_i, P_i\} = 1$$

for all i = 1, 2,...N. If this does not hold then the transformation is not canonical.

The Hamilton–Jacobi equation

By setting the canonically transformed Hamiltonian K = 0, and the type-2 generating function equal to Hamilton's principal function (also the action $\mathcal{S}$) plus an arbitrary constant C:

$$G_2(\mathrm{q},t) = \mathcal{S}(\mathrm{q},t) + C,$$

the generalized momenta become:

$$p = \frac{\partial S}{\partial q}$$

and P is constant, then the Hamiltonian-Jacobi equation (HJE) can be derived from the type-2 canonical transformation:

$$H = -\frac{\partial S}{\partial t}$$

where H is the Hamiltonian as before:

$$H = H(q, p, t) = H\left(q, \frac{\partial S}{\partial q}, t\right)$$

Another related function is Hamilton's characteristic function

$$W(q) = S(q, t) + Et$$

used to solve the HJE by additive separation of variables for a time-independent Hamiltonian H.

The study of the solutions of the Hamilton–Jacobi equations leads naturally to the study of symplectic manifolds and symplectic topology. In this formulation, the solutions of the Hamilton–Jacobi equations are the integral curves of Hamiltonian vector fields.

Routhian Mechanics

Routhian mechanics is a hybrid formulation of Lagrangian and Hamiltonian mechanics, not often used but especially useful for removing cyclic coordinates. If the Lagrangian of a system has s cyclic coordinates $q = q_1, q_2, \ldots q_s$ with conjugate momenta $p = p_1, p_2, \ldots p_s$, with the rest of the coordinates non-cyclic and denoted $\zeta = \zeta_1, \zeta_1, \ldots, \zeta_{N-s}$, they can be removed by introducing the *Routhian*:

$$R = p \cdot \dot{q} - L(q, p, \zeta, \dot{\zeta}),$$

which leads to a set of $2s$ Hamiltonian equations for the cyclic coordinates q,

$$\dot{\mathbf{q}} = +\frac{\partial R}{\partial p}, \quad \dot{p} = -\frac{\partial R}{\partial q},$$

and $N - s$ Lagrangian equations in the non cyclic coordinates ζ.

$$\frac{d}{dt}\frac{\partial R}{\partial \dot{\zeta}} = \frac{\partial R}{\partial \zeta}.$$

Set up in this way, although the Routhian has the form of the Hamiltonian, it can be thought of a Lagrangian with $N - s$ degrees of freedom.

The coordinates q do not have to be cyclic, the partition between which coordinates enter the Hamiltonian equations and those which enter the Lagrangian equations is arbitrary. It is simply convenient to let the Hamiltonian equations remove the cyclic coordinates, leaving the non cyclic coordinates to the Lagrangian equations of motion.

Appellian Mechanics

Appell's equation of motion involve generalized accelerations, the second time derivatives of the generalized coordinates:

$$\alpha_r = \ddot{q}_r = \frac{d^2 q_r}{dt^2},$$

as well as generalized forces mentioned above in D'Alembert's principle. The equations are

$$Q_r = \frac{\partial S}{\partial \alpha_r}, \quad S = \frac{1}{2}\sum_{k=1}^{N} m_k \mathrm{a}_k^2,$$

where

$$\mathrm{a}_k = \ddot{\mathrm{r}}_k = \frac{d^2 \mathrm{r}_k}{dt^2}$$

is the acceleration of the *k* particle, the second time derivative of its position vector. Each acceleration a_k is expressed in terms of the generalized accelerations α_r, likewise each r_k are expressed in terms the generalized coordinates q_r.

Extensions to Classical Field Theory

Lagrangian Field Theory

Generalized coordinates apply to discrete particles. For *N* scalar fields $\phi_i(\mathrm{r}, t)$ where $i = 1, 2, \ldots N$, the Lagrangian density is a function of these fields and their space and time derivatives, and possibly the space and time coordinates themselves:

$$\mathcal{L} = \mathcal{L}(\phi_1, \phi_2, \ldots \nabla\phi_1, \nabla\phi_2, \ldots \partial\phi_1/\partial t, \partial\phi_2/\partial t, \ldots \mathrm{r}, t).$$

and the Euler–Lagrange equations have an analogue for fields:

$$\partial_\mu\left(\frac{\partial\mathcal{L}}{\partial(\partial_\mu\phi_i)}\right) = \frac{\partial\mathcal{L}}{\partial\phi_i},$$

where ∂_μ denotes the 4-gradient and the summation convention has been used. For *N* scalar fields, these Lagranian field equations are a set of *N* second order partial differential equations in the fields, which in general will be coupled and nonlinear.

This scalar field formulation can be extended to vector fields, tensor fields, and spinor fields.

The Lagrangian is the volume integral of the Lagrangian density:

$$L = \int_V \mathcal{L} dV.$$

Originally developed for classical fields, the above formulation is applicable to all physical fields in classical, quantum, and relativistic situations: such as Newtonian gravity, classical electromagnetism, general relativity, and quantum field theory. It is a question of determining the correct Lagrangian density to generate the correct field equation.

Hamiltonian Field Theory

The corresponding "momentum" field densities conjugate to the N scalar fields $\phi_i(\mathrm{r}, t)$ are:

$$\pi_i(\mathrm{r},t) = \frac{\partial \mathcal{L}}{\partial \dot{\phi}_i} \quad \dot{\phi}_i \equiv \frac{\partial \phi_i}{\partial t}.$$

where in this context the overdot denotes a partial time derivative, not a total time derivative. The Hamiltonian density $\mathcal{H}$ is defined by analogy with mechanics:

$$\mathcal{H}(\phi_1, \phi_2, \ldots, \pi_1, \pi_2, \ldots, \mathrm{r}, t) = \sum_{i=1}^{N} \dot{\phi}_i(\mathrm{r},t)\pi_i(\mathrm{r},t) - \mathcal{L}.$$

The equations of motion are:

$$\dot{\phi}_i = +\frac{\delta \mathcal{H}}{\delta \pi_i}, \quad \dot{\pi}_i = -\frac{\delta \mathcal{H}}{\delta \phi_i},$$

where the variational derivative

$$\frac{\delta}{\delta \phi_i} = \frac{\partial}{\partial \phi_i} - \partial_\mu \frac{\partial}{\partial(\partial_\mu \phi_i)}$$

must be used instead of merely partial derivatives. For N fields, these Hamiltonian field equations are a set of $2N$ first order partial differential equations, which in general will be coupled and nonlinear.

Again, the volume integral of the Hamiltonian density is the Hamiltonian

$$H = \int_V \mathcal{H} dV.$$

Symmetry, Conservation, and Noether's Theorem

Each transformation can be described by an operator (i.e. function acting on the position r or momentum p variables to change them). The following are the cases when the operator does not change r or p, i.e. symmetries.

Transformation	Operator	Position	Momentum
Translational symmetry	$X(\mathrm{a})$	$\mathrm{r} \to \mathrm{r} + \mathrm{a}$	$\mathrm{p} \to \mathrm{p}$
Time translation	$U(t_0)$	$\mathrm{r}(t) \to \mathrm{r}(t + t_0)$	$\mathrm{p}(t) \to \mathrm{p}(t + t_0)$
Rotational invariance	$R(\hat{\mathrm{n}}, \theta)$	$\mathrm{r} \to R(\hat{\mathrm{n}}, \theta)\mathrm{r}$	$\mathrm{p} \to R(\hat{\mathrm{n}}, \theta)\mathrm{p}$
Galilean transformations	$G(\mathbf{v})$	$\mathrm{r} \to \mathrm{r} + \mathrm{v}t$	$\mathrm{p} \to \mathrm{p} + m\mathbf{v}$
Parity	P	$\mathrm{r} \to -\mathrm{r}$	$\mathrm{p} \to -\mathrm{p}$
T-symmetry	T	$\mathrm{r} \to \mathrm{r}(-t)$	$\mathrm{p} \to -\mathrm{p}(-t)$

where $R(\hat{\mathbf{n}}, \theta)$ is the rotation matrix about an axis defined by the unit vector $\hat{\mathbf{n}}$ and angle θ.

Noether's Theorem

Noether's theorem states that a continuous symmetry transformation of the action corresponds to a conservation law, i.e. the action (and hence the Lagrangian) doesn't change under a transformation parameterized by a parameter s:

$$L[q(s,t), \dot{q}(s,t)] = L[q(t), \dot{q}(t)]$$

the Lagrangian describes the same motion independent of s, which can be length, angle of rotation, or time. The corresponding momenta to q will be conserved.

Lagrangian Mechanics

"Lagrangian mechanics" is, fundamentally, just another way of looking at Newtonian mechanics. Newtonian mechanics, in a nutshell, says:

$$\text{(Second law) } F = \frac{d(mv)}{dt} = m\frac{dv}{dt} = m\ddot{x}$$

$$\text{(third law) } F_{1\to 2} = -F_{2\to 1} \text{ \{For any two interacting bodies\}}$$

The fundamental forces in the universe are all *conservative*, and many forces we deal with in everyday life are conservative as well (friction being one obvious exception). A conservative force can be represented as the gradient of a *potential;* when an object is being affected only by conservative forces, we can rewrite the second law as:

$$m\ddot{x} = -\frac{\partial}{\partial x}\phi$$

or, in vector form, using r as the object's position vector,

$$m\ddot{x} = -\nabla \phi$$

where ϕ is the potential function.

The "Lagrangian formulation" of Newtonian mechanics is based on equation, which, again, is just an alternate form of Newton's laws which is applicable in cases where the forces are conservative. Lagrangian mechanics adds no new "semantics" -- it's just a mathematical change, not a change in the physics.

So why use it? Because

Newtonian mechanics has a problem: It works very nicely in Cartesian coordinates, but it's difficult to switch to a different coordinate system. Something as simple as changing to polar coordinates is cumbersome; finding the equations of motion of a particle acting under a "central force" in polar coordinates is tedious. The Lagrangian formulation, in contrast, is *independent* of the coordinates, and the equations of motion for a non-Cartesian coordinate system can typically be found immediately using it. That's (most of) the point in "Lagrangian mechanics".

From Newtonian to Lagrangian Mechanics

Newton's Laws

For simplicity, Newton's laws can be illustrated for one particle without much loss of generality (for a system of N particles, all of these equations apply to each particle in the system). The equation of motion for particle of mass m is Newton's second law of 1687, in modern vector notation

$$\mathrm{F} = m\mathrm{a}$$

where a is its acceleration and F the resultant force acting *on* it. In three spatial dimensions, this is a system of three coupled second order ordinary differential equations to solve, since there are three components in this vector equation. The solutions are the position vectors r of the particles at time t, subject to the initial conditions of r and v when $t = 0$.

Isaac Newton

Newton's laws are easy to use in Cartesian coordinates, but Cartesian coordinates are not always convenient, and for other coordinate systems the equations of motion can become complicated. In a set of curvilinear coordinates $\xi = (\xi^1, \xi^2, \xi^3)$, the law in tensor index notation is the *"Lagrangian form"*

$$F^a = m\left(\frac{d^2\xi^a}{dt^2} + \Gamma^a{}_{bc}\frac{d\xi^b}{dt}\frac{d\xi^c}{dt}\right) = g^{ak}\left(\frac{d}{dt}\frac{\partial T}{\partial \dot{\xi}^k} - \frac{\partial T}{\partial \xi^k}\right), \quad \dot{\xi}^a \equiv \frac{d\xi^a}{dt},$$

where F^a is the ath contravariant components of the resultant force acting on the particle, $\Gamma^a{}_{bc}$ are the Christoffel symbols of the second kind,

$$T = \frac{1}{2} m g_{bc} \frac{d\xi^b}{dt}\frac{d\xi^c}{dt},$$

is the kinetic energy of the particle, and g_{bc} the covariant components of the metric tensor of the curvilinear coordinate system. All the indices a, b, c, each take the values 1, 2, 3. Curvilinear coordinates are not the same as generalized coordinates.

It may seem like an overcomplication to cast Newton's law in this form, but there are advantages. The acceleration components in terms of the Christoffel symbols can be avoided by evaluating derivatives of the kinetic energy instead. If there is no resultant force acting on the particle, F = 0, it does not accelerate, but moves with constant velocity in a straight line. Mathematically, the solutions of the differential equation are geodesics, the curves of extremal length between two points in space (These may end up being minimal so the shortest paths, but that is not necessary). In flat 3d real space the geodesics are simply straight lines. So for a free particle, Newton's second law coincides with the geodesic equation, and states free particles follow geodesics, the extremal trajectories it can move along. If the particle is subject to forces, F ≠ 0, the particle accelerates due to forces acting on it, and deviates away from the geodesics it would follow if free. With appropriate extensions of the quantities given here in flat 3d space to 4d curved spacetime, the above form of Newton's law also carries over to Einstein's general relativity, in which case free particles follow geodesics in curved spacetime that are no longer "straight lines" in the ordinary sense.

However, we still need to know the total resultant force F acting on the particle, which in turn requires the resultant non-constraint force N plus the resultant constraint force C,

$$\mathrm{F} = \mathrm{C} + \mathrm{N}.$$

The constraint forces can be complicated, since they will generally depend on time. Also, if there are constraints, the curvilinear coordinates are not independent but related by one or more constraint equations.

The constraint forces can either be eliminated from the equations of motion so only the non-constraint forces remain, or included by including the constraint equations in the equations of motion.

D'Alembert's Principle

Jean d'Alembert (1717—1783)

A fundamental result in analytical mechanics is D'Alembert's principle, introduced in 1708 by Jacques Bernoulli to understand static equilibrium, and developed by D'Alembert in 1743 to solve dynamical problems. The principle asserts for N particles

$$\sum_{k=1}^{N}(\mathrm{N}_k + \mathrm{C}_k - m_k \mathrm{a}_k)\cdot \delta \mathrm{r}_k = 0.$$

The δr_k are virtual displacements, by definition they are infinitesimal changes in the configuration of the system consistent with the constraint forces acting on the system *at an instant of time*, i.e. in such a way that the constraint forces maintain the constrained motion. They are not the same as the actual displacements in the system, which are caused by the resultant constraint and non-constraint forces acting on the particle to accelerate and move it. Virtual work is the work done along a virtual displacement for any force (constraint or non-constraint).

Since the constraint forces act perpendicular to the motion of each particle in the system to maintain the constraints, the total virtual work by the constraint forces acting on the system is zero;

$$\sum_{k=1}^{N}\mathrm{C}_k \cdot \delta \mathrm{r}_k = 0,$$

so that

$$\sum_{k=1}^{N}(\mathrm{N}_k - m_k \mathrm{a}_k)\cdot \delta \mathrm{r}_k = 0.$$

Thus D'Alembert's principle allows us to concentrate on only the applied non-constraint forces, and exclude the constraint forces in the equations of motion. The form shown is also independent of the choice of coordinates. However, it cannot be readily used to set up the equations of motion in an arbitrary coordinate system since the displacements δr_k might be connected by a constraint equation, which prevents us from setting the N individual summands to 0. We will therefore seek a system of mutually independent coordinates for which the total sum will be 0 if and only if the individual summands are 0. Setting each of the summands to 0 will eventually give us our separated equations of motion.

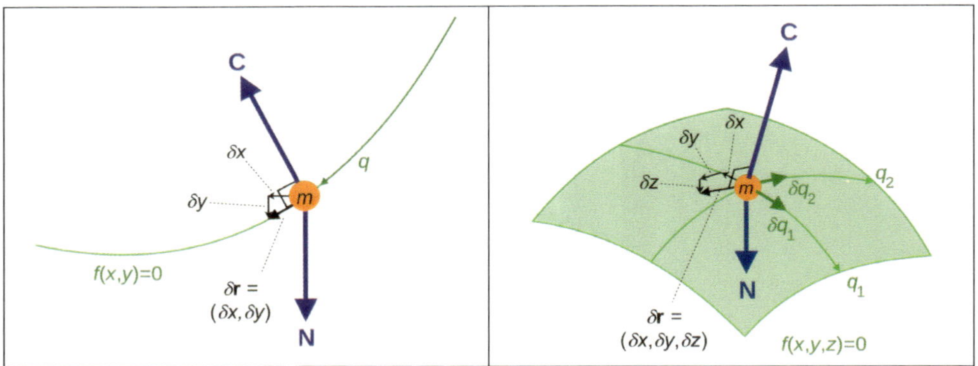

One degree of freedom.	Two degrees of freedom.

Constraint force **C** and virtual displacement $\delta\mathbf{r}$ for a particle of mass m confined to a curve. The resultant non-constraint force is **N**.

Equations of Motion from D'Alembert's Principle

If there are constraints on particle k, then since the coordinates of the position $\mathrm{r}_k = (x_k, y_k, z_k)$ are linked together by a constraint equation, so are those of the virtual displacements $\delta\mathrm{r}_k = (\delta x_k, \delta y_k, \delta z_k)$. Since the generalized coordinates are independent, we can avoid the complications with the $\delta\mathrm{r}_k$ by converting to virtual displacements in the generalized coordinates. These are related in the same form as a total differential,

$$\delta\mathrm{r}_k = \sum_{j=1}^{n} \frac{\partial \mathrm{r}_k}{\partial q_j} \delta q_j .$$

There is no partial time derivative with respect to time multiplied by a time increment, since this is a virtual displacement, one along the constraints in an *instant* of time.

The first term in D'Alembert's principle above is the virtual work done by the non-constraint forces N_k along the virtual displacements $\delta\mathrm{r}_k$, and can without loss of generality be converted into the generalized analogues by the definition of generalized forces

$$Q_j = \sum_{k=1}^{N} \mathrm{N}_k \cdot \frac{\partial \mathrm{r}_k}{\partial q_j},$$

so that

$$\sum_{k=1}^{N} \mathrm{N}_k \cdot \delta\mathrm{r}_k = \sum_{k=1}^{N} \mathrm{N}_k \cdot \sum_{j=1}^{n} \frac{\partial \mathrm{r}_k}{\partial q_j} \delta q_j = \sum_{j=1}^{n} Q_j \delta q_j .$$

This is half of the conversion to generalized coordinates. It remains to convert the acceleration term into generalized coordinates, which is not immediately obvious. Recalling the Lagrange form of Newton's second law, the partial derivatives of the kinetic energy with respect to the generalized coordinates and velocities can be found to give the desired result;

$$\sum_{k=1}^{N} m_k \mathrm{a}_k \cdot \frac{\partial \mathrm{r}_k}{\partial q_j} = \frac{\mathrm{d}}{\mathrm{d}t}\frac{\partial T}{\partial \dot{q}_j} - \frac{\partial T}{\partial q_j}.$$

Now D'Alembert's principle is in the generalized coordinates as required,

$$\sum_{j=1}^{n}\left[Q_j - \left(\frac{\mathrm{d}}{\mathrm{d}t}\frac{\partial T}{\partial \dot{q}_j} - \frac{\partial T}{\partial q_j}\right)\right]\delta q_j = 0,$$

and since these virtual displacements δq_j are independent and nonzero, the coefficients can be equated to zero, resulting in Lagrange's equations or the generalized equations of motion,

$$Q_j = \frac{\mathrm{d}}{\mathrm{d}t}\frac{\partial T}{\partial \dot{q}_j} - \frac{\partial T}{\partial q_j}$$

These equations are equivalent to Newton's laws *for the non-constraint forces*. The generalized forces in this equation are derived from the non-constraint forces only – the constraint forces have been excluded from D'Alembert's principle and do not need to be found. The generalized forces may be non-conservative, provided they satisfy D'Alembert's principle.

Euler–Lagrange Equations and Hamilton's Principle

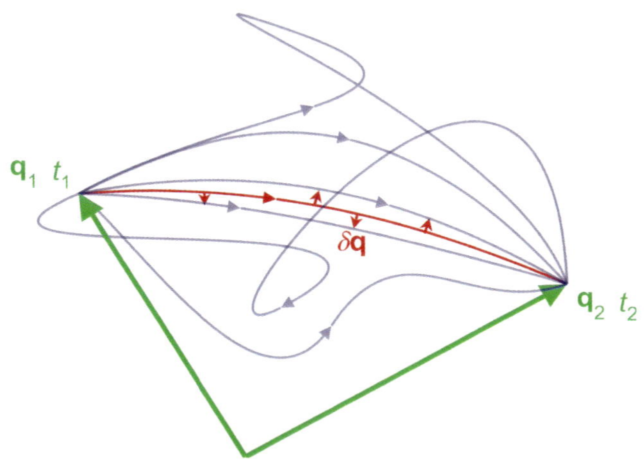

As the system evolves, **q** traces a path through configuration space (only some are shown). The path taken by the system (red) has a stationary action ($\delta S = 0$) under small changes in the configuration of the system ($\delta\mathbf{q}$).

For a non-conservative force which depends on velocity, it *may* be possible to find a potential energy function V that depends on positions and velocities. If the generalized forces Q_i can be derived from a potential V such that

$$Q_j = \frac{\mathrm{d}}{\mathrm{d}t}\frac{\partial V}{\partial \dot{q}_j} - \frac{\partial V}{\partial q_j},$$

equating to Lagrange's equations and defining the Lagrangian as $L = T - V$ obtains Lagrange's equations of the second kind or the Euler–Lagrange equations of motion

$$\frac{\partial L}{\partial q_j} - \frac{\mathrm{d}}{\mathrm{d}t}\frac{\partial L}{\partial \dot{q}_j} = 0.$$

However, the Euler–Lagrange equations can only account for non-conservative forces *if* a potential can be found as shown. This may not always be possible for non-conservative forces, and Lagrange's equations do not involve any potential, only generalized forces; therefore they are more general than the Euler–Lagrange equations.

The Euler–Lagrange equations also follow from the calculus of variations. The *variation* of the Lagrangian is

$$\delta L = \sum_{j=1}^{n}\left(\frac{\partial L}{\partial q_j}\delta q_j + \frac{\partial L}{\partial \dot{q}_j}\delta \dot{q}_j\right), \quad \delta \dot{q}_j \equiv \delta \frac{\mathrm{d}q_j}{\mathrm{d}t} \equiv \frac{\mathrm{d}(\delta q_j)}{\mathrm{d}t},$$

which has a similar form to the total differential of L, but the virtual displacements and their time derivatives replace differentials, and there is no time increment in accordance with the definition of the virtual displacements. An integration by parts with respect to time can transfer the time derivative of δq_j to the $\partial L/\partial(\mathrm{d}q_j/\mathrm{d}t)$, in the process exchanging $\mathrm{d}(\delta q_j)/\mathrm{d}t$ for δq_j, allowing the independent virtual displacements to be factorized from the derivatives of the Lagrangian,

$$\int_{t_1}^{t_2}\delta L \mathrm{d}t = \left[\sum_{j=1}^{n}\frac{\partial L}{\partial \dot{q}_j}\delta q_j\right]_{t_1}^{t_2} + \int_{t_1}^{t_2}\sum_{j=1}^{n}\left(\frac{\partial L}{\partial q_j} - \frac{\mathrm{d}}{\mathrm{d}t}\frac{\partial L}{\partial \dot{q}_j}\right)\delta q_j \mathrm{d}t.$$

Now, if the condition $\delta q_j(t_1) = \delta q_j(t_2) = 0$ holds for all j, the terms not integrated are zero. If in addition the entire time integral of δL is zero, then because the δq_j are independent, and the only way for a definite integral to be zero is if the integrand equals zero, each of the coefficients of δq_j must also be zero. Then we obtain the equations of motion. This can be summarized by Hamilton's principle;

$$\int_{t_1}^{t_2}\delta L \mathrm{d}t = 0.$$

The time integral of the Lagrangian is another quantity called the action, defined as

$$S = \int_{t_1}^{t_2} L\, dt,$$

which is a functional; it takes in the Lagrangian function for all times between t_1 and t_2 and returns a scalar value. Its dimensions are the same as [angular momentum], [energy]·[time], or [length]·[momentum]. With this definition Hamilton's principle is

$$\delta S = 0.$$

Thus, instead of thinking about particles accelerating in response to applied forces, one might think of them picking out the path with a stationary action, with the end points of the path in configuration space held fixed at the initial and final times. Hamilton's principle is sometimes referred to as the principle of least action, however the action functional need only be *stationary*, not necessarily a maximum or a minimum value. Any variation of the functional gives an increase in the functional integral of the action.

Historically, the idea of finding the shortest path a particle can follow subject to a force motivated the first applications of the calculus of variations to mechanical problems, such as the Brachistochrone problem solved by Jean Bernoulli in 1696, as well as Leibniz, Daniel Bernoulli, L'Hôpital around the same time, and Newton the following year. Newton himself was thinking along the lines of the variational calculus, but did not publish. These ideas in turn lead to the variational principles of mechanics, of Fermat, Maupertuis, Euler, Hamilton, and others.

Hamilton's principle can be applied to nonholonomic constraints if the constraint equations can be put into a certain form, a linear combination of first order differentials in the coordinates. The resulting constraint equation can be rearranged into first order differential equation. This will not be given here.

Lagrange Multipliers and Constraints

The Lagrangian L can be varied in the Cartesian r_k coordinates, for N particles,

$$\int_{t_1}^{t_1} \sum_{k=1}^{N} \left(\frac{\partial L}{\partial r_k} - \frac{d}{dt} \frac{\partial L}{\partial \dot{r} k} \right) . \delta r_k \, dt = 0.$$

Hamilton's principle is still valid even if the coordinates L is expressed in are not independent, here r_k, but the constraints are still assumed to be holonomic. As always the end points are fixed $\delta r_k(t_1) = \delta r_k(t_2) = 0$ for all k. What cannot be done is to simply equate the coefficients of δr_k to zero because the δr_k are not independent. Instead, the method of Lagrange multipliers can be used to include the constraints. Multiplying each constraint equation $f_i(r_k, t) = 0$ by a Lagrange multiplier λ_i for $i = 1, 2, \ldots, C$, and adding the results to the original Lagrangian, gives the new Lagrangian

$$L' = L(r_1, r_2, \ldots, \dot{r}_1, \dot{r}_2, \ldots, t) + \sum_{i=1}^{C} \lambda_i(t) f_i(r_k, t).$$

The Lagrange multipliers are arbitrary functions of time t, but not functions of the coordinates r_k, so the multipliers are on equal footing with the position coordinates. Varying this new Lagrangian and integrating with respect to time gives

$$\int_{t_1}^{t_2} \delta L' dt \int_{t_1}^{t_2} \sum_{k=1}^{N} \left(\frac{\partial L}{\partial \mathbf{r}_k} - \frac{d}{dt} \frac{\partial L}{\partial \dot{\mathbf{r}}_k} + \sum_{i=1}^{C} \lambda_i \frac{\partial f_i}{\partial \mathbf{r}_k} \right) \cdot \delta \mathbf{r}_k \, dt = 0 \, .$$

The introduced multipliers can be found so that the coefficients of δr_k are zero, even though the r_k are not independent. The equations of motion follow. From the preceding analysis, obtaining the solution to this integral is equivalent to the statement

$$\frac{\partial L'}{\partial \mathrm{r}_k} - \frac{\mathrm{d}}{\mathrm{d}t}\frac{\partial L'}{\partial \dot{r}_k} = 0 \quad \Rightarrow \quad \frac{\partial L}{\partial \mathrm{r}_k} - \frac{\mathrm{d}}{\mathrm{d}t}\frac{\partial L}{\partial \dot{r}_k} + \sum_{i=1}^{C} \lambda_i \frac{\partial f_i}{\partial \dot{r}_k} = 0,$$

which are Lagrange's equations of the first kind. Also, the λ_i Euler-Lagrange equations for the new Lagrangian return the constraint equations

$$\frac{\partial L'}{\partial \lambda_i} - \frac{\mathrm{d}}{\mathrm{d}t}\frac{\partial L'}{\partial \dot{\lambda}_i} = 0 \quad \Rightarrow \quad f_i(\mathrm{r}_k, t) = 0.$$

For the case of a conservative force given by the gradient of some potential energy V, a function of the r_k coordinates only, substituting the Lagrangian $L = T - V$ gives

$$\underbrace{\frac{\partial T}{\partial \mathrm{r}_k} - \frac{\mathrm{d}}{\mathrm{d}t}\frac{\partial T}{\partial \mathrm{r}_k}}_{-\mathrm{F}_k} + \underbrace{-\frac{\partial V}{\partial \mathrm{r}_k}}_{\mathrm{N}_k} + \sum_{i=1}^{C} \lambda_i \frac{\partial f_i}{\partial \mathrm{r}_k} = 0,$$

and identifying the derivatives of kinetic energy as the (negative of the) resultant force, and the derivatives of the potential equaling the non-constraint force, it follows the constraint forces are

$$\mathrm{C}_k = \sum_{i=1}^{C} \lambda_i \frac{\partial f_i}{\partial \mathrm{r}_k},$$

thus giving the constraint forces explicitly in terms of the constraint equations and the Lagrange multipliers.

Properties of the Euler–Lagrange Equation

In some cases, the Lagrangian has properties which can provide information about the system without solving the equations of motion. These follow from Lagrange's equations of the second kind.

Non Uniqueness

The Lagrangian of a given system is not unique. A Lagrangian L can be multiplied by a nonzero constant a, an arbitrary constant b can be added, and the new Lagrangian $aL + b$ will describe exactly the same motion as L. A less obvious result is that two Lagrangians describing the same system can differ by the total derivative (not partial) of some function $f(\mathrm{q}, t)$ with respect to time;

$$L' = L + \frac{\mathrm{d}f(\mathrm{q}, t)}{\mathrm{d}t}.$$

Each Lagrangian will obtain exactly the same equations of motion.

Invariance under Point Transformations

Given a set of generalized coordinates q, if we change these variables to a new set of generalized coordinates s according to a point transformation q = q(s, t), the new Lagrangian L' is a function of the new coordinates

$$L(\mathrm{q}(\mathrm{s},t),\dot{\mathrm{q}}(\mathrm{s},\dot{\mathrm{s}},t),t) = L'(\mathrm{s},\dot{\mathrm{s}},t),$$

and by the chain rule for partial differentiation, Lagrange's equations are invariant under this transformation;

$$\frac{\mathrm{d}}{\mathrm{d}t}\frac{\partial L'}{\partial \dot{s}_i} = \frac{\partial L'}{\partial s_i}.$$

This may simplify the equations of motion.

Cyclic Coordinates and Conserved Momenta

An important property of the Lagrangian is that conserved quantities can easily be read off from it. The *generalized momentum* "canonically conjugate to" the coordinate q_i is defined by

$$p_i = \frac{\partial L}{\partial \dot{q}_i}.$$

If the Lagrangian L does *not* depend on some coordinate q_i, it follows immediately from the Euler–Lagrange equations that

$$\dot{p}_i = \frac{\mathrm{d}}{\mathrm{d}t}\frac{\partial L}{\partial \dot{q}_i} = \frac{\partial L}{\partial q_i} = 0;$$

and integrating shows the corresponding generalized momentum equals a constant, a conserved quantity. This is a special case of Noether's theorem. Such coordinates are called "cyclic" or "ignorable".

For example, a system may have a Lagrangian

$$L(r,\theta,\dot{s},\dot{z},\dot{r},\dot{\theta},\dot{\phi},t),$$

where r and z are lengths along straight lines, s is an arc length along some curve, and θ and ϕ are angles. Notice z, s, and ϕ are all absent in the Lagrangian even though their velocities are not. Then the momenta

$$p_z = \frac{\partial L}{\partial \dot{z}}, \quad p_s = \frac{\partial L}{\partial \dot{s}}, \quad p_\phi = \frac{\partial L}{\partial \dot{\phi}},$$

are all conserved quantities. The units and nature of each generalized momentum will depend on the corresponding coordinate; in this case p_z is a translational momentum in the z direction, p_s is also a translational momentum along the curve s is measured, and p_ϕ is an angular momentum in the plane the angle ϕ is measured in. However complicated the motion of the system is, all the coordinates and velocities will vary in such a way that these momenta are conserved.

Energy Conservation

Taking the total derivative of the Lagrangian $L = T - V$ with respect to time leads to the general result

$$-\frac{\partial L}{\partial t} = \frac{\mathrm{d}}{\mathrm{d}t}\left(\sum_{i=1}^{n} \dot{q}_i \frac{\partial L}{\partial \dot{q}_i} - L\right).$$

If the entire Lagrangian is explicitly independent of time, it follows the partial time derivative of the Lagrangian is zero, $\partial L/\partial t = 0$, so the quantity under the total time derivative in brackets

$$E = \sum_{i=1}^{n} \dot{q}_i \frac{\partial L}{\partial \dot{q}_i} - L$$

must be a constant for all times during the motion of the system, and it also follows the kinetic energy is a homogenous function of degree 2 in the generalized velocities. If in addition the potential V is only a function of coordinates and independent of velocities, it follows by direct calculation, or use of Euler's theorem for homogenous functions, that

$$\sum_{i=1}^{n} \dot{q}_i \frac{\partial L}{\partial \dot{q}_i} = \sum_{i=1}^{n} \dot{q}_i \frac{\partial T}{\partial \dot{q}_i} = 2T.$$

Under all these circumstances, the constant

$$E = T + V$$

is the *total* conserved energy of the system. The kinetic and potential energies still change as the system evolves, but the motion of the system will be such that their sum, the total energy, is constant. This is a valuable simplification, since the energy E is a constant of integration that counts as an arbitrary constant for the problem, and it may be possible to integrate the velocities from this energy relation to solve for the coordinates. In the case the velocity or kinetic energy or both depends on time, then the energy is *not* conserved.

Mechanical Similarity

If the potential energy is a homogeneous function of the coordinates and independent of time, and all position vectors are scaled by the same nonzero constant α, $\mathrm{r}_k' = \alpha \mathrm{r}_k$, so that

$$V(\alpha \mathrm{r}_1, \alpha \mathrm{r}_2, \ldots, \alpha \mathrm{r}_N) = \alpha^N V(\mathrm{r}_1, \mathrm{r}_2, \ldots, \mathrm{r}_N)$$

and time is scaled by a factor β, $t' = \beta t$, then the velocities v_k are scaled by a factor of α/β and the kinetic energy T by $(\alpha/\beta)^2$. The entire Lagrangian has been scaled by the same factor if

$$\frac{\alpha^2}{\beta^2} = \alpha^N \quad \Rightarrow \quad \beta = \alpha^{1-N/2}.$$

Since the lengths and times have been scaled, the trajectories of the particles in the system follow geometrically similar paths differing in size. The length l traversed in time t in the original trajectory corresponds to a new length l' traversed in time t' in the new trajectory, given by the ratios

$$\frac{t'}{t} = \left(\frac{l'}{l}\right)^{1-N/2}.$$

Interacting Particles

For a given system, if two subsystems A and B are non-interacting, the Lagrangian L of the overall system is the sum of the Lagrangians L_A and L_B for the subsystems:

$$L = L_A + L_B.$$

If they do interact this is not possible. In some situations, it may be possible to separate the Lagrangian of the system L into the sum of non-interacting Lagrangians, plus another Lagrangian L_{AB} containing information about the interaction,

$$L = L_A + L_B + L_{AB}.$$

This may be physically motivated by taking the non-interacting Lagrangians to be kinetic energies only, while the interaction Lagrangian is the system's total potential energy. Also, in the limiting case of negligible interaction, L_{AB} tends to zero reducing to the non-interacting case above.

The extension to more than two non-interacting subsystems is straightforward – the overall Lagrangian is the sum of the separate Lagrangians for each subsystem. If there are interactions, then interaction Lagrangians may be added.

Examples

The following examples apply Lagrange's equations of the second kind to mechanical problems.

Conservative Force

A particle of mass m moves under the influence of a conservative force derived from the gradient ∇ of a scalar potential,

$$\mathrm{F} = -\nabla V(\mathrm{r}).$$

If there are more particles, in accordance with the above results, the total kinetic energy is a sum over all the particle kinetic energies, and the potential is a function of all the coordinates.

Cartesian Coordinates

The Lagrangian of the particle can be written

$$L(x, y, z, \dot{x}, \dot{y}, \dot{z}) = \frac{1}{2}m(\dot{x}^2 + \dot{y}^2 + \dot{z}^2) - V(x, y, z).$$

The equations of motion for the particle are found by applying the Euler–Lagrange equation, for the x coordinate

$$\frac{\mathrm{d}}{\mathrm{d}t}\left(\frac{\partial L}{\partial \dot{x}}\right) = \frac{\partial L}{\partial x},$$

with derivatives

$$\frac{\partial L}{\partial x} = -\frac{\partial V}{\partial x}, \quad \frac{\partial L}{\partial \dot{x}} = m\dot{x}, \quad \frac{\mathrm{d}}{\mathrm{d}t}\left(\frac{\partial L}{\partial \dot{x}}\right) = m\ddot{x},$$

hence

$$m\ddot{r} = -\frac{\partial V}{\partial x}$$

and similarly for the y and z coordinates. Collecting the equations in vector form we find

$$m\ddot{r} = -\nabla V$$

which is Newton's second law of motion for a particle subject to a conservative force.

Polar coordinates in 2d and 3d

The Lagrangian for the above problem in spherical coordinates, with a central potential, is

$$L = \frac{m}{2}(\dot{r}^2 + r^2\dot{\theta}^2 + r^2\sin^2\theta\dot{\varphi}^2) - V(r),$$

so the Euler–Lagrange equations are

$$m\ddot{r} - mr(\dot{\theta}^2 + \sin^2\theta\dot{\varphi}^2) + \frac{\partial V}{\partial r} = 0,$$

$$\frac{\mathrm{d}}{\mathrm{d}t}(mr^2\dot{\theta}) - mr^2\sin\theta\cos\theta\dot{\varphi}^2 = 0,$$

$$\frac{\mathrm{d}}{\mathrm{d}t}(mr^2\sin^2\theta\dot{\varphi}) = 0.$$

The ϕ coordinate is cyclic since it does not appear in the Lagrangian, so the conserved momentum in the system is the angular momentum

$$p_\varphi = \frac{\partial L}{\partial \dot{\varphi}} = mr^2 \sin^2 \theta \dot{\varphi},$$

in which r, θ and $d\phi/dt$ can all vary with time, but only in such a way that p_ϕ is constant.

Pendulum on a Movable Support

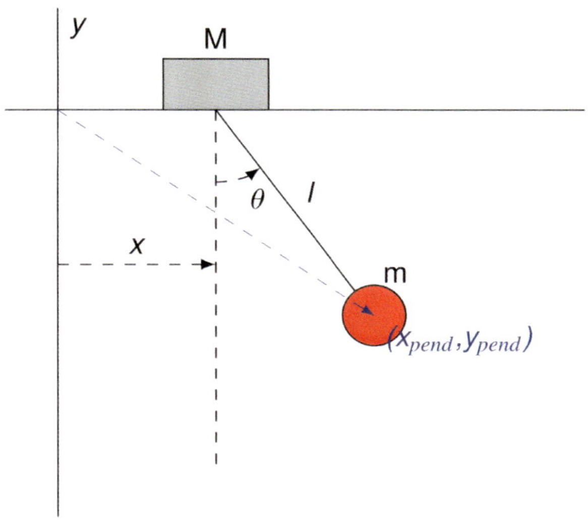

Sketch of the situation with definition of the coordinates

Consider a pendulum of mass m and length ℓ, which is attached to a support with mass M, which can move along a line in the x-direction. Let x be the coordinate along the line of the support, and let us denote the position of the pendulum by the angle θ from the vertical. The coordinates and velocity components of the pendulum bob are

$$x_{\text{pend}} = x + \ell \sin\theta \qquad \Rightarrow \quad \dot{x}_{\text{pend}} = \dot{x} + \ell\dot{\theta}\cos\theta$$
$$y_{\text{pend}} = -\ell\cos\theta \qquad \Rightarrow \quad \dot{y}_{\text{pend}} = \ell\dot{\theta}\sin\theta$$

The generalized coordinates can be taken to be x and θ. The kinetic energy of the system is then

$$T = \frac{1}{2}M\dot{x}^2 + \frac{1}{2}m\left(\dot{x}^2_{\text{pend}} + \dot{y}^2_{\text{pend}}\right)$$

and the potential energy is

$$V = mgy_{\text{pend}}$$

giving the Lagrangian

$$\begin{aligned} L &= T - V \\ &= \frac{1}{2}M\dot{x}^2 + \frac{1}{2}m\left[\left(\dot{x} + \ell\dot{\theta}\cos\theta\right)^2 + \left(\ell\dot{\theta}\sin\theta\right)^2\right] + mg\ell\cos\theta \\ &= \frac{1}{2}(M+m)\dot{x}^2 + m\dot{x}\ell\dot{\theta}\cos\theta + \frac{1}{2}m\ell^2\dot{\theta}^2 + mg\ell\cos\theta \end{aligned}$$

Since x is absent from the Lagrangian, it is a cyclic coordinate. The conserved momentum is

$$p_x = \frac{\partial L}{\partial \dot{x}} = (M+m)\dot{x} + m\ell\dot{\theta}\cos\theta;$$

and the Lagrange equation for the support coordinate x is

$$(M+m)\ddot{x} + m\ell\ddot{\theta}\cos\theta - m\ell\dot{\theta}^2\sin\theta = 0.$$

The Lagrange equation for the angle θ is

$$\frac{\mathrm{d}}{\mathrm{d}t}\left[m(\dot{x}\ell\cos\theta + \ell^2\dot{\theta})\right] + m\ell(\dot{x}\dot{\theta} + g)\sin\theta = 0;$$

and simplifying

$$\ddot{\theta} + \frac{\ddot{x}}{\ell}\cos\theta + \frac{g}{\ell}\sin\theta = 0.$$

These equations may look quite complicated, but finding them with Newton's laws would have required carefully identifying all forces, which would have been much more laborious and prone to errors. By considering limit cases, the correctness of this system can be verified: For example, $\ddot{x} \to 0$ should give the equations of motion for a simple pendulum that is at rest in some inertial frame, while $\ddot{\theta} \to 0$ should give the equations for a pendulum in a constantly accelerating system, etc. Furthermore, it is trivial to obtain the results numerically, given suitable starting conditions and a chosen time step, by stepping through the results iteratively.

Two-body Central Force Problem

Two bodies of masses m_1 and m_2 with position vectors r_1 and r_2 are in orbit about each other due to an attractive central potential V. We may write down the Lagrangian in terms of the position coordinates as they are, but it is an established procedure to convert the two-body problem into a one-body problem as follows. Introduce the Jacobi coordinates; the separation of the bodies $r = r_2 - r_1$ and the location of the center of mass $R = (m_1 r_1 + m_2 r_2)/(m_1 + m_2)$. The Lagrangian is then

$$L = \underbrace{\frac{1}{2}M\dot{\mathbf{R}}^2}_{L_{\mathrm{cm}}} + \underbrace{\frac{1}{2}\mu\dot{\mathbf{r}}^2 - V(|\mathbf{r}|)}_{L_{\mathrm{rel}}}$$

where $M = m_1 + m_2$ is the total mass, $\mu = m_1 m_2/(m_1 + m_2)$ is the reduced mass, and V the potential of the radial force, which depends only on the magnitude of the separation $|\mathrm{r}| = |\mathrm{r}_2 - \mathrm{r}_1|$. The Lagrangian splits into a *center-of-mass* term L_{cm} and a *relative motion* term L_{rel}.

The Euler–Lagrange equation for R is simply

$$M\ddot{\mathbf{R}} = 0,$$

which states the center of mass moves in a straight line at constant velocity.

Since the relative motion only depends on the magnitude of the separation, it is ideal to use polar coordinates (r, θ) and take $r = |\mathrm{r}|$,

$$L_{\mathrm{rel}} = \frac{1}{2}\mu(\dot{r}^2 + r^2\dot{\theta}^2) - V(r),$$

so θ is an ignorable coordinate with the corresponding conserved (angular) momentum

$$p_\theta = \frac{\partial L_{\mathrm{rel}}}{\partial \dot{\theta}} = \mu r^2 \dot{\theta} = \ell.$$

The radial coordinate r and angular velocity $\mathrm{d}\theta/\mathrm{d}t$ can vary with time, but only in such a way that ℓ is constant. The Lagrange equation for r is

$$\aleph\,\dot{}^{2} - \frac{dV}{dr} = \ddot{r}.$$

This equation is identical to the radial equation obtained using Newton's laws in a *co-rotating* reference frame, that is, a frame rotating with the reduced mass so it appears stationary. Eliminating the angular velocity $\mathrm{d}\theta/\mathrm{d}t$ from this radial equation,

$$\mu\ddot{r} \;\aleph\; \frac{\mathrm{d}V}{\mathrm{d}r} \quad \frac{\ell^2}{\mu r^3}.$$

which is the equation of motion for a one-dimensional problem in which a particle of mass μ is subjected to the inward central force $-\,\mathrm{d}V/\mathrm{d}r$ and a second outward force, called in this context the centrifugal force

$$F_{\mathrm{cf}} = \mu r\dot{\theta}^2 = \frac{\ell^2}{\mu r^3}.$$

Of course, if one remains entirely within the one-dimensional formulation, ℓ enters only as some imposed parameter of the external outward force, and its interpretation as angular momentum depends upon the more general two-dimensional problem from which the one-dimensional problem originated.

If one arrives at this equation using Newtonian mechanics in a co-rotating frame, the interpretation is evident as the centrifugal force in that frame due to the rotation of the frame itself. If one arrives at this equation directly by using the generalized coordinates (r, θ) and simply following the Lagrangian formulation without thinking about frames at all, the interpretation is that the centrifugal force is an outgrowth of *using polar coordinates*. As Hildebrand says:

> "Since such quantities are not true physical forces, they are often called *inertia forces*. Their presence or absence depends, not upon the particular problem at hand, but *upon the coordinate system chosen*." In particular, if Cartesian coordinates are chosen, the centrifugal force disappears, and the formulation involves only the central force itself, which provides the centripetal force for a curved motion.

This viewpoint, that fictitious forces originate in the choice of coordinates, often is expressed by users of the Lagrangian method. This view arises naturally in the Lagrangian approach, because the frame of reference is (possibly unconsciously) selected by the choice of coordinates. Unfortunately, this usage of "inertial force" conflicts with the Newtonian idea of an inertial force. In the Newtonian view, an inertial force originates in the acceleration of the frame of observation (the fact that it is not an inertial frame of reference), not in the choice of coordinate system. To keep matters clear, it is safest to refer to the Lagrangian inertial forces as *generalized* inertial forces, to distinguish them from the Newtonian vector inertial forces. That is, one should avoid following Hildebrand when he says "we deal *always* with *generalized* forces, velocities accelerations, and momenta. For brevity, the adjective "generalized" will be omitted frequently."

It is known that the Lagrangian of a system is not unique. Within the Lagrangian formalism the Newtonian fictitious forces can be identified by the existence of alternative Lagrangians in which the fictitious forces disappear, sometimes found by exploiting the symmetry of the system.

Electromagnetism

A test particle is a particle whose mass and charge are assumed to be so small that its effect on external system is insignificant. It is often a hypothetical simplified point particle with no properties other than mass and charge. Real particles like electrons and up quarks are more complex and have additional terms in their Lagrangians.

The Lagrangian for a charged particle with electrical charge q, interacting with an electromagnetic field, is the prototypical example of a velocity-dependent potential. The electric scalar potential $\phi = \phi(\mathrm{r}, t)$ and magnetic vector potential $\mathrm{A} = \mathrm{A}(\mathrm{r}, t)$ are defined from the electric field $\mathrm{E} = \mathrm{E}(\mathrm{r}, t)$ and magnetic field $\mathrm{B} = \mathrm{B}(\mathrm{r}, t)$ as follows;

$$\mathrm{E} = -\nabla\phi - \frac{\partial \mathrm{A}}{\partial t}, \quad \mathrm{B} = \nabla \times \mathrm{A}.$$

The Lagrangian of a massive charged test particle in an electromagnetic field is

$$L = \frac{m}{2}\dot{\mathrm{r}}^2 - q\phi + q\dot{\mathrm{r}}\cdot\mathrm{A},$$

which produces the Lorentz force law

$$m\ddot{r} = q\mathrm{E} + q\dot{r} \times \mathrm{B}.$$

An interesting detail in this example is the generalized momentum conjugate to r is the ordinary momentum plus a contribution from the A field,

$$\mathrm{p} = \frac{\partial L}{\partial r} = m\dot{r} + q\mathrm{A}.$$

If r is cyclic, which happens if the ϕ and A fields are uniform (independent of position), then this expression for p given here is the conserved momentum, while the usual quantity *mv* is not. This relation is also used in the minimal coupling prescription in quantum mechanics and quantum field theory.

Extensions to Include Non-conservative Forces

Dissipation (i.e. non-conservative systems) can also be treated with an effective Lagrangian formulated by a certain doubling of the degrees of freedom.

In a more general formulation, the forces could be both conservative and viscous. If an appropriate transformation can be found from the F_i, Rayleigh suggests using a dissipation function, *D*, of the following form:

$$D = \frac{1}{2}\sum_{j=1}^{m}\sum_{k=1}^{m} C_{jk}\dot{q}_j\dot{q}_k.$$

where C_{jk} are constants that are related to the damping coefficients in the physical system, though not necessarily equal to them. If *D* is defined this way, then

$$Q_j = -\frac{\partial V}{\partial q_j} - \frac{\partial D}{\partial \dot{q}_j}$$

and

$$\frac{\mathrm{d}}{\mathrm{d}t}\left(\frac{\partial L}{\partial \dot{q}_j}\right) - \frac{\partial L}{\partial q_j} + \frac{\partial D}{\partial \dot{q}_j} = 0.$$

Other Contexts and Formulations

The ideas in Lagrangian mechanics have numerous applications in other areas of physics, and can adopt generalized results from the calculus of variations.

Alternative Formulations of Classical Mechanics

A closely related formulation of classical mechanics is Hamiltonian mechanics. The Hamiltonian is defined by

$$H = \sum_{i=1}^{n} \dot{q}_i \frac{\partial L}{\partial \dot{q}_i} - L.$$

and can be obtained by performing a Legendre transformation on the Lagrangian, which introduces new variables canonically conjugate to the original variables. For example, given a set of generalized coordinates, the variables canonically conjugate are the generalized momenta. This doubles the number of variables, but makes differential equations first order. The Hamiltonian is a particularly ubiquitous quantity in quantum mechanics.

Routhian mechanics is a hybrid formulation of Lagrangian and Hamiltonian mechanics, which is not often used in practice but an efficient formulation for cyclic coordinates.

Momentum Space Formulation

The Euler–Lagrange equations can also be formulated in terms of the generalized momenta rather than generalized coordinates. Performing a Legendre transformation on the generalized coordinate Lagrangian L(q, dq/dt, t) obtains the generalized momenta Lagrangian L'(p, dp/dt, t) in terms of the original Lagrangian, as well the EL equations in terms of the generalized momenta. Both Lagrangians contain the same information, and either can be used to solve for the motion of the system. In practice generalized coordinates are more convenient to use and interpret than generalized momenta.

Higher Derivatives of Generalized Coordinates

There is no reason to restrict the derivatives of generalized coordinates to first order only. It is possible to derive modified EL equations for a Lagrangian containing higher order derivatives.

Optics

Lagrangian mechanics can be applied to geometrical optics, by applying variational principles to rays of light in a medium, and solving the EL equations gives the equations of the paths the light rays follow.

Relativistic Formulation

Lagrangian mechanics can be formulated in special relativity and general relativity. Some features of Lagrangian mechanics are retained in the relativistic theories but difficulties quickly appear in other respects. In particular, the EL equations take the same form, and the connection between cyclic coordinates and conserved momenta still applies, however the Lagrangian must be modified and is not simply the kinetic minus the potential energy of a particle. Also, it is not straightforward to handle multiparticle systems in a manifestly covariant way, it may be possible if a particular frame of reference is singled out.

Quantum Mechanics

In quantum mechanics, action and quantum-mechanical phase are related via Planck's constant, and the principle of stationary action can be understood in terms of constructive interference of wave functions.

In 1948, Feynman discovered the path integral formulation extending the principle of least action to quantum mechanics for electrons and photons. In this formulation, particles travel every possible path between the initial and final states; the probability of a specific final state is obtained by summing over all possible trajectories leading to it. In the classical regime, the path integral formulation cleanly reproduces Hamilton's principle, and Fermat's principle in optics.

Classical Field Theory

In Lagrangian mechanics, the generalized coordinates form a discrete set of variables that define the configuration of a system. In classical field theory, the physical system is not a set of discrete particles, but rather a continuous field $\phi(\mathrm{r}, t)$ defined over a region of 3d space. Associated with the field is a Lagrangian density;

$$\mathcal{L}(\phi, \nabla\phi, \partial\phi / \partial t, \mathrm{r}, t)$$

defined in terms of the field and its space and time derivatives at a location r and time t. Analogous to the particle case, for non-relativistic applications the Lagrangian density is also the kinetic energy density of the field, minus its potential energy density (this is not true in general, and the Lagrangian density has to be "reverse engineered"). The Lagrangian is then the volume integral of the Lagrangian density over 3d space;

$$L(t) = \int \mathcal{L} \mathrm{d}^3\mathrm{r}$$

where $\mathrm{d}^3\mathrm{r}$ is a 3d differential volume element. The Lagrangian is a function of time since the Lagrangian density has implicit space dependence via the fields, and may have explicit spatial dependence, but these are removed in the integral, leaving only time in as the variable for the Lagrangian.

Noether's Theorem

The action principle, and the Lagrangian formalism, are tied closely to Noether's theorem, which connects physical conserved quantities to continuous symmetries of a physical system.

If the Lagrangian is invariant under a symmetry, then the resulting equations of motion are also invariant under that symmetry. This characteristic is very helpful in showing that theories are consistent with either special relativity or general relativity.

Lagrangian (Field Theory)

Lagrangian mechanics is a powerful system for analyzing the motion of a system of particles. It can be extended to cover the dynamics of a field.

The Lagrangian function for a particle system is defined as the difference between its kinetic energy and its potential energy. The integral of the Lagrangian function over time is called the action of the system. It was initially believed that a system carried out a motion that minimized its action. This was called the Principle of Least Action. Later it was realized that the motion of a system was such that an infinitesimal variation in the path had no effect on the action of the system. The action was said to be stationary with respect to small variations in its motion.

The potential energy of a system is a function of only its space-time coordinates and the kinetic energy a function of only the time-derivatives of the coordinates. Where a system's motion is subject to constraints there exist a set of generalized coordinates that are compatible with those constraints and hence which simplify the analysis of the system's motion.

Stationarity implies that the system's motion is given by a set of equations known as the Euler-Lagrange equations for the system. For example, for a system of one particle whose potential energy depends upon only one variable x, the Euler-Lagrange equation is

$$\left(\left(d/dt\right)\left(\partial L/\partial v\right) - \left(\partial L/\partial x\right) = 0$$

$$\textit{where}$$

$$v = \left(dx/dt\right)$$

A beautiful result of Lagrangian mechanics is what is known as Noether's Theorem. This is that if there is a continuous transformation of the coordinates of the Lagrangian function then there exist a quantity that is conserved; i.e., is constant over time. If the Lagrangian function is unaffected by a change in the time coordinate then system energy is conserved. If it is unaffected by a change in a space coordinate then the momentum associated with that coordinate is conserved.

The momentum p associated with a coordinate x is given by

$$p = \left(\partial L/\partial v\right)$$

$$\text{where}$$

$$v = \left(dx/dt\right)$$

A comparison of this definition with the Euler-Lagrange equations reveals that

$$\left(dp/dt\right) = \left(\partial L/\partial x\right)$$

This leads to the Hamiltonian formulation of mechanics.

Fields

Let $q^\mu=(q^o, q^1, ..., q^n)$ be the vector of generalized position coordinates. Let ϕ be a field variable; i.e., a variable defined over all space. The field variable could be a scalar, a vector or a tensor. In order to reduce the cumbersomeness of the equations let the derivatives of ϕ with respect to the spatial coordinates be denoted as $\partial_\mu\phi$ where,

$$\partial_\mu\varphi = \left(\partial\varphi/\partial q^\mu\right)$$

The Lagrangian density L is given by

$$L=L\left(\varphi, \partial_\mu\varphi\right)$$

The integration of the Lagrangian density L over the spatial variables gives the Lagrangian function L for the system

$$L = K - V = \int L d^{\mu} q$$

The integration of the Lagrangian function over time gives the action J for the field; i.e.,

$$J = \int L dt$$

If time is included as one of the coordinates then the action can be represented as

$$J = \int L d^{\mu+1} q$$

Path Variation

The Lagrangian analysis of fields is constructed in analogy with the Lagrangian analysis of particle dynamics. Therefore it is necessary to consider the Lagrangian analysis of particle dynamics. It is therefore necessary to consider the Lagrangian analysis of a particle and the simplest case will suffice.

In particle mechanics a path x'(t) that coincides with another path x(t) at the end points but differs from it elsewhere is called a variation of the path x(t) The variation is

$$\delta x(t) = x'(t) - x(t)$$
$$for \quad 0 \le t \le T$$

For infinitesimal variations in the path the variation in the Lagrangian density is given by:

$$\delta L = (\partial L / \partial x)\delta x + (\partial L / \partial v)\delta v$$
$$where$$
$$v = dx / dt$$

However

$$\delta v = \delta(dx/dt) = d/dt\ (\delta x)$$

Thus

$$\delta L = (\partial L / \partial x)\delta x + (\partial L / \partial v) d / dt\ (\delta x) dt$$
$$and$$
$$\delta J = \int (\delta L) dt$$
$$and\ hence$$
$$\delta J = \int \left[(\partial L / \partial x)\delta x + (\partial L / \partial v) d / dt\ (\delta x) \right] dt$$

The term $\int (\partial L/\partial v) d/dt\ (\delta x)]dt$ can be integrated by parts ($\int UdV = UV - \int VdU$) to

$$(\partial L / \partial v)\delta x - \int (d/dt(\partial L / \partial v))\delta x dt$$

Since $\delta x(t)$ is equal to zero at the limits of the integration the first term vanishes.

Thus

$$\delta J = \int[\delta(\partial L/\partial v)\delta x - \int\left(d/dt(\partial L/\partial v)\right)\delta x]dt$$

or, equivalently

$$\delta J = \int\left[\delta\left(\partial L/\partial v\right) - \int\left(d/dt\left(\partial L/\partial v\right)\right)\right]\delta x dt$$

Stationarity of action requires that δJ is equal to zero. Thus

$$\delta J = \int[\delta(\partial L/\partial v) - \int(d/dt(\partial L/\partial v))]\delta x dt = 0$$

Since the variation δx(t) is arbitrary the only way for the integral to be zero is for the integrand to be zero everywhere on the interval; i.e.,

$$\partial L/\partial v) - \int(d/dt(\partial L/\partial v) = 0$$

This is the Euler-Lagrange equation for the particle's dynamics. The other important equation for the particle is the equation defining the momentum p of the particle. That equation is

$$p = \left(\partial L/\partial v\right)$$

where

$$v = \left(dx/dt\right)$$

The Lagrangian density of a field is a function of $\phi(q^\mu)$, which id the field variable as a function of the coordinates and the derivatives $\partial_\mu\phi$ of the field variable with respect to the coordinates. For example, the coordinates may be time and the spatial coordinates.

In symbols the relationships are

$$L = L\left(\varphi, \partial_\mu\varphi\right)$$

where $\partial_\mu\varphi$ represents the vector of partial derivatives with respect to the coordinates.

The Euler-Lagrange equation for a field is

$$\left(\partial L/\partial\varphi\right) - \partial_\mu\left(\partial L/\partial v^\mu\right) = 0$$

where

$$v^\mu = \partial_\mu\varphi$$

and where the repetition of the index μ in the second term indicates that there is a summation of that term with respect to that index.

Hamiltonian Mechanics

The Hamiltonian Formulation of Mechanics is equivalent to Newton's Laws and to the Lagrangian Formulation. Like the Lagrangian Formulation, one can use generalized coordinates with the Hamiltonian, however, the Hamiltonian is written in terms of coordinates and their conjugate momenta rather than the coordinates and their time derivatives as with the Lagrangian.

The coordinate and conjugate momentum plays an important role in the quantization of both mechanical and other problems so we will find that the Hamiltonian is very commonly used in Quantum Mechanics, where the commutation relations be coordinates and their conjugate momenta form the basis for quantization. In addition, for most problems, the Hamiltonian is the (conserved) total energy of the system.

The Hamiltonian formalism is another tool with which physicists can solve complex problems. An example of this is Liouville's theorem.

Derivation of the Hamilton Equations

The transformation of one set of variables, coordinates and velocities, to another, generalised coordinates and momenta, can be carried out via a Legendre transformation. Let us try to transform our Lagrange function into a quantity that is described by generalised coordinates and momenta.

The total differential of the Lagrangian is given by

$$dL = \sum_i \frac{\partial L}{\partial r_i} dr_i + \sum_i \frac{\partial L}{\partial \dot{r}_i} d\dot{r}_i = \sum_i \dot{p}_i dr_i + \sum_i p_i d\dot{r}_i$$

Since are $\frac{\partial L}{\partial \dot{r}_i}$ by definition the generalised momenta and $\frac{\partial L}{\partial r_i} \dot{p}_i$ by the Lagrange equation. Now we can rewrite the second term using the chain rule $\sum_i p_i d\dot{r}_i = d(p_i \dot{r}_i) - \sum_i \dot{r}_i dp_i$ to

$$dL = \sum_i \dot{p}_i dr_i + d(p_i \dot{r}_i) - \sum_i \dot{r}_i dp_i$$

Rearranging this we end up with

$$d(p_i \dot{r}_i - L) = -\sum_i \dot{p}_i dr_i + \dot{r}_i dp_i.$$

The argument of the total differential on the left side is (as we have seen before when talking about symmetries) the energy of the system, now only expressed in terms of coordinates and momenta. In this form we call it the "Hamiltonian" or "Hamilton's function" of the system

$$H(p, r, t) = p_i \dot{r}_i - L \cdot$$

From the second part of equation we get to know that the total differential of the Hamiltonian is given by

$$dH = -\sum_i \dot{p}_i dr_i + \sum_i \dot{r}_i dp_i \, .$$

Using this we can derive the required equations of motion in terms of p and r:

$$\dot{r}_i = \frac{\partial H}{\partial p_i} \text{ and } \dot{p}_i = -\frac{\partial H}{\partial r_i} \, .$$

These are the so-called "Hamilton's equations". They form a set of 2s first-order differential equations for the 2s unknown functions $p_i(t)$ and $r_i(t)$. Because of their simplicity and symmetry they are also called "canonical equations".

As a Reformulation of Lagrangian Mechanics

Starting with Lagrangian mechanics, the equations of motion are based on generalized coordinates

$$\{q^j \mid j = 1, \ldots, N\}$$

and matching generalized velocities

$$\{\dot{q}^j \mid j = 1, \ldots, N\}$$

We write the Lagrangian as

$$\mathcal{L}\left(q^j, \dot{q}^j, t\right)$$

with the subscripted variables understood to represent all *N* variables of that type. Hamiltonian mechanics aims to replace the generalized velocity variables with generalized momentum variables, also known as *conjugate momenta*. By doing so, it is possible to handle certain systems, such as aspects of quantum mechanics, that would otherwise be even more complicated.

For each generalized velocity, there is one corresponding conjugate momentum, defined as:

$$p_j = \frac{\partial \mathcal{L}}{\partial \dot{q}^j}$$

In Cartesian coordinates, the generalized momenta are precisely the physical linear momenta. In circular polar coordinates, the generalized momentum corresponding to the angular velocity is the physical angular momentum. For an arbitrary choice of generalized coordinates, it may not be possible to obtain an intuitive interpretation of the conjugate momenta.

One thing which is not too obvious in this coordinate dependent formulation is that different generalized coordinates are really nothing more than different coordinate patches on the same symplectic manifold.

The *Hamiltonian* is the Legendre transform of the Lagrangian:

$$\mathcal{H}\left(q^j, p_j, t\right) = \left(\sum_i \dot{q}^i p_i\right) - \mathcal{L}\left(q^j, \dot{q}^j, t\right).$$

If the transformation equations defining the generalized coordinates are independent of t, and the Lagrangian is a sum of products of functions (in the generalized coordinates) which are homogeneous of order 0, 1 or 2, then it can be shown that H is equal to the total energy $E = T + V$.

Each side in the definition of H produces a differential:

$$\begin{aligned} \mathrm{d}\mathcal{H} &= \sum_i \left[\left(\frac{\partial \mathcal{H}}{\partial q^i}\right)\mathrm{d}q^i + \left(\frac{\partial \mathcal{H}}{\partial p_i}\right)\mathrm{d}p_i\right] + \left(\frac{\partial \mathcal{H}}{\partial t}\right)\mathrm{d}t \\ &= \sum_i \left[\dot{q}^i\,\mathrm{d}p_i + p_i\,\mathrm{d}\dot{q}^i - \left(\frac{\partial \mathcal{L}}{\partial q^i}\right)\mathrm{d}q^i - \left(\frac{\partial \mathcal{L}}{\partial \dot{q}^i}\right)\mathrm{d}\dot{q}^i\right] - \left(\frac{\partial \mathcal{L}}{\partial t}\right)\mathrm{d}t. \end{aligned}$$

Substituting the previous definition of the conjugate momenta into this equation and matching coefficients, we obtain the equations of motion of Hamiltonian mechanics, known as the canonical equations of Hamilton:

$$\frac{\partial \mathcal{H}}{\partial q^j} = -\dot{p}_j \quad , \quad \frac{\partial \mathcal{H}}{\partial p_j} = \dot{q}^j \quad , \quad \frac{\partial \mathcal{H}}{\partial t} = -\frac{\partial \mathcal{L}}{\partial t}.$$

Hamilton's equations consist of $2n$ first-order differential equations, while Lagrange's equations consist of n second-order equations. However, Hamilton's equations usually don't reduce the difficulty of finding explicit solutions. They still offer some advantages, since important theoretical results can be derived because coordinates and momenta are independent variables with nearly symmetric roles.

Hamilton's equations have another advantage over Lagrange's equations: if a system has a symmetry, such that a coordinate does not occur in the Hamiltonian, the corresponding momentum is conserved, and that coordinate can be ignored in the other equations of the set. Effectively, this reduces the problem from n coordinates to $(n - 1)$ coordinates. In the Lagrangian framework, of course the result that the corresponding momentum is conserved still follows immediately, but all the generalized velocities still occur in the Lagrangian - we still have to solve a system of equations in n coordinates.

The Lagrangian and Hamiltonian approaches provide the groundwork for deeper results in the theory of classical mechanics, and for formulations of quantum mechanics.

Geometry of Hamiltonian Systems

A Hamiltonian system may be understood as a fiber bundle E over time R, with the fibers E_t, $t \in R$, being the position space. The Lagrangian is thus a function on the jet bundle J over E; taking the fiberwise Legendre transform of the Lagrangian produces a function on the dual bundle over time whose fiber at t is the cotangent space T^*E_t, which comes equipped with a natural symplectic form, and this latter function is the Hamiltonian.

Generalization to Quantum Mechanics through Poisson Bracket

Hamilton's equations above work well for classical mechanics, but not for quantum mechanics, since the differential equations discussed assume that one can specify the exact position and momentum of the particle simultaneously at any point in time. However, the equations can be further generalized to then be extended to apply to quantum mechanics as well as to classical mechanics, through the deformation of the Poisson algebra over p and q to the algebra of Moyal brackets.

Specifically, the more general form of the Hamilton's equation reads

$$\frac{\mathrm{d}f}{\mathrm{d}t} = \{f, \mathcal{H}\} + \frac{\partial f}{\partial t}$$

where f is some function of p and q, and H is the Hamiltonian. To find out the rules for evaluating a Poisson bracket without resorting to differential equations, a Poisson bracket is the name for the Lie bracket in a Poisson algebra. These Poisson brackets can then be extended to Moyal brackets comporting to an inequivalent Lie algebra, as proven by Hilbrand J. Groenewold, and thereby describe quantum mechanical diffusion in phase space. This more algebraic approach not only permits ultimately extending probability distributions in phase space to Wigner quasi-probability distributions, but, at the mere Poisson bracket classical setting, also provides more power in helping analyze the relevant conserved quantities in a system.

Mathematical Formalism

Any smooth real-valued function H on a symplectic manifold can be used to define a Hamiltonian system. The function H is known as the Hamiltonian or the energy function. The symplectic manifold is then called the phase space. The Hamiltonian induces a special vector field on the symplectic manifold, known as the Hamiltonian vector field.

The Hamiltonian vector field (a special type of symplectic vector field) induces a Hamiltonian flow on the manifold. This is a one-parameter family of transformations of the manifold (the parameter of the curves is commonly called the time); in other words an isotopy of symplectomorphisms, starting with the identity. By Liouville's theorem, each symplectomorphism preserves the volume form on the phase space. The collection of symplectomorphisms induced by the Hamiltonian flow is commonly called the Hamiltonian mechanics of the Hamiltonian system.

The symplectic structure induces a Poisson bracket. The Poisson bracket gives the space of functions on the manifold the structure of a Lie algebra.

Given a function f

$$\frac{\mathrm{d}}{\mathrm{d}t} f = \frac{\partial}{\partial t} f + \{f, \mathcal{H}\}$$

If we have a probability distribution, ρ, then (since the phase space velocity ($\dot{p}_i, \dot{q}_i$) has zero divergence, and probability is conserved) its convective derivative can be shown to be zero and so

$$\frac{\partial}{\partial t}\rho = -\{\rho, \mathcal{H}\}$$

This is called Liouville's theorem. Every smooth function G over the symplectic manifold generates a one-parameter family of symplectomorphisms and if $\{G, H\} = 0$, then G is conserved and the symplectomorphisms are symmetry transformations.

A Hamiltonian may have multiple conserved quantities G_i. If the symplectic manifold has dimension $2n$ and there are n functionally independent conserved quantities G_i which are in involution (i.e., $\{G_i, G_j\} = 0$), then the Hamiltonian is Liouville integrable. The Liouville-Arnold theorem says that locally, any Liouville integrable Hamiltonian can be transformed via a symplectomorphism into a new Hamiltonian with the conserved quantities G_i as coordinates; the new coordinates are called *action-angle coordinates*. The transformed Hamiltonian depends only on the G_i, and hence the equations of motion have the simple form

$$\dot{G}_i = 0\ , \quad \dot{\varphi}_i = F(G)$$

for some function F. There is an entire field focusing on small deviations from integrable systems governed by the KAM theorem.

The integrability of Hamiltonian vector fields is an open question. In general, Hamiltonian systems are chaotic; concepts of measure, completeness, integrability and stability are poorly defined.

Riemannian Manifolds

An important special case consists of those Hamiltonians that are quadratic forms, that is, Hamiltonians that can be written as

$$\mathcal{H}(q, p) = \tfrac{1}{2}\langle p, p\rangle_q$$

where $\langle\ ,\ \rangle_q$ is a smoothly varying inner product on the fibers T^*_qQ, the cotangent space to the point q in the configuration space, sometimes called a cometric. This Hamiltonian consists entirely of the kinetic term.

If one considers a Riemannian manifold or a pseudo-Riemannian manifold, the Riemannian metric induces a linear isomorphism between the tangent and cotangent bundles. Using this isomorphism, one can define a cometric. (In coordinates, the matrix defining the cometric is the inverse of the matrix defining the metric.) The solutions to the Hamilton–Jacobi equations for this Hamiltonian are then the same as the geodesics on the manifold. In particular, the Hamiltonian flow in this case is the same thing as the geodesic flow.

Sub-riemannian Manifolds

When the cometric is degenerate, then it is not invertible. In this case, one does not have a Riemannian manifold, as one does not have a metric. However, the Hamiltonian still exists. In the case where the cometric is degenerate at every point q of the configuration space manifold Q, so that the rank of the cometric is less than the dimension of the manifold Q, one has a sub-Riemannian manifold.

The Hamiltonian in this case is known as a sub-Riemannian Hamiltonian. Every such Hamiltonian uniquely determines the cometric, and vice versa. This implies that every sub-Riemannian manifold is uniquely determined by its sub-Riemannian Hamiltonian, and that the converse is true: every sub-Riemannian manifold has a unique sub-Riemannian Hamiltonian. The existence of sub-Riemannian geodesics is given by the Chow–Rashevskii theorem.

The continuous, real-valued Heisenberg group provides a simple example of a sub-Riemannian manifold. For the Heisenberg group, the Hamiltonian is given by

$$\mathcal{H}\left(x, y, z, p_x, p_y, p_z\right) = \tfrac{1}{2}\left(p_x^2 + p_y^2\right)$$

p_z is not involved in the Hamiltonian.

Poisson Algebras

Hamiltonian systems can be generalized in various ways. Instead of simply looking at the algebra of smooth functions over a symplectic manifold, Hamiltonian mechanics can be formulated on general commutative unital real Poisson algebras. A state is a continuous linear functional on the Poisson algebra (equipped with some suitable topology) such that for any element A of the algebra, A^2 maps to a nonnegative real number.

A further generalization is given by Nambu dynamics.

Charged Particle in an Electromagnetic Field

A good illustration of Hamiltonian mechanics is given by the Hamiltonian of a charged particle in an electromagnetic field. In Cartesian coordinates (i.e. $q_i = x_i$), the Lagrangian of a non-relativistic classical particle in an electromagnetic field is (in SI Units):

$$\mathcal{L} = \sum_i \tfrac{1}{2} m \dot{x}_i^2 + \sum_i e \dot{x}_i A_i - e\varphi$$

where e is the electric charge of the particle (not necessarily the elementary charge), ϕ is the electric scalar potential, and the A_i are the components of the magnetic vector potential (these may be modified through a gauge transformation). This is called minimal coupling.

The generalized momenta are given by:

$$p_i = \frac{\partial \mathcal{L}}{\partial \dot{x}_i} = m\dot{x}_i + eA_i$$

Rearranging, the velocities are expressed in terms of the momenta:

$$\underline{p_i \quad eA_i}$$

If we substitute the definition of the momenta, and the definitions of the velocities in terms of the momenta, into the definition of the Hamiltonian given above, and then simplify and rearrange, we get:

$$\mathcal{H} = \left\{ \sum_i \dot{x}_i p_i \right\} - \mathcal{L} = \sum_i \frac{(p_i - eA_i)^2}{2m} + e\varphi$$

This equation is used frequently in quantum mechanics.

Relativistic Charged Particle in an Electromagnetic Field

The Lagrangian for a relativistic charged particle is given by:

$$\mathcal{L}(t) = -mc^2 \sqrt{1 - \frac{\dot{\mathrm{x}}(t)^2}{c^2}} - e\varphi\left(\mathrm{x}(t), t\right) + e\dot{\mathrm{x}}(t) \cdot \mathrm{A}\left(\mathrm{x}(t), t\right)$$

Thus the particle's canonical (total) momentum is

$$\mathrm{P}(t) = \frac{\partial \mathcal{L}(t)}{\partial \dot{\mathrm{x}}(t)} = \frac{m\dot{\mathrm{x}}(t)}{\sqrt{1 - \frac{\dot{\mathrm{x}}(t)^2}{c^2}}} + e\mathrm{A}\left(\mathrm{x}(t), t\right)$$

that is, the sum of the kinetic momentum and the potential momentum.

Solving for the velocity, we get

$$\dot{\mathrm{x}}(t) = \frac{\mathrm{P}(t) - e\mathrm{A}(\mathrm{x}(t), t)}{\sqrt{m^2 + \frac{1}{c^2}\left(\mathrm{P}(t) - e\mathrm{A}\left(\mathrm{x}(t), t\right)\right)^2}}$$

So the Hamiltonian is

$$\mathcal{H}(t) = \dot{\mathrm{x}}(t) \cdot \mathrm{P}(t) - \mathcal{L}(t) = c\sqrt{m^2c^2 + \left(\mathrm{P}(t) - e\mathrm{A}(\mathrm{x}(t), t)\right)^2} + e\varphi(\mathrm{x}(t), t)$$

From this we get the force equation (equivalent to the Euler–Lagrange equation)

$$\dot{\mathrm{P}} = -\frac{\partial \mathcal{H}}{\partial \mathrm{x}} = e(\nabla \mathrm{A}) \cdot \dot{\mathrm{x}} - e\nabla\varphi$$

from which one can derive

$$\frac{\mathrm{d}}{\mathrm{d}t}\left(\frac{m\dot{\mathrm{x}}}{\sqrt{1 - \frac{\dot{\mathrm{x}}^2}{c^2}}}\right) = e\mathrm{E} + e\dot{\mathrm{x}} \times \mathrm{B}$$

An equivalent expression for the Hamiltonian as function of the relativistic (kinetic) momentum, $\mathrm{p} = \gamma m\dot{\mathrm{x}}(t)$, is

$$\mathcal{H}(t) = \dot{\mathrm{x}}(t) \cdot \mathrm{p}(t) + \frac{mc^2}{\gamma} + e\varphi(\mathrm{x}(t),t) = \gamma mc^2 + e\varphi(\mathrm{x}(t),t) = E + V \cdot$$

This has the advantage that p can be measured experimentally whereas P cannot. Notice that the Hamiltonian (total energy) can be viewed as the sum of the relativistic energy (kinetic+rest), $E = \gamma mc^2$, plus the potential energy, $V = e\phi$.

Canonical Transformation

Coordinate transformations $\{q, p\} \to \{Q, P\}$ and Hamiltonian functions $H(q, p, t)$, $H'(Q, P, t)$ such that the form of the canonical equations of motion are preserved:

$$\dot{q} = \frac{\partial H}{\partial p} \;;\; \dot{p} = -\frac{\partial H}{\partial q}$$

$$\dot{Q} = \frac{\partial H'}{\partial Q} \;;\; \dot{P} = -\frac{\partial H}{\partial Q}$$

These transformations are called "canonical transformations". Not every coordinate transformation is so special; we will now derive the conditions for transformations to be "canonical" (i.e.. preserve the form of the canonical equations of motion), and we will find out the differences between the "old" and "new" Hamiltonian functions.

We know that we can derive canonical equations of motion for $\{q, p\}$ from an action principle of the form $S = \int (p\dot{q} - H)dt$ We can also derive canonical equations of motion for $\{Q, P\}$ *from the action* $S = \int (P\dot{Q} - H')dt$. The solutions to the action principle for are unchanged if $S - S' = \int (df / dt)dt$, for $f = f(q, t)$ a function of coordinates and time. If we use a restricted version of the action principle, keeping both q and p fixed at the initial and final times, then the functions f can be a function of coordinates and momenta: $F = F(q, p, t)$. Each such choice of function F will "generate" a change of coordinates if

$$\int \frac{dF}{dt} dt = \int (p\dot{q} - H)dt - \int (P\dot{Q} - H')dt$$

$$\int dF = int(pdq - Hdt - (P\,dQ - H'dt)$$

$$dF = pdq - PdQ + (H' - H)dt$$

This is a restricting condition on the coordinates, since the combination $pdq - PdQ + (H' - H)dt$ will not in general be an exact differential (i.e., we would not be be able to find an exact integral). When integrated, this expression gives us a function $F(q, Q, t)$ such that

$$\frac{\partial F}{\partial q} = p; \; \frac{\partial F}{\partial Q} = -P; \; \frac{\partial F}{\partial t} = H' - H$$

A different, and more useful, way to view these equations is to think that any function F(qQ, t) will generate a coordinate transformation $\{q, p\} \to \{Q, P\}$, where the change of coordinates are given by the equations

$$\frac{\partial F}{\partial q} = p; \ \frac{\partial F}{\partial Q} = P.$$

The equations of motion in either coordinate set will have the form of canonical equationsof motion, if $H'(Q, P, t) = H(q, p, t) + \partial F / \partial t$ where in the right hand side we use $q = q(Q,P,t), F = F(q(Q, P, t), p(Q, P, t), t)$.

The formalism can be derived in a straightforward manner for n generalized coordinates. For example, consider the generating function $F = qQ$. The coordinate change will be

$$\frac{\partial F}{\partial q} = Q = p$$

$$\frac{\partial F}{\partial Q} = q = -P$$

The new coordinate q is the old momentum p, and the new momentum P is minus the old coordinate q. Just like in Lagrangian mechanics the generalized coordinates need not have units of length or be position coordinates, in Hamiltonian mechanics the canonical momenta need not be related to linear momenta or velocities; in fact the roles of coordinates and momenta are interchangeable.

The solutions to the new equations of motion should of course be the same physical solutions as for the "old" problem. T illustrate this point, let's consider the Hamiltonian of a simple harmonic oscillator,

$$H = \frac{p^2}{2m} + \frac{1}{2}kq^2.$$

The canonical equations of motion are $\dot{q} = \partial H / \partial p = p / m$ and $\dot{p} = -\partial H / \partial q = -kq$. Using the q equation into the p equation, we obtain the SHO equation;

$$\ddot{q} = -\frac{k}{m} q$$

with the usual solution $q(t) = q_0 \cos(\omega t + \varphi)$, with $\omega^2 = k / m$. If we perform the change of coordinates generated by $F = qQ$ the new Hamiltonian is the same as the old one because $\partial F \ \partial t = 0$. The change of coordinates is $q = -P, \ p = Q$, so the new Hamiltonian is

$$H' = \frac{Q^2}{2m} + \frac{1}{2}k(-P)^2 = \frac{Q^2}{2m} + \frac{1}{2} kP^2$$

The new equations of motion are

$$\dot{Q} = \frac{\partial H'}{\partial P} = kP$$

$$\dot{P} = -\frac{\partial H'}{\partial Q} = Q / m.$$

Again using the $\dot{Q}$ equation into the $\dot{P}$ equation, we obtain a SHO equation

$$\frac{\ddot{Q}}{k} = -\frac{Q}{m}$$

with the usual solution $Q(t) = Q_0 \cos(\omega t + \psi 0)$, and $P(t) = Q / \dot{k} = -(Q_0\omega / k) \sin(\omega t + \psi 0)$ and $\omega^2 = k / m$. We see that the "old solution"

$$q(t) = q_0 \cos(\omega t + \varphi)$$

is the same as the "new" solution

$$-P(t) = (Q_0\omega / k)\ \sin(\omega_t + \psi_0)$$

if $\psi_0 = \varphi_0 + \pi / 2$ *and* $Q_0 = q_0\omega / k$. The initial values of the coordinates are indeed the ones that tell us about differences in the choices for coordinates. If the oscillator is a mass on a spring, a physical statement of initial conditions could be "start from rest with the spring stretched by a length Δ from its rest length. From that physical description, we have to come up with values for our constants of integration, be those q_0, φ_0 and Q_0, ψ_0. In this case, the condition "at rest" is described by $\varphi_0 = 0$, but $\psi_0 = \pi / 2$; while the initial spring length tells us that $q_0 = \Delta$ and $Q_0 = \Delta\omega / k$.

We know the Hamiltonian functions are the same under the transformation generated by $F = qQ$ because $\partial F / \partial t = 0$, but the Lagrangian functions will not be the same:

$$L' - L = (P\dot{Q} - H') - (p\dot{q} - H) = P\dot{Q} - p\dot{q} = (-q)\dot{p} - p\dot{q} = -d(qp) / dt$$

The Lagrangian functions differ by a total time derivative of the function $-qp = -mq\dot{q}$. This is not the generic transformation we learned about in Goldstein's problem 1-8, since it is a function of q, $\dot{q}$, not just q: canonical transformations are more general. We can use Legendre transformations to find generating functions of different pairs of variables other than q, Q associated with $F(q, Q, t)$. For example, if we want generating functions of q, P (that is, using P instead of Q), we know that $\Phi(q, P) = F(q, Q) + PQ$, and

$$d\Phi = d(F + PQ) = pdq - PdQ + (H' - H)dt + PdQ + QdP = pdq + QdP + (H' - H)dt$$

so the equations for the coordinate transformations and the change in Hamiltonian is

$$\frac{\partial \Phi}{\partial q} = p; \quad \frac{\partial \Phi}{\partial P} = Q; \quad \frac{\partial \Phi}{\partial t} = H' - H$$

Let us consider, for example, a change of coordinates generated by $\Phi(q, P, t) = qP - f(q, t)$ The new and old Hamiltonian functions are not the same now, but

$$H' = H + \frac{\partial \Phi}{\partial t} = H - \frac{\partial f}{\partial t}$$

The coordinate transformation is

$$p = \frac{\partial \Phi}{\partial q} = P - \frac{\partial f}{\partial q}$$

$$Q = \frac{\partial \Phi}{\partial P} = q$$

Only momenta are transformed, while coordinates are the same.

The change in Lagrangian will be

$$L' = PQ - H' = \left(p + \frac{\partial f}{\partial q} \right)\dot{q} - \left(H - \frac{\partial f}{\partial t} \right) = (p\dot{q} - H) + \left(\frac{\partial f}{\partial q} + - \frac{\partial f}{\partial t} \right) = L + \frac{df}{dt}$$

This is the case we studied in Goldstein's problem 1-8 and revisited many times through the course. These transformation includes the case of the electromagnetic gauge transformation.

We have seen then that all of the transformations generated by arbitrary functions of coordinates $f(q, t)$, are canonical transformations generated by functions of the form $\Phi(q, P, t) = qP - f(q, t)$, a small set of the generic functions $\Phi(q, P, t)$ we could use.

Liouville's Theorem

The direct conditions allow us to prove Liouville's theorem, which states that the *volume* in phase space is conserved under canonical transformations, i.e.,

$$\int dqdp = \int dQdP$$

By calculus, the latter integral must equal the former times the Jacobian J

$$\int dQdP = \int J dqdp$$

where the Jacobian is the determinant of the matrix of partial derivatives, which we write as

$$J \equiv \frac{\partial(Q, P)}{\partial(q, p)}$$

Exploiting the "division" property of Jacobians yields

$$J \equiv \frac{\partial(Q,P)}{\partial(q,P)} \Big/ \frac{\partial(q,p)}{\partial(q,P)}$$

Eliminating the repeated variables gives

$$J \equiv \frac{\partial(Q)}{\partial(q)} \Big/ \frac{\partial(p)}{\partial(p)}$$

Application of the direct conditions above yields $J = 1$.

Appell's Equation of Motion

Ordinary differential equations which describe the motions of both holonomic and non-holonomic systems, established by P.E. Appell. They are sometimes referred to as Gibbs–Appell equations, since they were first proposed by J.W. Gibbs for holonomic systems. The Appell equations in independent Lagrange coordinates $q_s (s = 1,....,n)$ have the form of second-order equations

$$\frac{\partial S}{\partial \ddot{q}_i} = Q_i^*, i = 1,...,k \leq n.$$

Here

$$S = \frac{1}{2}\sum_{V=1}^{N} m_v w_v^2$$

(m_v and w_v are the masses and the accelerations of the N points of the system) is the energy of acceleration of the system, which is so expressed that it contains the second derivatives of the q_i $i = 1,...,k$ only, the variations of which are considered as independent; Q_i^* are the generalized forces corresponding to the coordinates q_i, obtained as coefficients in front of the independent variations δq_i in the expression for the work of the given active forces F_v corresponding to virtual displacements δr_v

$$\sum_{V=1}^{N} F_v \delta r_v = \sum_{i=1}^{k} Q_1^* \delta q_i$$

In evaluating S and Q_i^* the dependent variables $\dot{q}_j(\delta q_j)(j = k+1,....,n)$ are expressed in terms of the independent velocities (variations) by solving the $n-k$ non-holonomic constraint equations, expressed in the generalized coordinates q_s (and by solving the equations for δq_s obtained from them) Differentiation with respect to the time t of the expressions found for $\dot{q}_j$ yields expressions for $\ddot{q}_j$ yields expressions for $\ddot{q}_j$ in terms of $\ddot{q}_i$.

Equations (1), together with the $n-k$ equations of the non-integrable constraints, form a system (of order $n+k$) of n differential equations involving the n unknowns q_s. For a holonomic system $k = n$ all velocities q_s and variations δq_s are independent $Q_i* = Q_i$ and equations are a different notation for the Lagrange equations (in mechanics)) of the second kind.

Appell's equations in quasi-coordinates π_r where

$$\dot{\pi} r = \sum_{i=1}^{n} a_{r_i} \dot{q}_i, \qquad r = 1, \ldots., k,$$

Here the form

$$\frac{\partial S}{\partial \ddot{\pi} r} = \Pi_r, \quad r = 1, \ldots, k \leq n.$$

Here S is the energy of acceleration, expressed in terms of the second "derivatives" $\ddot{\pi} r$ (with respect to the time) of the quasi-coordinates, and Π_r are the generalized forces corresponding to the quasi-coordinates. Equations (3), together with the equations of the $n-k$ non-integrable constraints and the k equations (2), form a system of $n+k$ differential equations of the first order with the same number of unknowns $q_{s,} s = 1, \ldots,$ and $\dot{\pi} r, r = 1, \ldots., k$.

Appell's equations are the most general equations of motion of mechanical systems.

Derivation of the Gibbs-appell Equations of Motion

Consider a system consisting of N particles, which is subject to a set of known forces, M holonomic constraints, and L anholonomic constraints.

If we represent the configuration of the system by a single point $x \equiv x_1, x_2, \ldots, x_{3N}$ in a 3N-dimensional Cartesian configuration space, [5] and if, in this space, we let f be the 3N-dimensional resultant of the given forces acting on the system and F be the 3N-dimensional resultant of all the constraint forces, then Newton's equations of motion for the system are

$$m_i \ddot{x}_i = f_i + F_i \; i = 1, 2, \ldots, 3N.$$

If we let $q = q_1 q_2, \ldots, q_{3N-M}$ be any $3N - M$ coordinates which together with the holonomic constraint conditions uniquely determine the configuration of the system, then the effect of the holonomic constraints on the possible configurations of the system can be described by the following relations:

$$x_i = x_i(q, t), \; i = 1, 2, \ldots, 3N.$$

If we now let $r \equiv \dot{r}_1, \dot{r}_2, \ldots, \dot{r}_{3N-M-L}$ be any *3N- M- L* quantities linear in $\dot{q}_1, \dot{q}_2, \ldots, \dot{q}_{3N-M}$ which together with the anholonomic constraint conditions uniquely determine the values of $\dot{q}_1, \dot{q}_2, \ldots, \dot{q}_{3N-M}$ then the effect of the anholonomic constraints on the possible motions of the system can be described by the following relations:

$$q_i = \sum_j a_{ij}(q, t) \dot{r}_j + b_i(q, t), \; i = 1, 2, \ldots, 3N - M.$$

Equations define the restrictions imposed by the constraints on the configuration and the motion of the system. In addition to these restrictions, we shall assume as an added requirement that the components F_i of the constraint force satisfy the following relation:

$$\sum_i F_i s_{ij} = 0, \quad j = 1, 2, ..., 3N - M - L,$$

where

$$S_{ij} \equiv \frac{\partial \dot{x}_i(q, \dot{r}, t)}{\partial \dot{r}_j} = \frac{\partial \ddot{x}_i(q, \dot{r}, \ddot{r}, t)}{\partial \ddot{r}_j}$$

Equations provide us with 12N- 2M- L equations in the 12N -2M- L unknowns $x_1, x_2, ..., x_{3N}, q_1, q_2, ..., q_{3N-M}, \dot{r}_1, \dot{r}_2, ..., r_{3N-M-L}, F_1, F_2, ...$ and F_3N. By judicious combination of the above equations, the number of equations and unknowns we have to worry about can be dramatically decreased. If we multiply equation by s_{ij}, sum over *i*, and make use of equation, we obtain

$$\sum_i m_i \ddot{x}_i s_{ij} = \sum_i f_i s_{ij}.$$

If we now define

$$S = \frac{1}{2} \sum_i m_i \ddot{x}_i^2$$

and

$$Q_j = \sum_i f_i s_{ij},$$

then equation can be written in the following form:

$$\frac{\partial S(q, \dot{r}, \ddot{r}, t)}{\partial \ddot{r}_j} = Q_J.$$

The *3N- M- L* equations are the Gibbs-Appell equations of motion. These equations together with the *3N- M* equations provide us with *6N- 2M- L* equations in the *6N- 2M- L* unknowns $q_1, q_2, ..., q_{3N-M}, \dot{r}_1 - \dot{r}_2, \bullet\bullet\bullet\bullet \dot{r}_{3N-M-L}$.

An even simpler and inore elegant form of the GibbsAppell equations can be obtained if we define

$$U \equiv \sum_i f_i \ddot{x}_i$$

$$R \equiv S\text{-}U.$$

It then follows that

$$Q_j = \frac{\partial U(q, \dot{r}, \ddot{r}, t)}{\partial_{\ddot{r}j}}$$

and the Gibbs-Appell equations assume the form

$$\frac{\partial R(q,\dot{r},\ddot{r},t)}{\partial_{\dot{r}_j}} = 0$$

This form of the equations of motion was introduced by Appell6 but is generally ignored in most treatments of the Gibbs-Appell method. We have found it quite useful. When we wish to make a distinction, we shall refer to equation as the first form of the Gibbs-Appell equations of motion and to equation as the second form.

The Gibbs-Appell equations of motion are not affected by additive terms of the form $\varphi(q,\dot{r},t)$, which occur in R, S, or U; that is, terms that when expressed as a function of $q,\dot{r},\ddot{r},$ and t do not contain at least one of the accelerations r;. Hence, in what follows, we will assume that the equality of two different values of either R, S, or U means that they differ at most by additive terms of the above form, and we will freely drop from expressions for R, S, or U terms of this form. It is also interesting to note that we could multiply $R(q,\dot{r},\ddot{r},t)$ by any function of the above form without altering the equations of motion.

Lagrange's Principle

In this topic we consider the validity of equation, which as indicated earlier we refer to as Lagrange's principle. If at time *t* the given system is in a configuration q, then from equations and it can be shown that the possible subsequent displacements dx of the system are those that satisfy the equations

$$dx_i = \sum_j S_{ij} d_{rj} + \alpha_i dt$$

for some set of values of dr and dt, where s_{ij} is defined by equation, and α_i is a known function of q and *t*, the exact form of which we will not need. The actual displacement will of course depend on the applied forces and the initial conditions.

If we examine individually the various constraint forces that are responsible for limiting the possible displacements to the above set of displacements, we find in general that they satisfy Lagrange's principle. For example, if our system consists of a single particle that is confined to a smooth surface, which may or may not be moving, then from equation it follows that the vector *s(j)*, whose components are s_{1j} s_{2j}, and s_{3j}, will be tangent to the surface. Since the surface is smooth, the constraint force F must be perpendicular to the surface, hence F•s *(j)* = 0 and Lagrange's principle is satisfied. As a second example, if our system consists of a disk or sphere rolling without slipping on either a stationary or moving surface, then the constraint force responsible for keeping the system from slipping is applied to the particle in the system that is in contact with the surface; hence the only non vanishing components of the *3N*-dimensional constraint force Fare those associated with this particle. But for those components of x associated with this particle s_{ij}=0, it follows that Lagrange's principle is valid for this constraint force. In a similar fashion, it is possible to show that an action-reaction pair of forces that maintain a fixed distance between two particles, or the forces that smoothly hinge two bodies together, will satisfy Lagrange's principle. We can proceed in this fashion and confirm for each of the constraint forces acting on a given system the validity of Lagrange's principle. Though we can thus verify Lagrange's principle by considering the constraint forces one by one, there is no general proof of the principle.

Virtual Displacements

The process of verifying Lagrange's principle can frequently be simplified if we introduce the concept of a virtual displacement. We define a virtual displacement to be one of the possible displacements, as given by, for which $dt = 0$; or equivalently a displacement δx that satisfies the equations

$$\delta x_i = \sum_j s_{ij} \; \delta r_i$$

for some value of 8r. Thus, given a system in a configuration q at time t, the set of all virtual displacements is a particular subset of the set of all possible displacements. Physically such displacements are the displacements that would occur if the system was frozen in its motion at timet, and the system was then moved without violating any of the constraints operating on the system.

A definition of a virtual displacement is sometimes made which allows the displacement to violate one of the constraints. Our definition excludes such displacements. It is also possible to consider finite as well as infinitesimal virtual displacements. We restrict our consideration to infinitesimal virtual displacements. Hence, whenever the term "a virtual displacement" is used in this article, it is to be read by those who employ the term in the broader sense as "an infinitesimal virtual displacement that does not violate any of the constraints."

From the above definition, equation, it follows that the work done by the constraint force F in a virtual displacement is given by

$$\delta W = \sum_i F_i \delta x_i = \sum_i F_i \left(\sum_j S_{ij} \delta r_j \right) = \sum_j \left(\sum_i F_i s_{ij} \right) \delta r_j .$$

A necessary and sufficient condition that the above quantity vanishes for arbitrary values of is that Lagrange's principle is satisfied. Hence assuming Lagrange's principle to be true is equivalent to assuming that the work done by the constraint forces in an arbitrary virtual displacement is zero. If we are interested in determining whether or not Lagrange's principle is satisfied for a particular constraint, it is generally easier to simply confirm that the work done by the force in an arbitrary virtual displacement is zero rather than to try to prove Lagrange's principle directly. It also follows from equations and the definition of work that the work done by the given force fin a virtual displacement is given by

$$\delta W = \sum_i Q_i \delta r_i .$$

It is often much easier to determine the values of the Q; by exploiting the above equation than by directly using the definition given by equation.

It should be noted that the introduction of the concept of a virtual displacement is a convenience and not a necessity. We were very careful in our derivation of the Gibbs-Appell equations of motion not to use this concept nor to employ any variational principles. This has been done to counteract the impression conveyed by many mechanics texts that any advanced technique used in mechanics requires the introduction of virtual quantities and the employment of variational principles.

Lagrange's Equations of Motion

If there are no anholonomic constraints, and we let $\dot{r}_i = \dot{q}_i$ then it is relatively easy to show that

$$\frac{\partial S(q,\dot{q},\ddot{q},t)}{\partial \ddot{q}_i} = \frac{d}{dt}\left(\frac{\partial T(q,\dot{q},t)}{\partial \dot{q}_i}\right) - \frac{\partial T(q.\dot{q},t)}{\partial q_i}$$

and

$$Q_j \equiv \sum_i f_i \frac{\partial \dot{x}_i(q,\dot{q},t)}{\partial \dot{q}_j} = \sum_i f_i \frac{\partial x_i(q,t)}{\partial q_j}$$

where T is the kinetic energy of the system and the definition of Q_j goes over to the usual definition of the generalized component of the force associated with the generalized coordinate q_j. If we use Gibbs Appell equations, we obtain Lagrange's equations of motion. It follows that Lagrange's equations of motion can be considered as a special case of the Gibbs-Appell equations of motion.

It is possible starting from the Gibbs-Appell equations to obtain Lagrange's equations for quasi coordinates but these equations are usually harder to apply than the Gibbs Appell equations and will not be considered here.

Gibbs-appell Versus Lagrange

There are two features of the Gibbs-Appell method that in principle if not in practice make it superior to Lagrange's method: (1) The Gibbs-Appell method handles anholonomic constraints in a simple, direct fashion whereas Lagrange's method requires additional machinery to handle such constraints; (2) the Gibbs-Appell method allows the introduction of motional coordinates that are not simply the time derivatives of the configurational coordinates, thus giving it a versatility not shared by Lagrange's method without substantial modification. These features allow certain systems to be handled far more economically using the Gibbs-Appell method than using Lagrange's method. The problem of finding equations of motion for a disk rolling on a flat surface is a good illustrative example and will be considered later.

If we are dealing with holonomic systems and there is no immediate advantage to introducing motional coordinates that are other than the time derivatives of the configurational coordinates, then Lagrange's equations are usually simpler to use than the Gibbs-Appell equations. Frequently, however, we are unnecessarily deterred from using the Gibbs-Appell method for such problems by the fact that thoughtless evaluation of the function $S(\ q,\dot{q},\ddot{q},t)$ frequently leads to expressions containing a discouraging number of terms. Suppose, for instance, we wish to obtain the equations of motion of an unconstrained particle in terms of the spherical coordinates r, θ, and ϕ. A straightforward substitution of $x = r\ sin\theta\ cos\theta$, $y = r\ sin\theta\ cos\theta$, and $z = r\,cos\,\theta$ into the expression $S\ =\ \frac{1}{2}m\left(\ddot{x}^2\ +\ \ddot{y}^2\ +\ddot{z}^2\right)$ will result in 6 distinct terms in $.\dot{x}^2$, 36 distinct terms in $\dot{y}^2$, and 6 distinct terms in $\ddot{z}^2$. By contrast, Lagrange's method requires us to obtain the kinetic energy function $T(q,\dot{q},t)$ and, in the case considered above, a straightforward substitution of $x = r\ sin\theta\ cos\theta$,

$y = r\ sin\theta\ cos\theta$, and $z = r\,cos\,\theta$ into $T\ = \frac{1}{2}m\left(\dot{x}^2 + \dot{y}^2 + \dot{z}^2\right)$ the expression will result in 6 distinct terms in $\dot{x}^2$, 6 distinct terms in $\dot{y}^2$, and 3 distinct terms in $\ddot{z}^2$. Hence it would appear that for this problem the Lagrange method is considerably easier than the Gibbs-Appell method. The operational disparity between the two methods in problems such as this is not however as great as the above comments imply, for the following reasons: (1) Terms in S that do not contain accelerations can be dropped. (2) There are a number of ~nalytical relations, which are available or which frequent use of the Gibbs-Appell method would generate, that can be used to simplify the evaluation of S; for example, the following relation

$$\frac{\cdot\cdot}{(\alpha \sin\beta)^2} + \frac{\cdot\cdot}{(\alpha\cos\beta)^2}$$
$$= \ddot{\alpha}^2 + \ddot{\alpha}^2\ddot{\beta}^2 - 2\alpha\ddot{\alpha}\dot{\beta}^2 + 4\alpha\dot{\alpha}\dot{\beta}\ddot{\beta}$$
$$+ fcn(\alpha, \beta, \dot{\alpha}, \dot{\beta})$$

greatly simplifies the evaluation of Sin the above example. (3) Once we have obtained Sand T, it is easier to obtain the Gibbs-Appell equations from S than to obtain Lagrange's equations from T. (4) It is not always necessary or even advisable to complete the squares in Sin order to use it in the Gibbs-Appell equations of motion. In general, we will find that ifterms of the form $(\phi_1 + \phi_2 + \ ...\ )^2$ occur in S, then, depending on the problem, determination of as I ori may in one case be best achieved by completing the squares, combining terms, and then taking the derivative; while in another case it may best be achieved by not completing the square but rather working with the derivative in the form

$$2(\phi_1 + \phi_2 + \ ...\ )[\partial(\phi_1 + \phi_2 + ...)/\partial\ddot{r}].$$

Although certain classes of problems are generally better handled with one method rather than another, it is difficult to definitively determine that one method is clearly superior to another for all problems of a certain class. For example, using the Gibbs-Appell method, it is possible with limited analytical effort to obtain equations of motion for a multiple planar pendulum containing an arbitrary number of components, 7 apparently more readily than with Lagrange's method, despite the fact that the system is a holonomic system and the motional coordinates chosen are the time derivatives of the configurational coordinates.

Examples

Euler's Equations

Euler's equations provide an excellent illustration of Appell's formulation.

Consider a rigid body of N particles joined by rigid rods. The rotation of the body may be described by an angular velocity vector ω, and the corresponding angular acceleration vector

$$\alpha = \frac{d\omega}{dt}$$

The generalized force for a rotation is the torque N, since the work done for an infinitesimal rotation $\delta\phi$ is $dW = \mathrm{N}\cdot\delta\phi$. The velocity of the kth particle is given by

$$\mathrm{v}_k = \omega \times \mathrm{r}_k$$

where r_k is the particle's position in Cartesian coordinates; its corresponding acceleration is

$$\mathrm{a}_k = \frac{d\mathrm{v}_k}{dt} = \alpha \times \mathrm{r}_k + \omega \times \mathrm{v}_k$$

Therefore, the function S may be written as

$$S = \frac{1}{2}\sum_{k=1}^{N} m_k\left(\mathrm{a}_k \cdot \mathrm{a}_k\right) = \frac{1}{2}\sum_{k=1}^{N} m_k\left\{\left(\alpha\times\mathrm{r}_k\right)^2 + \left(\omega\times\mathrm{v}_k\right)^2 + 2\left(\alpha\times\mathrm{r}_k\right)\cdot\left(\omega\times\mathrm{v}_k\right)\right\}$$

Setting the derivative of S with respect to á equal to the torque yields Euler's equations

$$I_{xx}\alpha_x - \left(I_{yy} - I_{zz}\right)\omega_y\omega_z = N_x$$

$$I_{yy}\alpha_y - \left(I_{zz} - I_{xx}\right)\omega_z\omega_x = N_y$$

$$I_{zz}\alpha_z - \left(I_{xx} - I_{yy}\right)\omega_x\omega_y = N_z.$$

Udwadia–Kalaba Equation

The Fundamental Equation of Constrained Motion is a mathematical method for derivation of the equations of motion of a constrained system. This equation was originally developed by Firdaus E. Udwadia and Robert E. Kalaba in a series of papers, beginning in 1992. This equation has application in the field of analytical

The Central Problem of Constrained Motion

In the study of the dynamics of mechanical systems, the configuration of a given system S is, in general, completely described by n generalized coordinates so that its generalized coordinate n-vector is given by

$$\mathrm{q} := [q_1, q_2, \ldots, q_n]^{\mathrm{T}}.$$

where T denotes matrix transpose. Using Newtonian or Lagrangian dynamics, the unconstrained equations of motion of the system S under study can be derived as a matrix equation:

Udwadia–Kalaba equations of motion (*Unconstrained*)

$$\mathrm{M}(q,t)\ddot{\mathrm{q}}(t) = \mathrm{Q}(q,\dot{q},t),$$

where the dots represent derivatives with respect to time:

$$\dot{q}_i = \frac{dq_i}{dt}.$$

It is assumed that the initial conditions q(o) and $\dot{\text{q}}(0)$ are known. We call the system *S* unconstrained because $\dot{\text{q}}(0)$ may be arbitrarily assigned.

The *n*-vector Q denotes the total generalized force acted on the system by some external influence; it can be expressed as the sum of all the conservative forces as well as *non*-conservative forces.

The *n*-by-*n* matrix M is symmetric, and it can be positive definite $(\text{M} > 0)$ or semi-positive definite $(\text{M} \geq 0)$. Typically, it is assumed that M is positive definite; however, it is not uncommon to derive the unconstrained equations of motion of the system *S* such that M is only semi-positive definite; i.e., the mass matrix may be singular (it has no inverse matrix).

Constraints

We now assume that the unconstrained system *S* is subjected to a set of *m* consistent equality constraints given by

$$\text{A}(q,\dot{q},t)\ddot{\text{q}} = \text{b}(q,\dot{q},t),$$

where A is a known *m*-by-*n* matrix of rank *r* and b is a known *m*-vector. We note that this set of constraint equations encompass a very general variety of holonomic and non-holonomic equality constraints. For example, holonomic constraints of the form

$$\varphi(q,t) = 0$$

can be differentiated twice with respect to time while non-holonomic constraints of the form

$$\psi(q,\dot{q},t) = 0$$

can be differentiated once with respect to time to obtain the *m*-by-*n* matrix A and the *m*-vector b. In short, constraints may be specified that are

1. nonlinear functions of displacement and velocity,
2. explicitly dependent on time, and
3. functionally dependent.

As a consequence of subjecting these constraints to the unconstrained system *S*, an additional force is conceptualized to arise, namely, the force of constraint. Therefore, the constrained system S_c becomes

Udwadia–Kalaba equations of motion (*Constrained*)

$$\text{M}\ddot{\text{q}} = \text{Q} + \text{Q}_c(q,\dot{q},t),$$

where Q_c—the constraint force—is the additional force needed to satisfy the imposed constraints.

The central problem of constrained motion is now stated as follows:

1. given the unconstrained equations of motion of the system S,
2. given the generalized displacement $q(t)$ and the generalized velocity $\dot{q}(t)$ of the constrained system S_c at time t, and
3. given the constraints in the form $A\ddot{q} = b$ as stated above,

find the equations of motion for the *constrained* system—the acceleration—at time t, which is in accordance with the agreed upon principles of analytical dynamics.

Equation of Motion

The solution to this central problem is given by the Udwadia–Kaleba equation. When the matrix M is positive definite, the equation of motion of the constrained system S_c, at each instant of time, is

$$M\ddot{q} = Q + M^{1/2}\left(AM^{-1/2}\right)^{+}(b - AM^{-1}Q),$$

where the '+' symbol denotes the pseudoinverse of the matrix $AM^{-1/2}$. The force of constraint is thus given explicitly as

$$Q_c = M^{1/2}\left(AM^{-1/2}\right)^{+}(b - AM^{-1}Q),$$

and since the matrix M is positive definite the generalized acceleration of the constrained system S_c is determined explicitly by

$$\ddot{q} = M^{-1}Q + M^{-1/2}\left(AM^{-1/2}\right)^{+}(b - AM^{-1}Q).$$

In the case that the matrix M is semi-positive definite $(M \geq 0)$, the above equation cannot be used directly because M may be singular. Furthermore, the generalized accelerations may not be unique unless the $(n + m)$-by-n matrix

$$\widehat{M} = \begin{bmatrix} M \\ \mathbf{A} \end{bmatrix}$$

has full rank (rank = n). But since the observed accelerations of mechanical systems in nature are always unique, this rank condition is a necessary and sufficient condition for obtaining the uniquely defined generalized accelerations of the constrained system S_c at each instant of time. Thus, when $\widehat{M}$ has full rank, the equations of motion of the constrained system S_c at each instant of time are uniquely determined by (1) creating the auxiliary unconstrained system

$$M_{\mathbf{A}}\ddot{q} := (M + A^{+}A)\ddot{q} = Q + A^{+}b := Q_{\mathbf{b}},$$

and by (2) applying the fundamental equation of constrained motion to this auxiliary unconstrained system so that the auxiliary constrained equations of motion are explicitly given by

$$M_A\ddot{q} = Q_b + M_A^{1/2}(AM_A^{-1/2})^+(b - AM_A^{-1}Q_b).$$

Moreover, when the matrix $\widehat{M}$ has full rank, the matrix M_A is always positive definite. This yields, explicitly, the generalized accelerations of the constrained system S_c as

$$\ddot{q} = M_A^{-1}Q_b + M_A^{-1/2}(AM_A^{-1/2})^+(b - AM_A^{-1}Q_b).$$

This equation is valid when the matrix M is either positive definite *or* positive semi-definite! Additionally, the force of constraint that causes the constrained system S_c—a system that may have a singular mass matrix M—to satisfy the imposed constraints is explicitly given by

$$Q_c = M_A^{1/2}(AM_A^{-1/2})^+(b - AM_A^{-1}Q_b).$$

Non-ideal Constraints

At any time during the motion we may consider perturbing the system by a virtual displacement δr consistent with the constraints of the system. The displacement is allowed to be either reversible or irreversible. If the displacement is irreversible, then it performs virtual work. We may write the virtual work of the displacement as

$$W_c(t) = C^T(q,\dot{q},t)\delta r(t)$$

The vector $C(q,\dot{q},t)$ describes the non-ideality of the virtual work and may be related, for example, to friction or drag forces (such forces have velocity dependence). This is a generalized D'Alembert's principle, where the usual form of the principle has vanishing virtual work with $C(q,\dot{q},t) = 0$.

The Udwadia–Kalaba equation is modified by an additional non-ideal constraint term to

$$M\ddot{q} = Q + M^{1/2}\left(AM^{-1/2}\right)^+(b - AM^{-1}Q) + M^{1/2}\left[I - \left(AM^{-1/2}\right)^+ AM^{-1/2}\right]M^{1/2}C.$$

Examples

Inverse Kepler Problem

The method can solve the inverse Kepler problem of determining the force law that corresponds to the orbits that are conic sections. We take there to be no external forces (not even gravity) and instead constrain the particle motion to follow orbits of the form

$$r = \epsilon x + l$$

where $r = \sqrt{x^2 + y^2}$, ϵ is the eccentricity, and l is the semi-latus rectum. Differentiating twice with respect to time and rearranging slightly gives a constraint

$$(x - r\epsilon)\ddot{x} + y\ddot{y} = -\frac{(x\dot{y} - y\dot{x})^2}{r^2}$$

We assume the body has a simple, constant mass. We also assume that angular momentum about the focus is conserved as

$$m(x\dot{y} - y\dot{x}) = L$$

with time derivative

$$x\ddot{y} - y\ddot{x} = 0$$

We can combine these two constraints into the matrix equation

$$\begin{pmatrix} x - r\epsilon & y \\ y & -x \end{pmatrix}\begin{pmatrix} \ddot{x} \\ \ddot{y} \end{pmatrix} = \begin{pmatrix} -\dfrac{L^2}{m^2 r^2} \\ 0 \end{pmatrix}$$

The constraint matrix has inverse

$$\begin{pmatrix} x - r\epsilon & y \\ y & -x \end{pmatrix}^{-1} = \frac{1}{lr}\begin{pmatrix} x & y \\ y & -(x - r\epsilon) \end{pmatrix}$$

The force of constraint is therefore the expected, central inverse square law

$$\mathrm{F}_c = m\mathrm{A}^{-1}\mathrm{b} = \frac{m}{lr}\begin{pmatrix} x & y \\ y & -(x - r\epsilon) \end{pmatrix}\begin{pmatrix} -\dfrac{L^2}{m^2 r^2} \\ 0 \end{pmatrix} = -\frac{L^2}{mlr^2}\begin{pmatrix} \cos\theta \\ \sin\theta \end{pmatrix}$$

Inclined Plane with Friction

Consider a small block of constant mass on an inclined plane at an angle α above horizontal. The constraint that the block lie on the plane can be written as

$$y = x\tan\alpha$$

After taking two time derivatives, we can put this into a standard constraint matrix equation form

$$\begin{pmatrix} -\tan\alpha & 1 \end{pmatrix}\begin{pmatrix} \ddot{x} \\ \ddot{y} \end{pmatrix} = 0$$

The constraint matrix has pseudoinverse

$$\begin{pmatrix} -\tan\alpha & 1 \end{pmatrix}^{+} = \cos^2\alpha\begin{pmatrix} -\tan\alpha \\ 1 \end{pmatrix}$$

We allow there to be sliding friction between the block and the inclined plane. We parameterize this force by a standard coefficient of friction multiplied by the normal force

$$C = -\mu mg \cos\alpha \,\text{sgn}\, \dot{y} \begin{pmatrix} \cos\alpha \\ \sin\alpha \end{pmatrix}$$

Whereas the force of gravity is reversible, the force of friction is not. Therefore, the virtual work associated with a virtual displacement will depend on C. We may summarize the three forces (external, ideal constraint, and non-ideal constraint) as follows:

$$F_{ext} = Q = -mg \begin{pmatrix} 0 \\ y \end{pmatrix}$$

$$F_{c,i} = -A^{+}AQ = mg \cos^2\alpha \begin{pmatrix} -\tan\alpha \\ 1 \end{pmatrix} \begin{pmatrix} -\tan\alpha & 1 \end{pmatrix} \begin{pmatrix} 0 \\ y \end{pmatrix} = mg \begin{pmatrix} -\sin\alpha\cos\alpha \\ \cos^2\alpha \end{pmatrix}$$

$$\mathbf{F}_{c,ni} = (\mathbf{I} - \mathbf{A}^{+}\mathbf{A})\mathbf{C} = -\mu mg \cos\alpha \,\text{sgn}\, \dot{y} \left[\begin{pmatrix} 1 & 0 \\ 0 & 1 \end{pmatrix} - \cos^2\alpha \begin{pmatrix} -\tan\alpha \\ 1 \end{pmatrix} \begin{pmatrix} -\tan\alpha & 1 \end{pmatrix} \right] = -\mu mg \cos\alpha \,\text{sgn}\, \dot{y} \begin{pmatrix} \cos^2\alpha \\ \sin\alpha\cos\alpha \end{pmatrix}$$

Combining the above, we find that the equations of motion are

$$\begin{pmatrix} \ddot{x} \\ \ddot{y} \end{pmatrix} = \frac{1}{m}\left(F_{ext} + F_{c,i} + F_{c,ni}\right) = -g\left(\sin\alpha + \mu\cos\alpha \,\text{sgn}\, \dot{y}\right) \begin{pmatrix} \cos\alpha \\ \sin\alpha \end{pmatrix}$$

This is like a constant downward acceleration due to gravity with a slight modification. If the block is moving up the inclined plane, then the friction increases the downward acceleration. If the block is moving down the inclined plane, then the friction reduces the downward acceleration.

Euler–Lagrange Equation

In this topic, we will give necessary conditions for an extremum of a function of the type

$$I(x) = \int_a^b F(x(t), x'(t), (t)dt,$$

with various types of boundary conditions. The necessary condition is in the form of a dierential equation that the extremal curve should satisfy, and this differential equation is called the Euler-Lagrange equation.

We begin with the simplest type of boundary conditions, where the curves are allowed to vary between two fixed points.

The Simplest Optimisation Problem

The simplest optimisation problem can be formulated as follows:

Let $F(\alpha,\ \beta,\ \gamma)$ be a function with continuous _rst and second partial derivatives with respect to $(\alpha,\ \beta,\ \gamma)$. Then Find $x \in C^1[a,\ b]$ such that $x(a) = y_a$ and $x(b) = y_b$, and which is an extremum for the function

$$I(x) = \int_a^b F\ (x(t),\ x'(t), t)dt.$$

In other words, the simplest optimisation problem consists of finding an extremum of a function of the form, where the class of admissible curves comprises all smooth curves joining two fixed points in figure. We will apply the necessary condition for an extremum (established in Theorem) to the solve the simplest optimisation problem.

Theorem: *Let* $S = \{x \in C^1[a,\ b] \mid x(a) = y_a$ and $x(b)\ = y_b\}$, and *let* $I\ :\ S\ \rightarrow\ R$ be a function of the form

$$I(x) = \int_a^b F\ (x(t),\ x'(t), t)dt.$$

If I has an extremum at $x_0\ \in\ S$, then x_0 satisfies the Euler-Lagrange equation:

$$\frac{\partial F}{\partial \alpha}\left(x_0(t),\ x_0'(t), t\right) - \frac{d}{dt}\left(\frac{\partial F}{\partial \beta}\left(x_0(t), x_0'(t), t\right)\right) = 0,\ t \in [a,\ b].$$

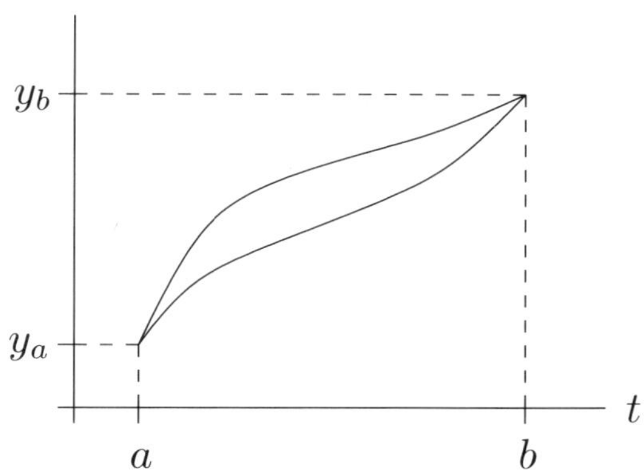

Possible paths joining the two fixed points (a, y_a) and (b, y_b).

Proof: The proof is long and so we divide it into several steps.

Step 1. First of all we note that the set S is not a vector space (unless $y_a = 0 = y_b$). So theorem is not applicable directly. Hence we introduce a new linear space X, and consider a new function $\tilde{I} : X \rightarrow \mathbb{R}$ which is defined in terms of the old function I.

Introduce the linear space

$$X = \{x\ \in\ C^1\ [a,\ b] \mid x(a) = x(b)\ =\ 0\},$$

with the induced norm from $C^1[a, b]$. Then for all $h \in X$, $x_0 + h$ satisfies $(x_0 + h)(a) = y_a$ and $(x_0 + h)(b) = y_b$. Defining $\tilde{I}(h) = I(x_0\ +\ h)$, for $h \in X$, we note that $\tilde{I} : X \rightarrow \mathbb{R}$ has a local extremum at 0.

It follows from theorem that $I(x_0)=\int_0^1 (x'_0(t)-1)^2\,dt=\int_0^1 (1-1)^2\,dt=\int_0^1 0dt=0.$

Step 2. We now calculate $D\tilde{I}(0)=0$. We have

$$\tilde{I}(h)-\tilde{I}(0)=\int_a^b F((x_0+h)(t),(x_0+h)'(t),t)dt-\int_a^b F(x_0(t),x_0'(t),(t)dt$$
$$=\int_a^b [F(x_0(t)+h(t),x_0'(t)+h'(t),t)dt-F(x_0(t),x_0'(t),t)]dt.$$

Recall that from Taylor's theorem, if F possesses partial derivatives of order 2 in a ball B of radius r around the point (αo, βo, γo) in $\mathbb{R}^3$, then for all (α, β, γ) ∈ B, there exists a Θ ∈ [0, 1] such that:

$$F(\alpha,\beta,\gamma)=F(\alpha_0,\beta_0,\gamma_0)+\left((\alpha-\alpha_0)\frac{\partial}{\partial\alpha}+(\beta-\beta_0)\frac{\partial}{\partial\beta}+(\gamma-\gamma_0)\frac{\partial}{\partial\gamma}\right)F\Big|_{(\alpha 0,\beta 0,\gamma 0)}+$$
$$\frac{1}{2!}\left((\alpha-\alpha_0)\frac{\partial}{\partial\alpha}+(\beta-\beta_0)\frac{\partial}{\partial\beta}+(\gamma-\gamma_0)\frac{\partial}{\partial\gamma}\right)^2 F\Big|_{(\alpha_0,\beta_0,\gamma_0)+\Theta((\alpha,\beta,\gamma)-(\alpha_0,\beta_0,\gamma_0))}$$

Hence for h ∈ X such that $\|h\|$ is small enough,

$$\tilde{I}(h)-\tilde{I}(0)=\int_a^b\left[\frac{\partial F}{\partial\alpha}(x_0(t),x_0'(t),t)h(t)+\frac{\partial F}{\partial\beta}(x_0(t),x_0'(t),t)\,h'(t)\right]dt+$$
$$\frac{1}{2!}\int_a^b\left(h(t)\frac{\partial}{\partial\alpha}+h'(t)\frac{\partial}{\partial\beta}\right)^2 F\Big|_{(x_0(t)+\Theta(t)h(t),x_0'(t)+\Theta(t)h'(t),t)}dt.$$

It can be checked that there exists a M > 0 such that

$$\left|\frac{1}{2!}\int_a^b\left(h(t)\frac{\partial}{\partial\alpha}+h'(t)\frac{\partial}{\partial\beta}\right)^2 F\Big|_{(x_0(t)+\Theta(t)h(t),x_0'(t)+\Theta(t)h'(t),t)}dt\right|\le M\|h\|^2,$$

And so $D\tilde{I}(0)$ is the Map.

Step 3. Next we show that if the map in is the zero map, then this implies that holds. Define

$$A(t)=\int_a^t\frac{\partial F}{\partial\alpha}(x0(\tau),x\,0\,0\,(\tau),\tau)\,d\tau.$$

Integrating by parts, we find that

$$\int_a^b\frac{\partial F}{\partial\alpha}(x_0(t),x'_0(t),t)h(t)dt=-\int_a^b A(t)h'(t)dt,$$

and so from, it follows that $D\tilde{I}(0)=0$ implies that

$$\int_a^b \left[-A(t) + \frac{\partial F}{\partial \beta}\left(x_0(t),\, x'_0(t), t\right) \right] h'(t)\, dt = 0 \textit{ for all } h \in X.$$

Step 4. Finally, using Lemma 1.4.5, we obtain

$$-A(t) + \frac{\partial F}{\partial \beta}\left(x_0(t),\, x'_0(t), t\right) = k \textit{ for all } t \in [a,\, b].$$

Differentiating with respect to t, we obtain (2.3). This completes the proof of theorem.

Note that the Euler-Lagrange equation is only a necessary condition for the existence of an extremum. However, in many cases, the Euler-Lagrange equation by itself is enough to give a complete solution of the problem. In fact, the existence of an extremum is sometimes clear from the context of the problem. If in such scenarios, there exists only one solution to the Euler-Lagrange equation, then this solution must a fortiori be the point for which the extremum is achieved.

Example. Let $S = \{x \in C^1[0,1] \;\; x(0) = 0 \text{ and } x(1) = 1\}$. Consider the function $I : S \to \mathbb{R}$ given by

$$I(x) = \int_0^1 \left(\frac{d}{dt} x(t) - 1 \right)^2 dt.$$

We wish to find $x_0 \in S$ that minimizes I. We proceed as follows:

Step 1. We have

$$F(\alpha,\, \beta,\, \gamma) = (\beta - 1)^2, \text{ and } so \; \frac{\partial F}{\partial \alpha} = 0 \text{ and } \frac{\partial F}{\partial \beta} = 2(\beta - 1).$$

Step 2. The Euler-Lagrange equation is now given by

$$0 - \frac{d}{dt}\left(2\left(x'_0(t) - 1\right)\right) = 0 \;\; for\ all\ t \in [0,\, 1].$$

Step 3. Integrating, we obtain $2\left(x'\ (t) - 1\right) = C$, for some constant C, and so $x'_0 = \frac{C}{2} + 1 =: A$ Integrating again, we have $x_0(t) = At + B$, where A and B are suitable constants.

Step 4. The constants A and B can be determined by using that fact that $x_0 \in S$, and so $x_0(0) = 0$ and $x_0(a) = 1$. Thus we have

A0 + B = 0,

A1 + B = 1,

which yield A = 1 and B = 0.

So the unique solution x_0 of the Euler-Lagrange equation in S is $x_0(t) = t$, $t \in [0, 1]$ in figure.

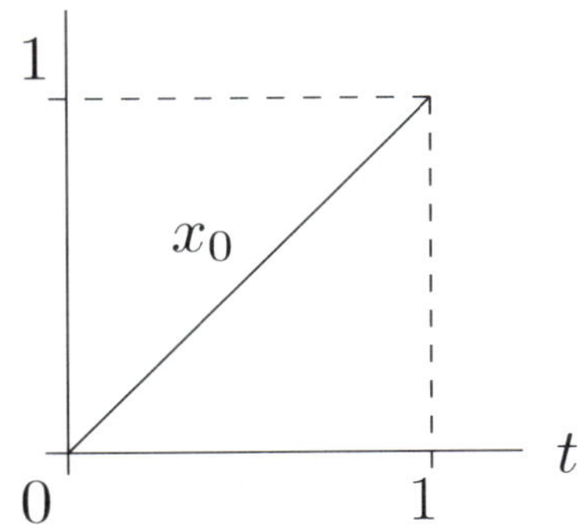

Minimizer for I.

Now we argue that the solution x_0 indeed minimizes I. Since $(x'(t) - 1)^2 \geq 0$ for all $t \in [0, 1]$, it follows that $I(x) \geq 0$ for all $x \in C^1 [0, 1]$. But

$$I(x_0) = \int_0^1 (x'_0(t) - 1)^2 \, dt = \int_0^1 (1 - 1)^2 \, dt = \int_0^1 0 dt = 0.$$

As $I(x) \geq 0 = I(x_0)$ for all $x \in S$, it follows that x_0 minimizes I.

References

- Leach, Dr Andrew (2001-01-30). Molecular Modelling: Principles and Applications (2 edition ed.). Harlow: Prentice Hall. pp. 369–370. ISBN 9780582382107
- von Neumann, J. (1932). "Zur Operatorenmethode In Der Klassischen Mechanik". Annals of Mathematics. 33 (3): 587–642. doi:10.2307/1968537. JSTOR 1968537
- Mauro, D. (2002). "On Koopman–Von Neumann Waves". International Journal of Modern Physics A. 17 (9): 1301. arXiv:quant-ph/0105112. Bibcode:2002IJMPA..17.1301M. doi:10.1142/S0217751X02009680
- Relativity, Gravitation, and Cosmology, R.J.A. Lambourne, Open University, Cambridge University Press, 2010, ISBN 978-0-521-13138-4
- Transtrum, M. K.; Van Huele, J. F. O. S. (2005). "Commutation relations for functions of operators". Journal of Mathematical Physics. 46 (6): 063510. Bibcode:2005JMP....46f3510T. doi:10.1063/1.1924703
- Transtrum, M. K.; Van Huele, J. F. O. S. (2005). "Commutation relations for functions of operators". Journal of Mathematical Physics. 46 (6): 063510. Bibcode:2005JMP....46f3510T. doi:10.1063/1.1924703
- Lanczos, Cornelius (1986). "II §5 Auxiliary conditions: the Lagrangian λ-method". The variational principles of mechanics (Reprint of University of Toronto 1970 4th ed.). Courier Dover. p. 43. ISBN 0-486-65067-7
- Kosyakov, B. P. (2007). Introduction to the classical theory of particles and fields. Berlin, Germany: Springer. doi:10.1007/978-3-540-40934-2

5

Understanding Special Relativity

Special relativity is a theory in physics, which defines the relationship of space and time. This chapter discusses in detail the principal topics, such as relativity of simultaneity, relativistic mass, Galilean invariance and frame of reference, in order to provide an extensive understanding of this important theory.

The theory of special relativity explains how space and time are linked for objects that are moving at a consistent speed in a straight line. One of its most famous aspects concerns objects moving at the speed of light.

Simply put, as an object approaches the speed of light, its mass becomes infinite and it is unable to go any faster than light travels. This cosmic speed limit has been a subject of much discussion in physics, and even in science fiction, as people think about how to travel across vast distances.

The theory of special relativity was developed by Albert Einstein in 1905, and it forms part of the basis of modern physics. After finishing his work in special relativity, Einstein spent a decade pondering what would happen if one introduced acceleration. This formed the basis of his general relativity, published in 1915.

The Special Theory of Relativity tells us that a moving object measures shorter in its direction of motion as its velocity increases until, at the speed of light, it disappears. It also tells us that moving clocks run more slowly as their velocity increases until, at the speed of light, they stop running altogether. In fact, it also tells us that the mass of a moving object measures more as its velocity increases until, at the speed of light, it becomes infinite.

Thus, one person's interval of space is not the same as another person's, and time runs at different rates for different observers travelling at different speeds. To some extent, the faster you go, the slower you age and the slimmer you are! The reason this is not obvious in everyday situations is that the differences at everyday speeds are infinitesimally small, and only really become apparent at speeds approaching that of light itself ("relativistic" speeds). The closer the speed of an objects approaches to the speed of light, the more warped lengths and time intervals become.

The amount of length contraction and time dilation is given by the Lorentz factor, named after the Dutch physicist Hendrik Lorentz, who had been exploring such transformation equations since as early as 1895, long before Einstein began his work (indeed some would claim that Lorentz and Henri Poincaré between them anticipated almost everything in Einstein's Special Theory of Relativity). The Lorentz factor, γ (gamma) is given by the equation $\gamma \equiv \frac{1}{\sqrt{1-\frac{v^2}{c^2}}}$, so that the effect increases exponentially as the object's velocity v approaches the speed of light c. Thus, the calculations show that at 25% of the speed of light, the effect is just 1.03 (a mere 3% slowing of time or contraction of length); at 50% of the speed of light, it is just 1.15; at 99% of the speed of light, time is slowed by a factor of about 7; and at 99.999, the factor is 224. So, if it were possible to travel in a spaceship at, say, 99.5% of the speed of light, a hypothetical observer looking in would see the clock moving about 10 times slower than normal and the astronaut inside moving in slow-motion, as though through treacle.

Reference Frames, Coordinates, and the Lorentz Transformation

Reference Frames and Relative Motion

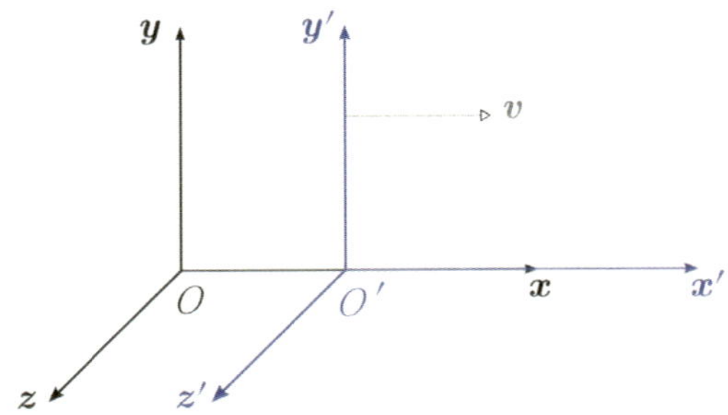

The primed system is in motion relative to the unprimed system with constant velocity v only along the x-axis, from the perspective of an observer stationary in the unprimed system.

By the principle of relativity, an observer stationary in the primed system will view a likewise construction except that the velocity they record will be $-v$. The changing of the speed of propagation of interaction from infinite in non-relativistic mechanics to a finite value will require a modification of the transformation equations mapping events in one frame to another.

Reference frames play a crucial role in relativity theory. The term reference frame as used here is an observational perspective in space which is not undergoing any change in motion (acceleration), from which a position can be measured along 3 spatial axes. In addition, a reference frame has the ability to determine measurements of the time of events using a 'clock' (any reference device with uniform periodicity).

An event is an occurrence that can be assigned a single unique time and location in space relative to a reference frame: it is a "point" in spacetime. Since the speed of light is constant in relativity in each and every reference frame, pulses of light can be used to unambiguously measure distances and refer back the times that events occurred to the clock, even though light takes time to reach the clock after the event has transpired.

For example, the explosion of a firecracker may be considered to be an "event". We can completely specify an event by its four spacetime coordinates: The time of occurrence and its 3-dimensional spatial location define a reference point. Let's call this reference frame S.

In relativity theory we often want to calculate the position of a point from a different reference point.

Suppose we have a second reference frame S', whose spatial axes and clock exactly coincide with that of S at time zero, but it is moving at a constant velocity v with respect to S along the x-axis.

Since there is no absolute reference frame in relativity theory, a concept of 'moving' doesn't strictly exist, as everything is always moving with respect to some other reference frame. Instead, any two frames that move at the same speed in the same direction are said to be *comoving*. Therefore, S and S' are not *comoving*.

Lorentz Transformation

Define the event to have spacetime coordinates (t,x,y,z) in system S and (t',x',y',z') in a reference frame moving at a velocity v with respect to that frame, S'. Then the Lorentz transformation specifies that these coordinates are related in the following way:

$$\begin{aligned} t' &= \gamma(t - vx/c^2) \\ x' &= \gamma(x - vt) \\ y' &= y \\ z' &= z, \end{aligned}$$

where

$$\gamma = \frac{1}{\sqrt{1 - \frac{v^2}{c^2}}}$$

is the Lorentz factor and c is the speed of light in vacuum, and the velocity v of S' is parallel to the x-axis. The y and z coordinates are unaffected; only the x and t coordinates are transformed. These Lorentz transformations form a one-parameter group of linear mappings, that parameter being called rapidity.

There is nothing special about the x-axis, the transformation can apply to the y- or z-axis, or indeed in any direction, which can be done by directions parallel to the motion (which are warped by the γ factor) and perpendicular.

A quantity invariant under Lorentz transformations is known as a Lorentz scalar.

Writing the Lorentz transformation and its inverse in terms of coordinate differences, where for instance one event has coordinates (x_1,t_1) and (x'_1,t'_1), another event has coordinates (x_2,t_2) and (x'_2,t'_2), and the differences are defined as

$$\Delta x' = x'_2 - x'_1, \quad \Delta x = x_2 - x_1,$$
$$\Delta t' = t'_2 - t'_1, \quad \Delta t = t_2 - t_1,$$

we get

$$\Delta x' = \gamma(\Delta x - v\Delta t), \quad \Delta x = \gamma(\Delta x' + v\Delta t'),$$
$$\Delta t' = \gamma\left(\Delta t - \frac{v\Delta x}{c^2}\right), \quad \Delta t = \gamma\left(\Delta t' + \frac{v\Delta x'}{c^2}\right).$$

These effects are explicitly related to our way of measuring *time intervals* between events which occur at the same place in a given coordinate system (called "co-local" events). These time intervals will be *different* in another coordinate system moving with respect to the first, unless the events are also simultaneous. Similarly, these effects also relate to our measured distances between separated but simultaneous events in a given coordinate system of choice. If these events are not co-local, but are separated by distance (space), they will *not* occur at the same *spatial distance* from each other when seen from another moving coordinate system. However, the spacetime interval will be the same for all observers.

Measurement Versus Visual Appearance

Time dilation and length contraction are not optical illusions, but genuine effects. Measurements of these effects are not an artifact of Doppler shift, nor are they the result of neglecting to take into account the time it takes light to travel from an event to an observer.

For many years, the distinction between the two had not been generally appreciated. For example, it had generally been thought that a length contracted object passing by an observer would in fact actually be *seen* as length contracted. In 1959, James Terrell and Roger Penrose independently pointed out that differential time lag effects in signals reaching the observer from the different parts of a moving object result in a fast moving object's visual appearance being quite different from its measured shape. For example, a receding object would *appear* contracted, an approaching object would *appear* elongated, and a passing object would have a skew appearance that has been likened to a rotation. A sphere in motion retains the appearance of a sphere, although images on the surface of the sphere will appear distorted.

Galaxy M87 streams out a black-hole-powered jet of electrons and other sub-atomic particles traveling at nearly the speed of light.

Another example where visual appearance is at odds with measurement comes from the observation of apparent superluminal motion in various radio galaxies, BL Lac objects, quasars, and other astronomical objects that eject relativistic-speed jets of matter at narrow angles with respect to the viewer. An optical illusion results giving the appearance of faster than light travel. In figure, galaxy M87 streams out a high-speed jet of subatomic particles almost directly towards us.

Consequences Derived from the Lorentz Transformation

The consequences of special relativity can be derived from the Lorentz transformation equations. These transformations, and hence special relativity, lead to different physical predictions than those of Newtonian mechanics when relative velocities become comparable to the speed of light. The speed of light is so much larger than anything humans encounter that some of the effects predicted by relativity are initially counterintuitive.

Relativity of Simultaneity

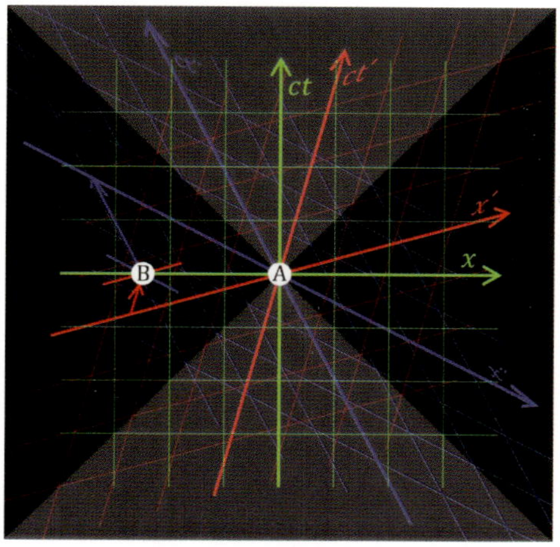

Event B is simultaneous with A in the green reference frame, but it occurs before A in the blue frame, and occurs after A in the red frame.

Two events happening in two different locations that occur simultaneously in the reference frame of one inertial observer, may occur non-simultaneously in the reference frame of another inertial observer (lack of absolute simultaneity).

From the first equation of the Lorentz transformation in terms of coordinate differences

$$\Delta t' = \gamma\left(\Delta t - \frac{v\Delta x}{c^2}\right),$$

it is clear that two events that are simultaneous in frame S (satisfying $\Delta t = 0$), are not necessarily simultaneous in another inertial frame S' (satisfying $\Delta t' = 0$). Only if these events are additionally co-local in frame S (satisfying $\Delta x = 0$), will they be simultaneous in another frame S'.

Time Dilation

The time lapse between two events is not invariant from one observer to another, but is dependent on the relative speeds of the observers' reference frames (e.g., the twin paradox which concerns a twin who flies off in a spaceship traveling near the speed of light and returns to discover that his or her twin sibling has aged much more).

Suppose a clock is at rest in the unprimed system S. The location of the clock on two different ticks is then characterized by $\Delta x = 0$. To find the relation between the times between these ticks as measured in both systems, the first equation can be used to find:

$$\Delta t' = \gamma \Delta t \quad \text{for events satisfying } \Delta x = 0.$$

This shows that the time ($\Delta t'$) between the two ticks as seen in the frame in which the clock is moving (S'), is *longer* than the time (Δt) between these ticks as measured in the rest frame of the clock (S). Time dilation explains a number of physical phenomena; for example, the lifetime of muons produced by cosmic rays impinging on the Earth's atmosphere is measured to be greater than the lifetimes of muons measured in the laboratory.

Length Contraction

The dimensions (e.g., length) of an object as measured by one observer may be smaller than the results of measurements of the same object made by another observer (e.g., the ladder paradox involves a long ladder traveling near the speed of light and being contained within a smaller garage).

Similarly, suppose a measuring rod is at rest and aligned along the x-axis in the unprimed system S. In this system, the length of this rod is written as Δx. To measure the length of this rod in the system S', in which the rod is moving, the distances $'$ to the end points of the rod must be measured simultaneously in that system S'. In other words, the measurement is characterized by $\Delta t' = 0$, which can be combined with the fourth equation to find the relation between the lengths Δx and $\Delta x'$:

$$\Delta x' = \frac{\Delta x}{\gamma} \quad \text{for events satisfying } \Delta t' = 0.$$

This shows that the length ($\Delta x'$) of the rod as measured in the frame in which it is moving (S'), is *shorter* than its length (Δx) in its own rest frame (S).

Composition of Velocities

Velocities (speeds) do not simply add. If the observer in S measures an object moving along the x axis at velocity u, then the observer in the S' system, a frame of reference moving at velocity v in the x direction with respect to S, will measure the object moving with velocity u' where (from the Lorentz transformations):

$$u' = \frac{dx'}{dt'} = \frac{\gamma(dx - vdt)}{\gamma\left(dt - \frac{vdx}{c^2}\right)} = \frac{\frac{dx}{dt} - v}{1 - \left(\frac{v}{c^2}\right)\left(\frac{dx}{dt}\right)} = \frac{u - v}{1 - \frac{uv}{c^2}}.$$

The other frame S will measure:

$$u = \frac{dx}{dt} = \frac{\gamma(dx' + vdt')}{\gamma\left(dt' + \frac{vdx'}{c^2}\right)} = \frac{\frac{dx'}{dt'} + v}{1 + \left(\frac{v}{c^2}\right)\left(\frac{dx'}{dt'}\right)} = \frac{u' + v}{1 + \frac{u'v}{c^2}}.$$

Notice that if the object were moving at the speed of light in the S system (i.e. $u = c$), then it would also be moving at the speed of light in the S' system. Also, if both u and v are small with respect to the speed of light, we will recover the intuitive Galilean transformation of velocities

$$u' \approx u - v.$$

The usual example given is that of a train (frame S' above) traveling due east with a velocity v with respect to the tracks (frame S). A child inside the train throws a baseball due east with a velocity u' with respect to the train. In nonrelativistic physics, an observer at rest on the tracks will measure the velocity of the baseball (due east) as $u = u' + v$, while in special relativity this is no longer true; instead the velocity of the baseball (due east) is given by the second equation:

$$u = \frac{u' + v}{1 + \frac{u'v}{c^2}}.$$

Again, there is nothing special about the x or east directions. This formalism applies to any direction by considering parallel and perpendicular components of motion to the direction of relative velocity v.

Theories of Relativity and Quantum Mechanics

Special relativity can be combined with quantum mechanics to form relativistic quantum mechanics and quantum electrodynamics. It is an unsolved problem in physics how *general* relativity and quantum mechanics can be unified; quantum gravity and a "theory of everything", which require a unification including general relativity too, are active and ongoing areas in theoretical research.

The early Bohr–Sommerfeld atomic model explained the fine structure of alkali metal atoms using both special relativity and the preliminary knowledge on quantum mechanics of the time.

In 1928, Paul Dirac constructed an influential relativistic wave equation, now known as the Dirac equation in his honour,that is fully compatible both with special relativity and with the final version of quantum theory existing after 1926. This equation explained not only the intrinsic angular momentum of the electrons called *spin*, it also led to the prediction of the antiparticle of the electron (the positron), and fine structure could only be fully explained with special relativity. It was the first foundation of *relativistic quantum mechanics*. In non-relativistic quantum mechanics, spin is phenomenological and cannot be explained.

On the other hand, the existence of antiparticles leads to the conclusion that relativistic quantum mechanics is not enough for a more accurate and complete theory of particle interactions. Instead, a theory of particles interpreted as quantized fields, called *quantum field theory*, becomes necessary; in which particles can be created and destroyed throughout space and time.

Galilean Invariance

Solutions to the fluid equations should be Galilean invariant, that is the same in every inertial frame. Recently, it has been reported that solutions generated by grid codes (like Athena) are not Galilean invariant. In fact, however, it is the test problems themselves that are flawed. They involve unresolved interfaces, so that solutions are dominated by truncation error, and it is this truncation error which violates Galilean invariance.

KH Instability Test

Initial Conditions

One of the problems used to test for Galilean invariance is the growth of the Kelvin-Helmholtz instability between two fluids of different densities. The interface between the fluids is an unresolved discontinuity (a *slip-surface*). The total velocity difference is equal to the sound speed C_s.

The test is based on the KH instability test in the Athena test suite; the figure below sketches the initial conditions.

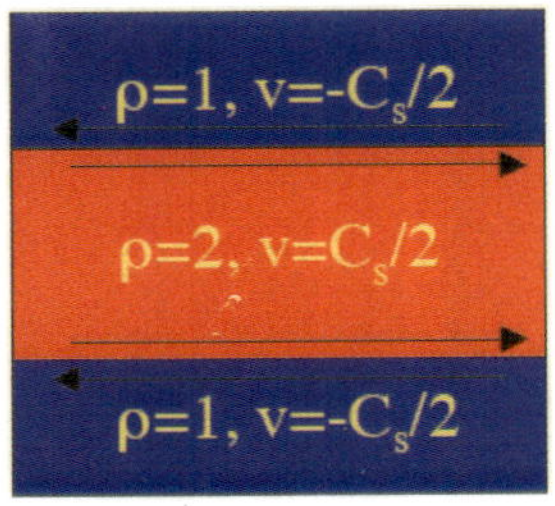

Results with Unresolved Interfaces

The images below show results typical of a grid code when the solution is computed in the rest frame (*left*), in a frame moving at the sound speed to the right (*center*), and a frame moving at ten times the sound speed to the right (*right*). Note the instability seems to disappear in the moving frames. This is because numerical diffusion at the interface damps the unstable modes.

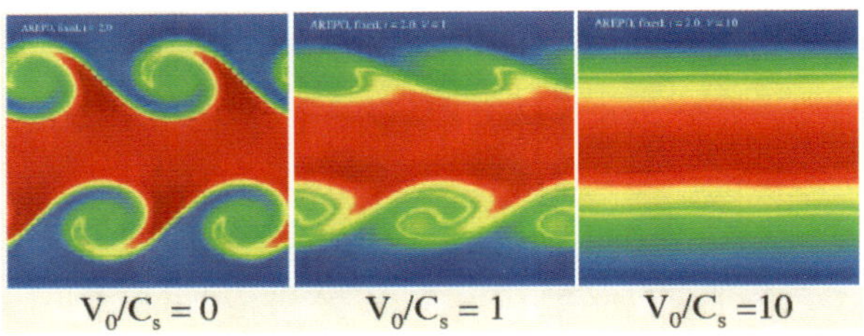

Of course, numerical diffusion is not only a problem in the moving frames. It also dominates the solution in the rest frame. The images below show the same test run at different resolutions *in the rest frame*. Note the solution is vastly different at each resolution, and the instability seems to disappear at the lowest resolution (where the truncation error is largest).

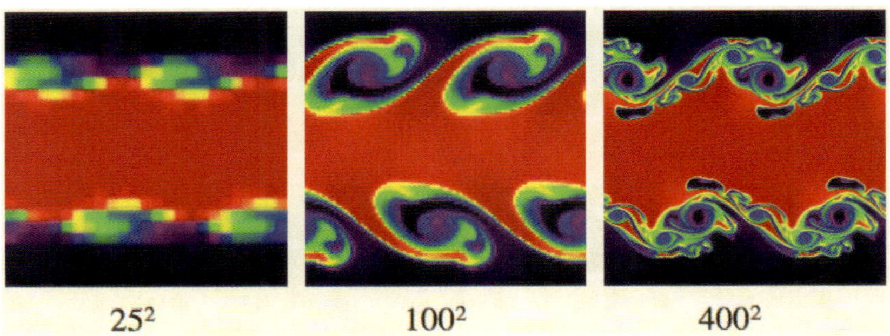

So which solution is correct? That is, *which solution should the code reproduce in any inertial frame?*. Of course, this question cannot be answered unless the test is based on a resolved solution.

Results with a Resolved Interface

The images on the left below show the streamwise (x-direction), and spanwise (y-direction) velocity fluctuations from a test using uniform density, and in which the interface is spread out using a hyperbolic tangent function, with a characteristic scale a equal to twice the grid cell size. This replaces the discontinuity at the interface by a very sharp but smooth profile, and introduces a minimum lengthscale *a* in the problem that can be resolved. The images on the right are from a calculation in the rest frame, and on the left from a calculation moving at $100C_s$.

An even more quantitative test is to measure the kinetic energy associated with motion in the spanwise (y-) direction. There are two lines shown in the figure at left below, one taken from each calculation. It only appears there is one line, because the results are very nearly identical. In both cases, exponential growth in the linear regime is captured correctly.

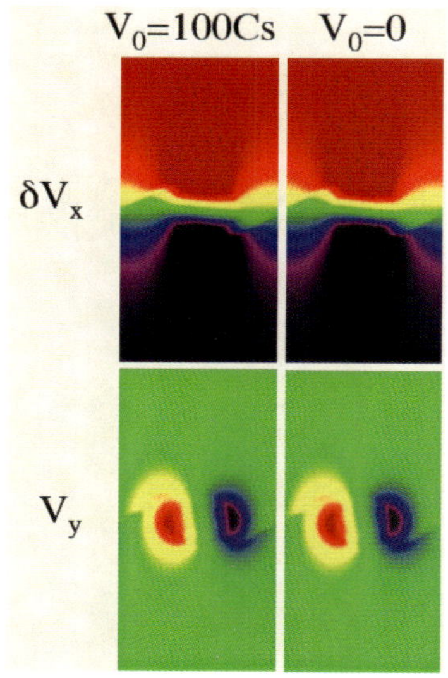

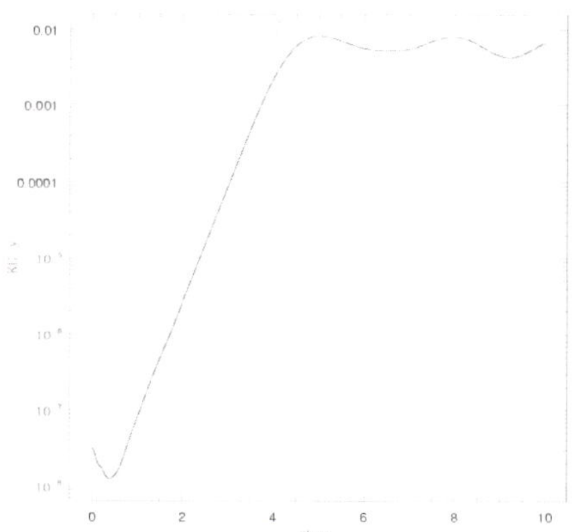

Formulation

Specifically, the term *Galilean invariance* today usually refers to this principle as applied to Newtonian mechanics, that is, Newton's laws hold in all frames related to one another by a Galilean transformation. In other words, all frames related to one another by such a transformation are inertial (meaning, Newton's equation of motion is valid in these frames). In this context it is sometimes called *Newtonian relativity*.

Among the axioms from Newton's theory are:

1. There exists an *absolute space*, in which Newton's laws are true. An inertial frame is a reference frame in relative uniform motion to absolute space.

2. All inertial frames share a *universal time*.

Galilean relativity can be shown as follows. Consider two inertial frames S and S'. A physical event in S will have position coordinates $r = (x,y,z)$ and time t in S, and $r' = (x',y',z')$ and time t' in S'. By the second axiom above, one can synchronize the clock in the two frames and assume $t = t'$. Suppose S' is in relative uniform motion to S with velocity v. Consider a point object whose position is given by functions $r'(t)$ *in S' and* $r(t)$ in S. We see that

$$r'(t) = r(t) - vt.$$

The velocity of the particle is given by the time derivative of the position:

$$u'(t) = \frac{d}{dt}r'(t) = \frac{d}{dt}r(t) - v = u(t) - v.$$

Another differentiation gives the acceleration in the two frames:

$$a'(t) = \frac{d}{dt}u'(t) = \frac{d}{dt}u(t) - 0 = a(t).$$

It is this simple but crucial result that implies Galilean relativity. Assuming that mass is invariant in all inertial frames, the above equation shows Newton's laws of mechanics, if valid in one frame, must hold for all frames. But it is assumed to hold in absolute space, therefore Galilean relativity holds.

Newton's Theory Versus Special Relativity

A comparison can be made between Newtonian relativity and special relativity.

Some of the assumptions and properties of Newton's theory are:

1. The existence of infinitely many inertial frames. Each frame is of infinite size (the entire universe may be covered by many linearly equivalent frames). Any two frames may be in relative uniform motion. (The relativistic nature of mechanics derived above shows that the absolute space assumption is not necessary.)

2. The inertial frames may move in *all* possible relative forms of uniform motion.

3. There is a universal, or absolute, notion of time.

4. Two inertial frames are related by a Galilean transformation.

5. In all inertial frames, Newton's laws, and gravity, hold.

In comparison, the corresponding statements from special relativity are as follows:

1. The existence, as well, of infinitely many non-inertial frames, each of which referenced to (and physically determined by) a unique set of spacetime coordinates. Each frame may be

of infinite size, but its definition is always determined locally by contextual physical conditions. Any two frames may be in relative non-uniform motion (as long as it is assumed that this condition of relative motion implies a relativistic dynamical effect -and later, mechanical effect in general relativity- between both frames).

2. Rather than freely allowing all conditions of relative uniform motion between frames of reference, the relative velocity between two inertial frames becomes bounded above by the speed of light.
3. Instead of universal time, each inertial frame possesses its own notion of time.
4. The Galilean transformations are replaced by Lorentz transformations.
5. In all inertial frames, *all* laws of physics are the same.

Notice both theories assume the existence of inertial frames. In practice, the size of the frames in which they remain valid differ greatly, depending on gravitational tidal forces.

In the appropriate context, a *local Newtonian inertial frame*, where Newton's theory remains a good model, extends to, roughly, 10^7 light years.

In special relativity, one considers *Einstein's cabins*, cabins that fall freely in a gravitational field. According to Einstein's thought experiment, a man in such a cabin experiences (to a good approximation) no gravity and therefore the cabin is an approximate inertial frame. However, one has to assume that the size of the cabin is sufficiently small so that the gravitational field is approximately parallel in its interior. This can greatly reduce the sizes of such approximate frames, in comparison to Newtonian frames. For example, an artificial satellite orbiting around earth can be viewed as a cabin. However, reasonably sensitive instruments would detect "microgravity" in such a situation because the "lines of force" of the Earth's gravitational field converge.

In general, the convergence of gravitational fields in the universe dictates the scale at which one might consider such (local) inertial frames. For example, a spaceship falling into a black hole or neutron star would (at a certain distance) be subjected to tidal forces so strong that it would be crushed in width and torn apart in length. In comparison, however, such forces might only be uncomfortable for the astronauts inside (compressing their joints, making it difficult to extend their limbs in any direction perpendicular to the gravity field of the star). Reducing the scale further, the forces at that distance might have almost no effects at all on a mouse. This illustrates the idea that all freely falling frames are locally inertial (acceleration and gravity-free) if the scale is chosen correctly.

Electromagnetism

Maxwell's equations governing electromagnetism possess a different symmetry, Lorentz invariance, under which lengths and times *are* affected by a change in velocity, which is then described mathematically by a Lorentz transformation.

Albert Einstein's central insight in formulating special relativity was that, for full consistency with electromagnetism, mechanics must also be revised such that Lorentz invariance replaces Galilean invariance. At the low relative velocities characteristic of everyday life, Lorentz invariance and Galilean invariance are nearly the same, but for relative velocities close to that of lightthey are very different.

Work, Kinetic Energy, and Momentum

Because the distance covered while applying a force to an object depends on the inertial frame of reference, so does the workdone. Due to Newton's law of reciprocal actions there is a reaction force; it does work depending on the inertial frame of reference in an opposite way. The total work done is independent of the inertial frame of reference.

Correspondingly the kinetic energy of an object, and even the change in this energy due to a change in velocity, depends on the inertial frame of reference. The total kinetic energy of an isolated system also depends on the inertial frame of reference: it is the sum of the total kinetic energy in a center of momentum frame and the kinetic energy the total mass would have if it were concentrated in the center of mass. Due to the conservation of momentum the latter does not change with time, so changes with time of the total kinetic energy do not depend on the inertial frame of reference.

By contrast, while the momentum of an object also depends on the inertial frame of reference, its change due to a change in velocity does not.

Frame of Reference

Reference frame, also called frame of reference, in dynamics, system of graduated lines symbolically attached to a body that serve to describe the position of points relative to the body. The position of a point on the surface of the Earth, for example, can be described by degrees of latitude, measured north and south from the Equator, and degrees of longitude, measured east and west from the great circle passing through Greenwich, England, and the poles.

The reference frames used in dynamics are known as coordinate systems with axes (lines) emanating from a point known as the origin. The position of a point moving parallel to a plane (plane motion) can be described by two numbers: (1) either the distances of the point from two lines at right angles to one another on the plane (rectangular coordinates), or (2) the length of a line with one end fixed at the origin and the other end at the moving point and the angle that the line makes with a fixed axis (polar coordinates). Motion in three dimensions can be described by three rectangular coordinates or by the length of a line emanating from the origin and two angles (spherical coordinates); one of these angles is equivalent to degrees of longitude and the other to degrees of latitude. In all cases a line from the origin to the point is known as the position vector for the point. As the point moves, the position vector changes in both magnitude and direction, and the velocity of the point is defined in terms of these changes.

Strictly speaking, Newton's laws of motion are valid only in a coordinate system at rest with respect to the "fixed" stars. Such a system is known as a Newtonian, or inertial reference, frame. The laws are also valid in any set of rigid axes moving with constant velocity and without rotation relative to the inertial frame; this concept is known as the principle of Newtonian or Galilean relativity. A coordinate system attached to the Earth is not an inertial reference frame because the Earth rotates and is accelerated with respect to the Sun. Although the solutions to most engineering problems can be obtained to a satisfactory degree of accuracy by assuming that an Earth-based reference frame is an inertial one, there are some applications in which the rotation of the Earth cannot be neglected; among these is the operation of a gyroscopic compass.

To describe the position of a point that moves relative to a body that is moving relative to the Earth, it is usually convenient to use a reference frame attached to the moving body. The motion of the piston in the engine of a moving automobile, for example, is of more interest relative to the block than relative to the ground, whereas the motion of the moving parts of the engine relative to one another may be more important than their motions relative to the block.

Different Aspects of Frame of Reference

The need to distinguish between the various meanings of "frame of reference" has led to a variety of terms. For example, sometimes the type of coordinate system is attached as a modifier, as in *Cartesian frame of reference*. Sometimes the state of motion is emphasized, as in *rotating frame of reference*. Sometimes the way it transforms to frames considered as related is emphasized as in *Galilean frame of reference*. Sometimes frames are distinguished by the scale of their observations, as in *macroscopic* and *microscopic frames of reference*.

In this section, the term *observational frame of reference* is used when emphasis is upon the *state of motion* rather than upon the coordinate choice or the character of the observations or observational apparatus. In this sense, an observational frame of reference allows study of the effect of motion upon an entire family of coordinate systems that could be attached to this frame. On the other hand, a *coordinate system* may be employed for many purposes where the state of motion is not the primary concern. For example, a coordinate system may be adopted to take advantage of the symmetry of a system. In a still broader perspective, the formulation of many problems in physics employs *generalized coordinates*, *normal modes* or *eigenvectors*, which are only indirectly related to space and time. It seems useful to divorce the various aspects of a reference frame for the discussion below. We therefore take observational frames of reference, coordinate systems, and observational equipment as independent concepts, separated as below:

- An observational frame (such as an inertial frame or non-inertial frame of reference) is a physical concept related to state of motion.
- A coordinate system is a mathematical concept, amounting to a choice of language used to describe observations.Consequently, an observer in an observational frame of reference can choose to employ any coordinate system (Cartesian, polar, curvilinear, generalized) to describe observations made from that frame of reference. A change in the choice of this coordinate system does not change an observer's state of motion, and so does not entail a change in the observer's *observational* frame of reference. This viewpoint can be found elsewhere as well. Which is not to dispute that some coordinate systems may be a better choice for some observations than are others.
- Choice of what to measure and with what observational apparatus is a matter separate from the observer's state of motion and choice of coordinate system.

Coordinate Systems

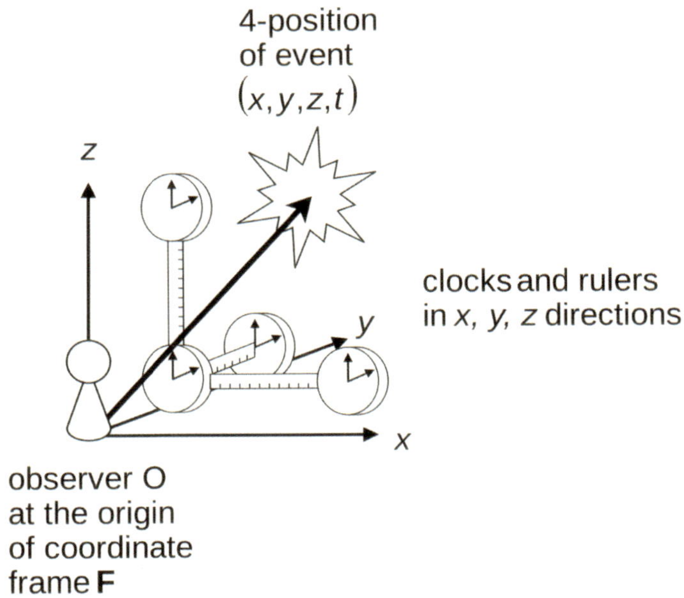

An observer O, situated at the origin of a local set of coordinates – a frame of reference F. The observer in this frame uses the coordinates $(x,\ y,\ z,\ t)$ to describe a spacetime event, shown as a star.

Although the term "coordinate system" is often used (particularly by physicists) in a nontechnical sense, the term "coordinate system" does have a precise meaning in mathematics, and sometimes that is what the physicist means as well.

A coordinate system in mathematics is a facet of geometry or of algebra,in particular, a property of manifolds (for example, in physics, configuration spaces or phase spaces). The coordinates of a point r in an n-dimensional space are simply an ordered set of n numbers:

$$\mathbf{r} = [x^1, x^2, \ldots, x^n].$$

In a general Banach space, these numbers could be (for example) coefficients in a functional expansion like a Fourier series. In a physical problem, they could be spacetime coordinates or normal mode amplitudes. In a robot design, they could be angles of relative rotations, linear displacements, or deformations of joints.Here we will suppose these coordinates can be related to a Cartesian coordinatesystem by a set of functions:

$$x^j = x^j(x, y, z, \ldots), \qquad j = 1, \ldots, n$$

where x, y, z, *etc.* are the n Cartesian coordinates of the point. Given these functions, coordinate surfaces are defined by the relations:

$$x^j(x, y, z, \ldots) = \text{constant}, \qquad j = 1, \ldots, n.$$

The intersection of these surfaces define coordinate lines. At any selected point, tangents to the intersecting coordinate lines at that point define a set of basis vectors $\{\mathbf{e}_1, \mathbf{e}_2, \ldots, \mathbf{e}_n\}$ at that point. That is:

$$\mathbf{e}_i(\mathbf{r}) = \lim_{\epsilon \to 0} \frac{\mathbf{r}\left(x^1, \ldots, x^i + \epsilon, \ldots, x^n\right) - \mathbf{r}\left(x^1, \ldots, x^i, \ldots, x^n\right)}{\epsilon}, \quad i = 1, \ldots, n$$

which can be normalized to be of unit length.

Coordinate surfaces, coordinate lines, and basis vectors are components of a coordinate system. If the basis vectors are orthogonal at every point, the coordinate system is an orthogonal coordinate system.

An important aspect of a coordinate system is its metric tensor g_{ik}, which determines the arc length ds in the coordinate system in terms of its coordinates:

$$(ds)^2 = g_{ik}\, dx^i\, dx^k ,$$

where repeated indices are summed over.

As is apparent from these remarks, a coordinate system is a mathematical construct, part of an axiomatic system. There is no necessary connection between coordinate systems and physical motion (or any other aspect of reality). However, coordinate systems can include time as a coordinate, and can be used to describe motion. Thus, Lorentz transformations and Galilean transformations may be viewed as coordinate transformations.

Observational Frames of Reference

An observational frame of reference, often referred to as a *physical frame of reference*, a *frame of reference*, or simply a *frame*, is a physical concept related to an observer and the observer's state of motion. Here we adopt the view expressed by Kumar and Barve: an observational frame of reference is characterized *only by its state of motion*. However, there is lack of unanimity on this point. In special relativity, the distinction is sometimes made between an *observer* and a *frame*. According to this view, a *frame* is an *observer* plus a coordinate lattice constructed to be an orthonormal right-handed set of spacelike vectors perpendicular to a timelike vector. This restricted view is not used here, and is not universally adopted even in discussions of relativity. In general relativity the use of general coordinate systems is common.

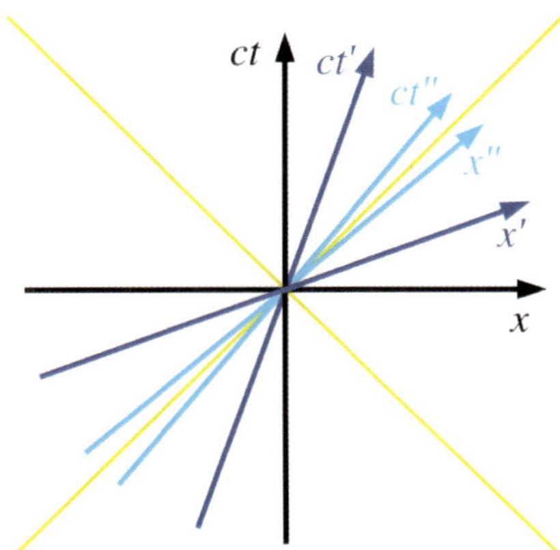

Three frames of reference in special relativity. The black frame is at rest. The primed frame moves at 40% of light speed, and the double primed frame at 80%. Note the scissors-like change as speed increases.

There are two types of observational reference frame: inertial and non-inertial. An inertial frame of reference is defined as one in which all laws of physics take on their simplest form. In special relativity these frames are related by Lorentz transformations, which are parametrized by rapidity. In Newtonian mechanics, a more restricted definition requires only that Newton's first law holds true; that is, a Newtonian inertial frame is one in which a free particle travels in a straight line at constant speed, or is at rest. These frames are related by Galilean transformations. These relativistic and Newtonian transformations are expressed in spaces of general dimension in terms of representations of the Poincaré group and of the Galilean group.

In contrast to the inertial frame, a non-inertial frame of reference is one in which fictitious forces must be invoked to explain observations. An example is an observational frame of reference centered at a point on the Earth's surface. This frame of reference orbits around the center of the Earth, which introduces the fictitious forces known as the Coriolis force, centrifugal force, and gravitational force. (All of these forces including gravity disappear in a truly inertial reference frame, which is one of free-fall.)

Examples of Inertial Frames of Reference

Simple Example

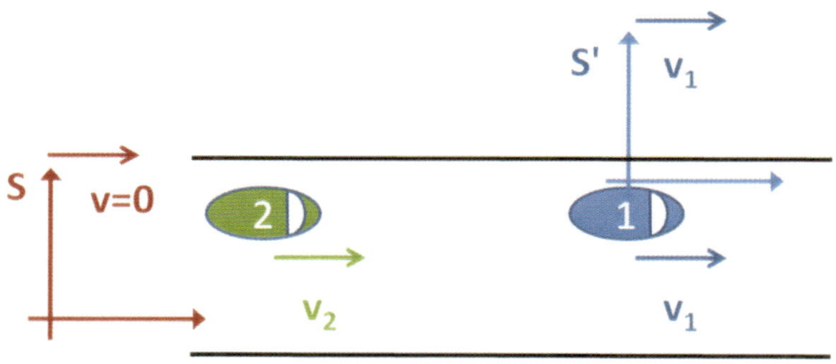

Figure: Two cars moving at different but constant velocities observed from stationary inertial frame *S*attached to the road and moving inertial frame *S'*attached to the first car.

Consider a situation common in everyday life. Two cars travel along a road, both moving at constant velocities in figure. At some particular moment, they are separated by 200 metres. The car in front is travelling at 22 metres per second and the car behind is travelling at 30 metres per second. If we want to find out how long it will take the second car to catch up with the first, there are three obvious "frames of reference" that we could choose.

First, we could observe the two cars from the side of the road. We define our "frame of reference" *S* as follows. We stand on the side of the road and start a stop-clock at the exact moment that the second car passes us, which happens to be when they are a distance d = 200 m apart. Since neither of the cars is accelerating, we can determine their positions by the following formulas, where $x_1(t)$ is the position in meters of car one after time *t* in seconds and $x_2(t)$ is the position of car two after time *t*.

$$x_1(t) = d + v_1 t = 200 + 22t \; ; \quad x_2(t) = v_2 t = 30t$$

Notice that these formulas predict at $t = 0$ s the first car is 200 m down the road and the second car is right beside us, as expected. We want to find the time at which $x_1 = x_2$. Therefore, we set $x_1 = x_2$ and solve for t, that is:

$$200 + 22t = 30t$$

$$8t = 200$$

$$t = 25 \quad \text{seconds}$$

Alternatively, we could choose a frame of reference S' situated in the first car. In this case, the first car is stationary and the second car is approaching from behind at a speed of $v_2 - v_1 = 8\ m\ /\ s$. In order to catch up to the first car, it will take a time of $\frac{d}{v_2 - v_1} = \frac{200}{8}\ s$, that is, 25 seconds, as before. Note how much easier the problem becomes by choosing a suitable frame of reference. The third possible frame of reference would be attached to the second car. That example resembles the case just discussed, except the second car is stationary and the first car moves backward towards it at 8m/s.

It would have been possible to choose a rotating, accelerating frame of reference, moving in a complicated manner, but this would have served to complicate the problem unnecessarily. It is also necessary to note that one is able to convert measurements made in one coordinate system to another. For example, suppose that your watch is running five minutes fast compared to the local standard time. If you know that this is the case, when somebody asks you what time it is, you are able to deduct five minutes from the time displayed on your watch in order to obtain the correct time. The measurements that an observer makes about a system depend therefore on the observer's frame of reference (you might say that the bus arrived at 5 past three, when in fact it arrived at three).

Non-inertial Frames

Here the relation between inertial and non-inertial observational frames of reference is considered. The basic difference between these frames is the need in non-inertial frames for fictitious forces.

An accelerated frame of reference is often delineated as being the "primed" frame, and all variables that are dependent on that frame are notated with primes, $e.g. x', y', a'$.

The vector from the origin of an inertial reference frame to the origin of an accelerated reference frame is commonly notated as R. Given a point of interest that exists in both frames, the vector from the inertial origin to the point is called r, and the vector from the accelerated origin to the point is called r′. From the geometry of the situation, we get

$$\mathbf{r} = \mathbf{R} + \mathbf{r}'$$

Taking the first and second derivatives of this with respect to time, we obtain

$$\mathbf{v} = \mathbf{V} + \mathbf{v}'$$

$$\mathbf{a} = \mathbf{A} + \mathbf{a}'$$

where V and A are the velocity and acceleration of the accelerated system with respect to the inertial system and v and a are the velocity and acceleration of the point of interest with respect to the inertial frame.

These equations allow transformations between the two coordinate systems; for example, we can now write Newton's second law as

$$\mathbf{F} = m\mathbf{a} = m\mathbf{A} + m\mathbf{a}'$$

When there is accelerated motion due to a force being exerted there is manifestation of inertia. If an electric car designed to recharge its battery system when decelerating is switched to braking, the batteries are recharged, illustrating the physical strength of manifestation of inertia. However, the manifestation of inertia does not prevent acceleration (or deceleration), for manifestation of inertia occurs in response to change in velocity due to a force. Seen from the perspective of a rotating frame of reference the manifestation of inertia appears to exert a force (either in centrifugal direction, or in a direction orthogonal to an object's motion, the Coriolis effect).

A common sort of accelerated reference frame is a frame that is both rotating and translating (an example is a frame of reference attached to a CD which is playing while the player is carried). This arrangement leads to the equation:

$$\mathbf{a} = \mathbf{a}' + \dot{\omega}\times\mathbf{r}' + 2\omega\times\mathbf{v}' + \omega\times(\omega\times\mathbf{r}') + \mathbf{A}_0$$

or, to solve for the acceleration in the accelerated frame,

$$\mathbf{a}' = \mathbf{a} - \dot{\omega}\times\mathbf{r}' - 2\omega\times\mathbf{v}' - \omega\times(\omega\times\mathbf{r}') - \mathbf{A}_0$$

Multiplying through by the mass m gives

$$\mathbf{F}' = \mathbf{F}_{\text{physical}} + \mathbf{F}'_{\text{Euler}} + \mathbf{F}'_{\text{Coriolis}} + \mathbf{F}'_{\text{centripetal}} - m\mathbf{A}_0$$

where

$$\mathbf{F}'_{\text{Euler}} = -m\dot{\omega}\times\mathbf{r}' \quad \text{(Euler force)}$$

$$\mathbf{F}'_{\text{Coriolis}} = -2m\omega\times\mathbf{v}' \quad \text{(Coriolis force)}$$

$$\mathbf{F}'_{\text{centrifugal}} = -m\omega\times(\omega\times\mathbf{r}') = m(\omega^2\mathbf{r}' - (\omega\cdot\mathbf{r}')\omega) \quad \text{(centrifugal force)}$$

Inertial Frame of Reference

A frame of reference that remains at rest or moves with constant velocity with respect to other frames of reference is called Inertial Frame of Reference. An inertial frame of reference has a constant velocity. That is, it is moving at a constant speed in a straight line, or it is standing still. Newton's laws of motion are valid in all inertial frames of reference. Here, a body does not change due to external forces. All inertial frames of reference are equivalent for the measurement of physical phenomena.

There are several ways to imagine this motion:

- Our earth.
- A space shuttle moving with constant velocity relative to the earth.
- A rocket moving with constant velocity relative to the earth.

For example, knowing the universal gravitational constant from experiment (and the masses involved), it is possible from Newton's Second Law,

$$force = mass \times acceleration,$$

to predict future planetary motions with great accuracy.

Suppose we know from experiment that these laws of mechanics are true in one frame of reference. How do they look in another frame, moving with respect to the first frame? To find out, we have to figure out how to get from position, velocity and acceleration in one frame to the corresponding quantities in the second frame.

Obviously, the two frames must have a constant relative velocity, otherwise the law of inertia won't hold in both of them. Let's choose the coordinates so that this velocity is along the x -axis of both of them.

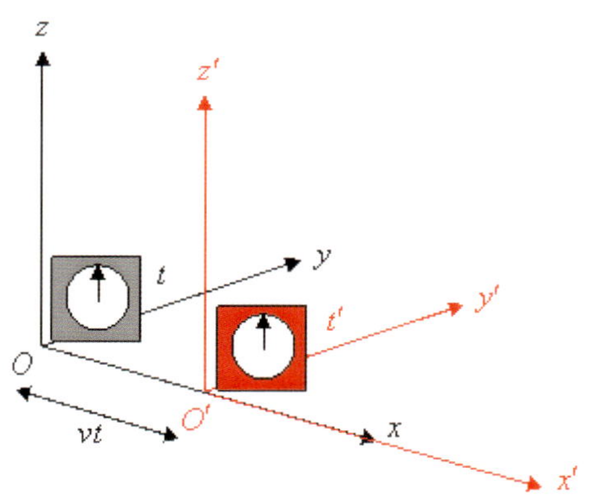

Two frames of reference relatively displaced along the x-axis

Notice we also throw in a clock with each frame.

Suppose S' is proceeding relative to S at speed v along the x -axis. For convenience, let us label the moment when O' passes O as the zero point of time keeping.

Newton's Inertial Frame of Reference

Frame *S'* has an arbitrary but fixed rotation with respect to frame *S*. They are both *inertial frames* provided a body not subject to forces appears to move in a straight line. If that motion is seen in one frame, it will also appear that way in the other.

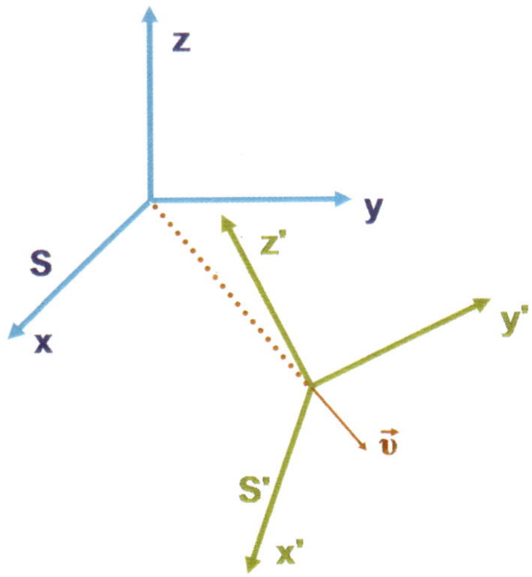

Figure: Two frames of reference moving with relative velocity $\vec{v}$.

Within the realm of Newtonian mechanics, an inertial frame of reference, or inertial reference frame, is one in which Newton's first law of motion is valid. However, the principle of special relativity generalizes the notion of inertial frame to include all physical laws, not simply Newton's first law.

Newton viewed the first law as valid in any reference frame that is in uniform motion relative to the fixed stars; that is, neither rotating nor accelerating relative to the stars. Today the notion of "absolute space" is abandoned, and an inertial frame in the field of classical mechanics is defined as:

> An inertial frame of reference is one in which the motion of a particle not subject to forces is in a straight line at constant speed.

Hence, with respect to an inertial frame, an object or body accelerates only when a physical force is applied, and (following Newton's first law of motion), in the absence of a net force, a body at rest will remain at rest and a body in motion will continue to move uniformly—that is, in a straight line and at constant speed. Newtonian inertial frames transform among each other according to the Galilean group of symmetries.

If this rule is interpreted as saying that straight-line motion is an indication of zero net force, the rule does not identify inertial reference frames because straight-line motion can be observed in a variety of frames. If the rule is interpreted as defining an inertial frame, then we have to be able to determine when zero net force is applied. The problem was summarized by Einstein:

> The weakness of the principle of inertia lies in this, that it involves an argument in a circle: a mass moves without acceleration if it is sufficiently far from other bodies; we know that it is sufficiently far from other bodies only by the fact that it moves without acceleration.

—*Albert Einstein*

There are several approaches to this issue. One approach is to argue that all real forces drop off with distance from their sources in a known manner, so we have only to be sure that a body is far enough away from all sources to ensure that no force is present. A possible issue with this approach is the historically long-lived view that the distant universe might affect matters (Mach's principle). Another approach is to identify all real sources for real forces and account for them. A possible issue with this approach is that we might miss something, or account inappropriately for their influence, perhaps, again, due to Mach's principle and an incomplete understanding of the universe. A third approach is to look at the way the forces transform when we shift reference frames. Fictitious forces, those that arise due to the acceleration of a frame, disappear in inertial frames, and have complicated rules of transformation in general cases. On the basis of universality of physical law and the request for frames where the laws are most simply expressed, inertial frames are distinguished by the absence of such fictitious forces.

Newton enunciated a principle of relativity himself in one of his corollaries to the laws of motion:

> The motions of bodies included in a given space are the same among themselves, whether that space is at rest or moves uniformly forward in a straight line.
>
> —*Isaac Newton*

This principle differs from the special principle in two ways: first, it is restricted to mechanics, and second, it makes no mention of simplicity. It shares with the special principle the invariance of the form of the description among mutually translating reference frames. The role of fictitious forces in classifying reference frames is:

Separating Non-inertial from Inertial Reference Frames

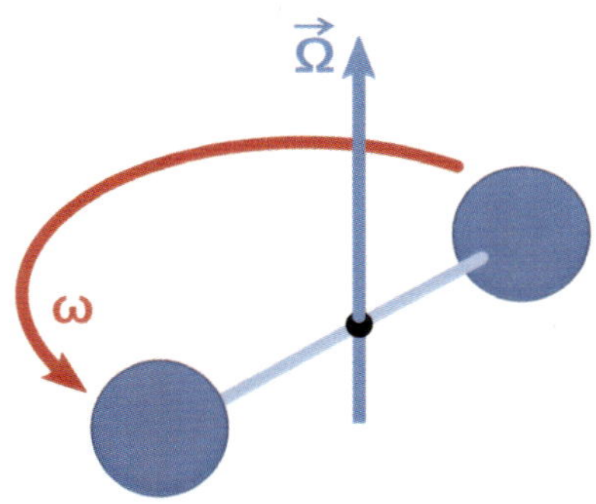

Figure: Two spheres tied with a string and rotating at an angular rate ω. Because of the rotation, the string tying the spheres together is under tension.

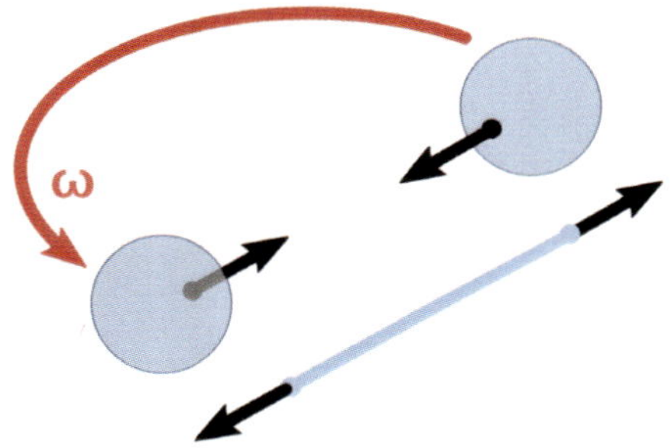

Figure: Exploded view of rotating spheres in an inertial frame of reference showing the centripetal forces on the spheres provided by the tension in the tying string.

Inertial and non-inertial reference frames can be distinguished by the absence or presence of fictitious forces, as explained shortly.

> The effect of this being in the noninertial frame is to require the observer to introduce a fictitious force into his calculations.
>
> —*Sidney Borowitz and Lawrence A Bornstein*

The presence of fictitious forces indicates the physical laws are not the simplest laws available so, in terms of the special principle of relativity, a frame where fictitious forces are present is not an inertial frame:

> The equations of motion in a non-inertial system differ from the equations in an inertial system by additional terms called inertial forces. This allows us to detect experimentally the non-inertial nature of a system.
>
> —*V. I. Arnol'd*

Bodies in non-inertial reference frames are subject to so-called *fictitious* forces (pseudo-forces); that is, forces that result from the acceleration of the reference frame itself and not from any physical force acting on the body. Examples of fictitious forces are the centrifugal force and the Coriolis force in rotating reference frames.

How then, are "fictitious" forces to be separated from "real" forces? It is hard to apply the Newtonian definition of an inertial frame without this separation. For example, consider a stationary object in an inertial frame. Being at rest, no net force is applied. But in a frame rotating about a fixed axis, the object appears to move in a circle, and is subject to centripetal force (which is made up of the Coriolis force and the centrifugal force). How can we decide that the rotating frame is a non-inertial frame? There are two approaches to this resolution: one approach is to look for the origin of the fictitious forces (the Coriolis force and the centrifugal force). We will find there are no sources for these forces, no associated force carriers, no originating bodies. A second approach is to look at a variety of frames of reference. For any inertial frame, the Coriolis force and the centrifugal force disappear, so application of the principle of special relativity would identify these frames where the forces disappear as sharing the same and the simplest physical laws, and hence rule that the rotating frame is not an inertial frame.

Newton examined this problem himself using rotating spheres, as shown in figures. He pointed out that if the spheres are not rotating, the tension in the tying string is measured as zero in every frame of reference. If the spheres only appear to rotate (that is, we are watching stationary spheres from a rotating frame), the zero tension in the string is accounted for by observing that the centripetal force is supplied by the centrifugal and Coriolis forces in combination, so no tension is needed. If the spheres really are rotating, the tension observed is exactly the centripetal force required by the circular motion. Thus, measurement of the tension in the string identifies the inertial frame: it is the one where the tension in the string provides exactly the centripetal force demanded by the motion as it is observed in that frame, and not a different value. That is, the inertial frame is the one where the fictitious forces vanish.

So much for fictitious forces due to rotation. However, for linear acceleration, Newton expressed the idea of undetectability of straight-line accelerations held in common:

> If bodies, any how moved among themselves, are urged in the direction of parallel lines by equal accelerative forces, they will continue to move among themselves, after the same manner as if they had been urged by no such forces.
>
> —*Isaac Newton*

This principle generalizes the notion of an inertial frame. For example, an observer confined in a free-falling lift will assert that he himself is a valid inertial frame, even if he is accelerating under gravity, so long as he has no knowledge about anything outside the lift. So, strictly speaking, inertial frame is a relative concept. With this in mind, we can define inertial frames collectively as a set of frames which are stationary or moving at constant velocity with respect to each other, so that a single inertial frame is defined as an element of this set.

For these ideas to apply, everything observed in the frame has to be subject to a base-line, common acceleration shared by the frame itself. That situation would apply, for example, to the elevator example, where all objects are subject to the same gravitational acceleration, and the elevator itself accelerates at the same rate.

Applications

Inertial navigation systems used a cluster of gyroscopes and accelerometers to determine accelerations relative to inertial space. After a gyroscope is spun up in a particular orientation in inertial space, the law of conservation of angular momentum requires that it retain that orientation as long as no external forces are applied to it. Three orthogonal gyroscopes establish an inertial reference frame, and the accelerators measure acceleration relative to that frame. The accelerations, along with a clock, can then be used to calculate the change in position. Thus, inertial navigation is a form of dead reckoning that requires no external input, and therefore cannot be jammed by any external or internal signal source.

A gyrocompass, employed for navigation of seagoing vessels, finds the geometric north. It does so, not by sensing the Earth's magnetic field, but by using inertial space as its reference. The outer casing of the gyrocompass device is held in such a way that it remains aligned with the local plumb line. When the gyroscope wheel inside the gyrocompass device is spun up, the way the gyroscope wheel is suspended causes the gyroscope wheel to gradually align its spinning axis with the Earth's axis. Alignment with the Earth's axis is the only direction for which the gyroscope's spinning axis can be stationary with respect to the Earth and not be required to change direction with respect to inertial space. After being spun up, a gyrocompass can reach the direction of alignment with the Earth's axis in as little as a quarter of an hour.

Newtonian Mechanics

Classical theories that use the Galilean transformation postulate the equivalence of all inertial reference frames. Some theories may even postulate the existence of a privileged frame which provides absolute space and absolute time. The Galilean transformation transforms coordinates from one inertial reference frame, s , to another, ′, by simple addition or subtraction of corrdinates:

$$\mathbf{r}' = \mathbf{r} - \mathbf{r}_0 - \mathbf{v}t$$

$$t' = t - t_0$$

where r_0 and t_0 represent shifts in the origin of space and time, and v is the relative velocity of the two inertial reference frames. Under Galilean transformations, the time $t_2 - t_1$ between two events is the same for all reference frames and the distance between two simultaneous events (or, equivalently, the length of any object, $|r_2 - r_1|$) is also the same.

Special Relativity

Einstein's theory of special relativity, like Newtonian mechanics, postulates the equivalence of all inertial reference frames. However, because special relativity postulates that the speed of light in free space is invariant, the transformation between inertial frames is the Lorentz transformation, not the Galilean transformation which is used in Newtonian mechanics. The invariance of the speed of light leads to counter-intuitive phenomena, such as time dilation and length contraction, and the relativity of simultaneity, which have been extensively verified experimentally. The Lorentz transformation reduces to the Galilean transformation as the speed of light approaches infinity or as the relative velocity between frames approaches zero.

General Relativity

General relativity is based upon the principle of equivalence:

> There is no experiment observers can perform to distinguish whether an acceleration arises because of a gravitational force or because their reference frame is accelerating.
>
> —*Douglas C. Giancoli*

This idea was introduced by Einstein's 1907 "Principle of Relativity and Gravitation" and later developed in 1911. Support for this principle is found in the Eötvös experiment, which determines whether the ratio of inertial to gravitational mass is the same for all bodies, regardless of size or composition. To date no difference has been found to a few parts in 10^{11}. For some discussion of the subtleties of the Eötvös experiment, such as the local mass distribution around the experimental site (including a quip about the mass of Eötvös himself).

Einstein's general theory modifies the distinction between nominally "inertial" and "noninertial" effects by replacing special relativity's "flat" Minkowski Space with a metric that produces non-zero curvature. In general relativity, the principle of inertia is replaced with the principle of geodesic motion, whereby objects move in a way dictated by the curvature of spacetime. As a consequence of this curvature, it is not a given in general relativity that inertial objects moving at a particular rate with respect to each other will continue to do so. This phenomenon of geodesic deviation means that inertial frames of reference do not exist globally as they do in Newtonian mechanics and special relativity.

However, the general theory reduces to the special theory over sufficiently small regions of spacetime, where curvature effects become less important and the earlier inertial frame arguments can come back into play. Consequently, modern special relativity is now sometimes described as only a "local theory". "Local" can encompass, for example, the entire Milky Way galaxy: The astronomer

Karl Schwarzschild observed the motion of pairs of stars orbiting each other. He found that the two orbits of the stars of such a system lie in a plane, and the perihelion of the orbits of the two stars remains pointing in the same direction with respect to the solar system. Schwarzschild pointed out that that was invariably seen: the direction of the angular momentum of all observed double star systems remains fixed with respect to the direction of the angular momentum of the Solar System. These observations allowed him to conclude that inertial frames inside the galaxy do not rotate with respect to one another, and that the space of the Milky Way is approximately Galilean or Minkowskian.

Non-inertial Reference Frame

A frame of reference is said to be a non-inertial frame of reference when a body, not acted upon by an external force, is accelerated. In a non-inertial frame of reference, Newton's laws of motion are not valid. It also does not have a constant velocity and is accelerating. There are several ways to imagine this motion:

- The frame could be travelling in a straight line, but be speeding up or slowing down.
- The frame could be travelling along a curved path at a steady speed.
- The frame could be travelling along a curved path and also speeding up or slowing down.

Example

The uniformly moving railway car, it was stated to be moving uniformly, so that any observations of a moving object would have an error, relative to the everyday world, but an error that was as constant as the motion of the railway car, so that uniform motion still appeared uniform, and non-uniform motion still appeared nonuniform. But what if the motion of the railway car is *not* constant? Then the error introduced into measurements of the moving object would be changing with time, which would make a constant motion look nonuniform, and therefore accelerated, which would suggest that a force is acting on the moving object, which is obviously incorrect, if it is really moving with constant motion.

Relativity of Simultaneity

One of the most fundamental deductions Albert Einstein made from the finite speed of light in his theory of special relativity is the relativity of simultaneity—because light takes a finite time to traverse a distance in space, it is not possible to define simultaneity with respect to a universal clock shared by all observers. In fact, purely due to their locations in space, two observers may disagree about the *order* in which two spatially separated events occurred. It is only because the speed of light is so great compared to distances we are familiar with in everyday life that this effect seems unfamiliar to us. Note that the relativity of simultaneity can be purely due to the finite speed of light; while it is usually discussed in conjunction with special relativity and moving observers, it can be observed in situations where none of the other relativistic effects are present.

Einstein's Train

Einstein's version of the experiment presumed that one observer was sitting midway inside a speeding traincar and another was standing on a platform as the train moved past. The train is struck by two bolts of lightning simultaneously, but at different positions along the axis of train movement (back and front of the train car). In the inertial frame of the standing observer, there are three events which are spatially dislocated, but simultaneous: standing observer facing the moving observer (i.e., the center of the train), lightning striking the front of the train car, and lightning striking the back of the car.

Since the events are placed along the axis of train movement, their time coordinates become projected to different time coordinates in the moving train's inertial frame. Events which occurred at space coordinates in the direction of train movement, happen *earlier* than events at coordinates opposite to the direction of train movement. In the moving train's inertial frame, this means that lightning will strike the front of the train car *before* the two observers align (face each other).

The Train-and-platform

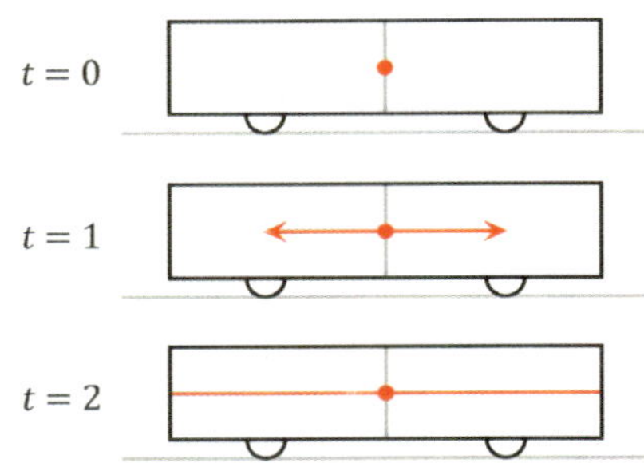

The train-and-platform experiment from the reference frame of an observer on board the train

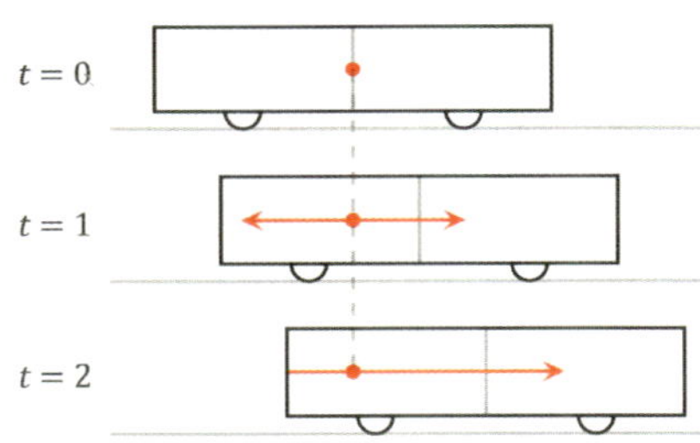

Reference frame of an observer standing on the platform (length contraction not depicted)

A popular picture for understanding this idea is provided by a thought experiment similar to those suggested by Daniel Frost Comstock in 1910 and Einstein in 1917. It also consists of one observer midway inside a speeding traincar and another observer standing on a platform as the train moves past.

A flash of light is given off at the center of the traincar just as the two observers pass each other. For the observer on board the train, the front and back of the traincar are at fixed distances from the light source and as such, according to this observer, the light will reach the front and back of the traincar at the same time.

For the observer standing on the platform, on the other hand, the rear of the traincar is moving (catching up) toward the point at which the flash was given off, and the front of the traincar is moving away from it. As the speed of light is finite and the same in all directions for all observers, the light headed for the back of the train will have less distance to cover than the light headed for the front. Thus, the flashes of light will strike the ends of the traincar at different times.

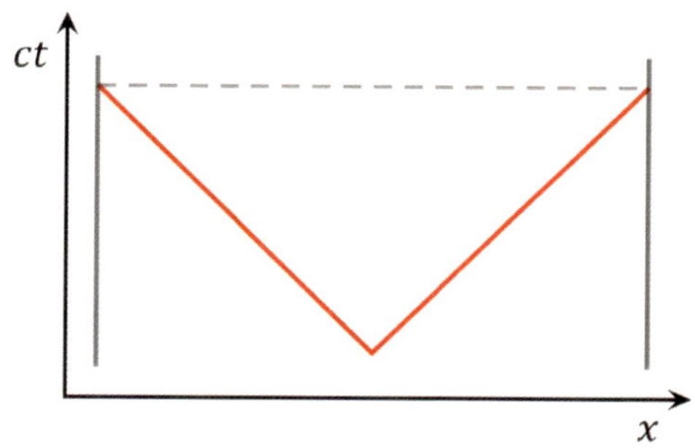

The spacetime diagram in the frame of the observer on the train.

Spacetime Diagrams

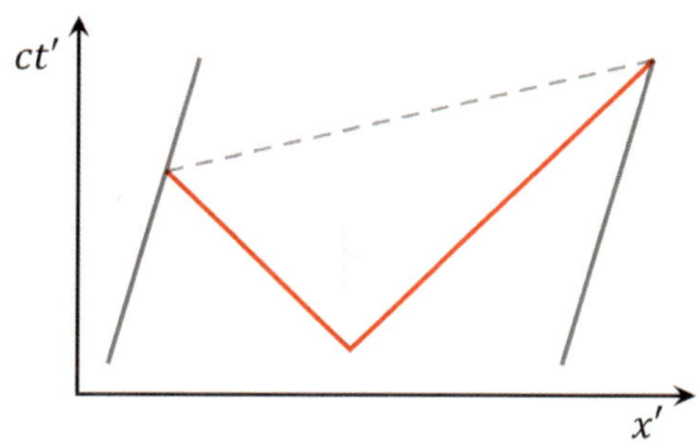

The same diagram in the frame of an observer who sees the train moving to the right.

It may be helpful to visualize this situation using spacetime diagrams. For a given observer, the t-axis is defined to be a point traced out in time by the origin of the spatial coordinate x, and is drawn vertically. The x-axis is defined as the set of all points in space at the time $t = 0$, and is drawn horizontally. The statement that the speed of light is the same for all observers is represented by drawing a light ray as a 45° line, regardless of the speed of the source relative to the speed of the observer.

In the first diagram, the two ends of the train are drawn as grey lines. Because the ends of the train are stationary with respect to the observer on the train, these lines are just vertical lines, showing their motion through time but not space. The flash of light is shown as the 45° red lines. The points at which the two light flashes hit the ends of the train are at the same level in the diagram. This means that the events are simultaneous.

In the second diagram, the two ends of the train moving to the right, are shown by parallel lines. The flash of light is given off at a point exactly halfway between the two ends of the train, and again form two 45° lines, expressing the constancy of the speed of light. In this picture, however, the points at which the light flashes hit the ends of the train are *not* at the same level; they are *not* simultaneous.

Lorentz Transformation

The relativity of simultaneity can be demonstrated using the Lorentz transformation, which relates the coordinates used by one observer to coordinates used by another in uniform relative motion with respect to the first.

Assume that the first observer uses coordinates labeled t, x, y, and z, while the second observer uses coordinates labeled t', x', y', and z'. Now suppose that the first observer sees the second moving in the x-direction at a velocity v. And suppose that the observers' coordinate axes are parallel and that they have the same origin. Then the Lorentz transformation expresses how the coordinates are related:

$$t' = \frac{t - vx/c^2}{\sqrt{1 - v^2/c^2}},$$

$$x' = \frac{x - vt}{\sqrt{1 - v^2/c^2}},$$

$$y' = y,$$

$$z' = z,$$

where c is the speed of light. If two events happen at the same time in the frame of the first observer, they will have identical values of the t-coordinate. However, if they have different values of the x-coordinate (different positions in the x-direction), they will have different values of the t' coordinate, so they will happen at different times in that frame. The term that accounts for the failure of absolute simultaneity is the $v\,x/c^2$.

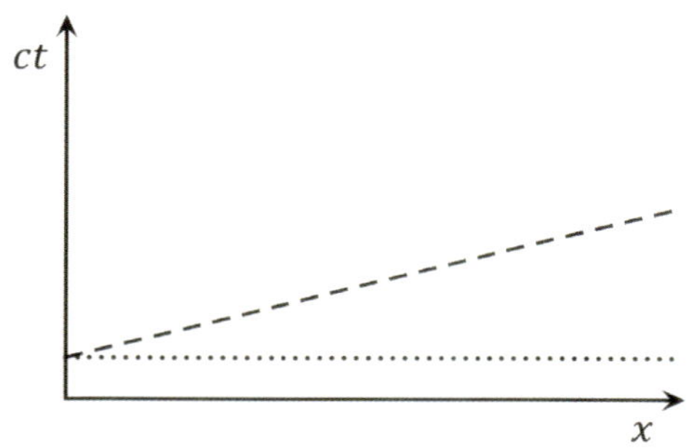

A spacetime diagram showing the set of points regarded as simultaneous by a stationary observer (horizontal dotted line) and the set of points regarded as simultaneous by an observer moving at v = 0.5c (dashed line).

The equation t' = constant defines a "line of simultaneity" in the (x', t') coordinate system for the second (moving) observer, just as the equation t = constant defines the "line of simultaneity" for the first (stationary) observer in the (x, t) coordinate system. From the above equations for the Lorentz transform it can be seen that t' is constant if and only if $t - v\,x/c^2$ = constant. Thus the set of points that make t constant are different from the set of points that makes t' constant. That is, the set of events which are regarded as simultaneous depends on the frame of reference used to make the comparison.

Graphically, this can be represented on a space-time diagram by the fact that a plot of the set of points regarded as simultaneous generates a line which depends on the observer. In the space-time diagram, the dashed line represents a set of points considered to be simultaneous with the origin by an observer moving with a velocity v of one-quarter of the speed of light. The dotted horizontal line represents the set of points regarded as simultaneous with the origin by a stationary observer. This diagram is drawn using the (x, t) coordinates of the stationary observer, and is scaled so that the speed of light is one, i.e., so that a ray of light would be represented by a line with a 45° angle from the x axis. From our previous analysis, given that $v = 0.25$ and $c = 1$, the equation of the dashed line of simultaneity is $t - 0.25x = 0$ and with $v = 0$, the equation of the dotted line of simultaneity is $t = 0$.

In general the second observer traces out a worldline in the spacetime of the first observer described by $t = x/v$, and the set of simultaneous events for the second observer (at the origin) is described by the line $t = vx$. Note the multiplicative inverse relation of the slopes of the worldline and simultaneous events, in accord with the principle of hyperbolic orthogonality.

Accelerated Observers

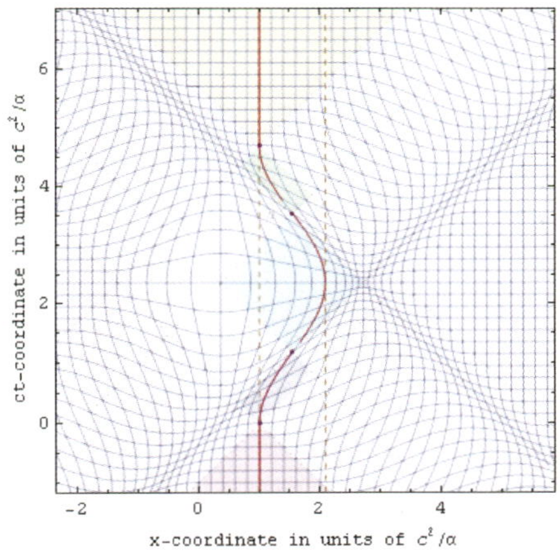

Roundtrip radar-time isocontours.

The Lorentz-transform calculation above uses a definition of extended-simultaneity (i.e. of when and where events occur *at which you were not present*) that might be referred to as the co-moving or "tangent free-float-frame" definition. This definition is naturally extrapolated to events in gravitationally-curved spacetimes, and to accelerated observers, through use of a radar-time/distance definition that (unlike the tangent free-float-frame definition for accelerated frames) assigns a unique time and position to any event.

The radar-time definition of extended-simultaneity further facilitates visualization of the way that acceleration curves spacetime for travelers in the absence of any gravitating objects. This is illustrated in the figure, which shows radar time/position isocontours for events in flat spacetime as experienced by a traveler (red trajectory) taking a constant proper-acceleration roundtrip. One caveat of this approach is that the time and place of remote events are not fully defined until light from such an event is able to reach our traveler.

Relativistic Mass

The concept of mass has always been fundamental to physics. It was present in the earliest days of the subject, and its importance has only grown as physics has diversified over the centuries. Its definition goes back to Galileo and Newton, for whom mass was that property of a body that enables it to resist externally imposed changes to its motion. Newton used mass to define momentum and force vectors: he defined a body's momentum as $p = mv$ (where v is its velocity), and he defined force to be the rate of increase of the body's momentum: $F = \mathrm{d}p/\mathrm{d}t$. When a body's mass is constant (as it usually is, except when we are analysing the motion of e.g. a rocket), the force law becomes $F = m\,\mathrm{d}v/\mathrm{d}t = ma$, where a is the body's acceleration.

This definition of mass was applied in a straightforward way for almost two centuries. Then Einstein arrived on the scene and, in his theory of motion known as *special relativity*, the situation became more complicated. The above definition of mass still holds for a body at rest, and so has come to be called the body's *rest mass*, denoted m_o if we wish to stress that we're dealing with rest mass. But when the body is moving we find that its force–acceleration relationship now depends on two quantities: the body's speed, and the angle between its direction of motion and the applied force. When we relate the force to the resulting acceleration along each of three mutually perpendicular spatial axes, we find that in each of the three expressions a factor of γm_o appears, where the *gamma factor* $\gamma = (1-v^{2/c2})^{-1/2}$ occurs frequently in special relativity.

The idea of a speed-dependent mass actually dates back to Lorentz's work. His 1904 paper *Electromagnetic Phenomena in a System Moving With Any Velocity Less Than That of Light* introduced the "longitudinal" and "transverse" electromagnetic masses of the electron. With these he could write the equations of motion for an electron in an electromagnetic field in the newtonian form, provided the electron's mass was allowed to increase with its speed. Between 1905 and 1909, the relativistic theory of force, momentum, and energy was developed by Planck, Lewis, and Tolman. It turned out that a single mass dependence could be used for any acceleration, thus enabling mass to retain its independence of the body's direction of acceleration, if a speed-dependent "relativistic mass" m was understood as present in Newton's original expression $p = mv$.

So a body moving with speed v and whose momentum has magnitude p has a relativistic mass given by $m = p/v$, and (it turns out) a total energy of mc^2. A body with rest mass m_o turns out to have relativistic mass γm_o. But the definition $m = p/v$ now also neatly defines a relativistic mass for a photon: this moves with speed c and has energy E, and electromagnetic theory gives it a momentum of magnitude $p = E/c$, so it has relativistic mass $p/v = E/c^2$. The expression $m = \gamma m_o$ doesn't apply to a photon, for which γ is infinite. But on the other hand, writing $m = \gamma m_o$ won't lead to any contradictions for a photon if we define the photon's rest mass to be zero.

It seems to have been Lewis who introduced the appropriate speed dependence of mass in 1908, but the term "relativistic mass" appeared later. (Gilbert Lewis was a chemist whose other claim to fame in physics was naming the photon in 1926.) Relativistic mass came into common usage in the relativity texts of the early 1920s written by Pauli, Eddington, and Born. But whereas rest mass is routinely used in many areas of physics, relativistic mass is mostly restricted to the dynamics of special relativity. Because of this, a body's rest mass tends to be called simply its "mass".

The quantities that a moving observer measures as scaled by γ in special relativity are not confined to mass. Two others commonly encountered in the subject are a body's *length in the direction of motion* and its *ageing rate*, both of which get reduced by a factor of γ when measured by a passing observer. So, a ruler has a *rest length*, being the length it was given on the production line, and a *relativistic* or *contracted* length in the direction of its motion, which is the length we measure it to have as it moves past us. Likewise, a stationary clock ages normally, but when it moves it ages slowly by the gamma factor so that its "factory tick rate" is reduced by γ. Lastly, an object has a rest mass, being the mass it "came off the production line with", and a relativistic mass, being defined as above. When at rest, the object's rest mass equals its relativistic mass. When it moves, its acceleration is determined by both its relativistic mass (or its rest mass, of course) and its velocity.

The use of these γ-scaled quantities is governed only by the extent to which they are useful. While contracted length and time intervals are used—or not—insofar as they simplify special relativity analyses, relativistic mass has found itself at the centre of much debate in recent years about whether it is necessary in a physics curriculum. All physicists use rest mass, but not all physicists would have relativistic mass appear in textbooks, preferring instead always to write it in terms of rest mass when it is used (although this can't be done for photons). So, if all physicists agree that rest mass is a very fundamental concept, then why use relativistic mass at all?

When particles are moving, relativistic mass provides a very economical description that absorbs the particles' motion naturally. For example, suppose we put an object on a set of scales that are capable of measuring incredibly small increases in weight. Now heat the object. As its temperature rises causing its constituents' thermal motion to increase, the reading on the scales will increase. If we prefer to maintain the usual idea that mass is proportional to weight—assuming we don't step onto an elevator or change our home planet midway through the experiment—then it follows that the object's mass has increased. If we define mass in such a way that the object's mass does not increase as it heats up, then we will have to give up the idea that mass is proportional to weight.

Another many-particle example occurs in pre-relativistic physics, in which the centre of mass of an object is calculated by "weighting" the position vector r_i of each of its particles by their mass m_i:

$$\text{Centre of mass} = \frac{\sum_i m_i r_i}{\sum_i m_i}$$

The same expression will hold relativistically *if* each of the above masses is now a particle's *relativistic* mass. If we prefer to use only rest mass then we must replace the m_i in the above expression by $\gamma_i\, m_i$ where m_i is *rest mass*, but now the expression has lost a certain economy. Similarly, if two objects with *relativistic* masses m_1 and m_2 collide and stick together in such a way that the resulting object is at rest, then its (relativistic = rest) mass will be $m_1 + m_2$. This accords with our intuition, and intuition is mostly what good conventions are about. In contrast, a rest-mass-only analysis describes the interaction by saying that the objects have (rest) masses of M_1 and M_2, with a combined (rest) mass of $\gamma_1 M_1 + \gamma_2 M_2$. Whether our intuition has anything to gain from this new expression is not clear.

Another place where the idea of relativistic mass surfaces is when describing the *cyclotron*, a device that accelerates charged particles in circles within a constant magnetic field. The cyclotron works by applying a varying electric field to the particles, and the frequency of this variation must be tuned to the natural orbital frequency that the particles acquire as they move in the magnetic field. But in practice we find that as the particles accelerate, they begin to get out of step with the applied electric field and can no longer be accelerated further. This can be described as a consequence of their masses increasing, which changes their orbital frequency in the magnetic field.

Lastly, the energy E of an object, whether moving or at rest, is given by Einstein's famous relation $E = mc^2$, where m is its *relativistic* mass. Because, for example, the photon has no rest mass but does have relativistic mass, the use of relativistic mass makes it much easier to describe the mass changes that happen when light interacts with matter. See the FAQ article What is the mass of a photon?.

While relativistic mass is useful in the context of special relativity, it is rest mass that appears most often in the modern language of relativity, which centres on "invariant quantities" to build a geometrical description of relativity. Geometrical objects are useful for unifying scenarios that can be described in different coordinate systems. Because there are multiple ways of describing scenarios in relativity depending on which frame we are in, it is useful to focus on whatever invariances we can find. This is, for example, one reason why vectors (i.e. arrows) are so useful in maths and physics; everyone can use the *same* arrow to express e.g. a velocity, even though they might each quantify the arrow using different components because each observer is using different coordinates. So the reason rest mass, rest length, and proper time find their way into the tensor language of relativity is that *all* observers agree on their values. (These invariants then join with other quantities in relativity: thus, for example, the *four-force* acting on a body equals its *rest* mass times its *four-acceleration*.) Some physicists cite this view to maintain that rest mass is the only way in which mass should be understood.

As with many things, the use of relativistic mass can be a matter of taste, but it seems that at least some physicists who vehemently oppose the use of relativistic mass believe, mistakenly, that pro-relativistic mass physicists are against the idea of *rest* mass. It's not clear just why there should be this perennial confusion about preferences, and why some of those who dislike the idea of relativistic mass show such fundamentalist opposition to a choice of formalism that can never produce wrong results. The world of physics and its language is full of useful alternative notations and ways of approaching things, and different choices of notation and language can shed light on the physics involved. Selecting one of the other of relativistic versus rest mass will never lead to problems for practitioners of the subject.

A debate of the subject surfaced in *Physics Today* in 1989 when Lev Okun wrote an article urging that relativistic mass should no longer be taught. Wolfgang Rindler responded with a letter to the editor defending its continued use. In 1991 Tom Sandin wrote an article in the *American Journal of Physics* that argued in favour of relativistic mass.

A commonly heard argument against the use of relativistic mass runs as follows: "The equation $E = mc^2$ says that a body's relativistic mass is proportional to its total energy, so why should we use two terms for what is essentially the same quantity? We should just stay with energy, and use the word ‹mass' to refer only to rest mass." The first difficulty with this line of reasoning is that it is quite selective; after all, it should surely rule out the use of *rest* mass as well, since within special relativity, rest mass is proportional to a body's rest energy. On that note, a second difficulty of the line of reasoning is more technical: equating energy and relativistic mass cannot be done more generally. In general relativity, it's natural to consider quantities that are conserved for a system moving on a geodesic. But $\gamma\, m$ is *not* generally conserved along geodesics. (Actually, $\gamma\, m$ is called p^t in the language of general relativity. It turns out that a closely related quantity, p_t, *will*be conserved along a geodesic if the metric is time independent.) Note, though, that while relativistic mass $\gamma\, m$ is not a body's total energy in general relativity, it's also not simply the source of gravity within the same theory. Finally, a third difficulty with the above commonly heard argument is that, in the interests of consistency, it should surely be applied to rule out either the "momentum density" or the "energy flux density" of light, since these also are simply related by a factor of c^2. Yet, and quite rightly, these last two terms co-exist in modern literature; no one ever suggests that either of these terms should be dropped in favour of the other, because they both have their uses and are fundamentally different quantities: a volume density and an areal density. The fact that they *can* be equated is the interesting thing physically, but one must not thereafter trivialise the physics by insisting that one of the two concepts be dropped.

So likewise can the concepts of mass and energy coexist. The above argument that $E = mc^2$ demotes mass in favour of energy—or rather, that it selectively demotes *relativistic* mass, but not *rest* mass—also neglects the very *definitions* of mass and energy. Mass is a property of a body that we have an intuitive feel for; its definition as a resistance to acceleration is very fundamental. Energy, on the other hand, is defined in physics in a technical way that involves the concept of a system's time evolution; this is not something that bears any obvious similarity to the concept of an object's resistance to being accelerated. If the concept of mass exists in some sense "prior" to that of energy, and if energy itself is defined in a different way to mass, then it does not seem reasonable to drop the idea of mass in favour of energy. Rather, $E = mc^2$ becomes an expression that tells us how much energy a given mass has; it also tells us how much a body will resist being accelerated depending on its energy content. And, perhaps best of all, it reminds us that Einstein's equation is a triumph of relating two disparate quantities—and this is one of the great aims of physics.

Another argument sometimes put forward for dropping the use of relativistic mass is that since e.g. all electrons have the same rest mass (whereas their relativistic masses depend on their speeds), then their rest mass is the only quantity able to be tabulated, and so we should discard the very idea of relativistic mass. But when we say without qualification that "the height of the Eiffel Tower is 324 metres", we clearly mean its rest length; but that doesn't mean the idea of contracted length should be discarded. Similarly, it's okay to say that the mass of an electron is about 10^{-30} kg without having to specify that we are referring to the rest mass; everyone knows we mean rest mass when we tabulate a particle's mass. That's purely a useful linguistic convention, and it does not imply that we have discarded the idea of relativistic mass, or that it should be discarded at all.

Everyone agrees that a moving train's rest mass is a fixed property inherent to it, just as its rest length is a fixed property inherent to it. And yet, strangely, many of the same physicists who insist that a moving train's mass does not scale by γ are quite happy to say that its length *does* scale by γ. There is no argument in the literature about the uses of rest length versus moving length, so why should there be any argument about the uses of rest mass versus moving mass?

Another mass concept that everyone agrees on is the idea of *reduced mass* in non-relativistic mechanics. When the mechanics of e.g. a sun–satellite system or a mass oscillating on a spring is analysed, a mass term appears that combines the two masses in a particular way. As far as the maths goes, it's *as if* we are replacing the two original bodies by two new ones: the first new body has *infinite* mass, and the second new body has a mass equal to the system's reduced mass, which has this name because it's smaller than either of the two original masses that gave rise to it. This is a fruitful way to view the original system, and it's completely standard. No one gets confused into thinking that we actually have an infinite mass and a reduced mass in our system. No one worries that the new, infinite, mass is somehow going to become a black hole, or that the reduced mass lost some of its atoms somewhere. Everyone knows the realm of applicability of the concept of reduced mass and how useful it is. Why then, do so many physicists criticise relativistic mass by squeezing it into realms where it was never intended to be used? They presumably don't do the same thing with reduced mass.

An optimistic view would hold that it's a measure of the richness of physics that focussing on different aspects of concepts like mass produces different insights: intuition in the case of relativistic mass in special relativity, and the also-intuitive notion of invariance and geometrical quantities in the case of rest mass within the tensor language of special and general relativity. The two aspects do not contradict each other, and there is room enough in the world of physics to accommodate them both.

Abandoning the use of relativistic mass is sometimes validated by quoting select physicists who are or were against the term, or by exhaustively tabulating which textbooks use the term. But real science isn't done this way. In the final analysis, the history of relativity, with its quotations from those in favour of relativistic mass and those against, has no real bearing on whether the idea itself has value. The question to ask is not whether relativistic mass is fashionable or not, or who likes the idea and who doesn't; rather, as in any area of physics notation and language, we should always ask "Is it *useful*?". And relativistic mass is certainly a useful concept. There can be little doubt that *some* of its vocal opponents even use it to gain intuition when analysing a scenario in special relativity.

Relativistic Version of F = ma

The concept of relativistic mass is neatly encapsulated in the expression $F = \mathrm{d}(mv)/\mathrm{d}t$, where m is relativistic mass. This says that an impulse $F\,\mathrm{d}t$ causes an infinitesimal increase in a body's *relativistic momentum* mv.

Besides this definition and use of relativistic mass, we wish here to write down the relativistic version of Newton's second law, $F = ma$. In Newton's mechanics, this equation relates vectors F and a via the mass m of the object being accelerated, which is invariant in Newton's theory. Because m is just a number, in Newton's theory the force on an object is always parallel to the resulting acceleration.

The corresponding equation in special relativity is a little more complicated. It turns out that the force F is not always parallel to the acceleration a. We can express this fact using matrix notation. Let m be the *rest* mass, and v be the velocity as a column vector, whose entries are expressed as fractions of c and whose magnitude v is the speed as a fraction of c. Let v^t be the velocity as a row vector, and let 1 be the 3×3 identity matrix. As usual, set $\gamma = (1 - v^2)^{-1/2}$. The relativistic version of $F = ma$ turns out to be

$$F = (1 + \gamma^2 \, v \, v^t)\, \gamma \, m \, a$$

and

$$a = \frac{(1 - v\,v^t)\,\mathrm{F}}{\gamma\,m}$$

So defining mass via force and acceleration isn't as simple as it was for Newton (although it *is* simple, in principle, to define the mass as relating impulse and momentum increase, as mentioned a few lines up). Nevertheless, the three components of the two expressions above share a factor of $\gamma\, m$, and the rest mass m only ever appears in both expressions accompanied by γ. The acceleration is not necessarily parallel to the force that produced it, and it's not hard to see from the above equations that it's easier to accelerate a mass sideways to its motion than it is to accelerate it in the direction of its motion. This is how relativity reproduces Lorentz's original concepts of longitudinal and transverse masses; they are actually contained in these equations. The directional dependence that the newtonian meaning of mass has now taken on is neatly contained in the matrices $1 + \gamma^2 v\, v^t$ and $1 - v\, v^t$, and the remaining factor $\gamma\, m$ is the relativistic mass. Taking our cue from the equations like this, to isolate quantities that might prove useful, is a powerful tool in mathematical physics.

References

- Albert Einstein (2001). Relativity: The Special and the General Theory (Reprint of 1920 translation by Robert W. Lawson ed.). Routledge. p. 48. ISBN 0-415-25384-5
- Jeroen van Dongen (2009). "On the role of the Michelson–Morley experiment: Einstein in Chicago". Archive for History of Exact Sciences. 63: 655–663. arXiv:0908.1545. doi:10.1007/s00407-009-0050-5
- Abraham Loeb; Mark J. Reid; Andreas Brunthaler; Heino Falcke (2005). "Constraints on the proper motion of the Andromeda galaxy based on the survival of its satellite M33" (PDF). The Astrophysical Journal. 633 (2): 894–898. arXiv:astro-ph/0506609. Bibcode:2005ApJ...633..894L. doi:10.1086/491644
- Baglio, Julien (26 May 2007). "Acceleration in special relativity: What is the meaning of "uniformly accelerated movement" ?" (PDF). Physics Department, ENS Cachan. Retrieved 22 January 2016
- Jump up^ Richard Phillips Feynman (1998). Six Not-so-easy Pieces: Einstein's relativity, symmetry, and space–time (Reprint of 1995 ed.). Basic Books. p. 68. ISBN 0-201-32842-9
- Penrose, Roger (24 October 2008). "The Apparent Shape of a Relativistically Moving Sphere". Mathematical Proceedings of the Cambridge Philosophical Society. 55 (01): 137. Bibcode:1959PCPS...55..137P. doi:10.1017/S0305004100033776
- Dolby, Carl E.; Gull, Stephen F. (December 2001). "On radar time and the twin "paradox"". American Journal of Physics. 69 (12): 1257–1261. arXiv:gr-qc/0104077. Bibcode:2001AmJPh..69.1257D. doi:10.1119/1.1407254
- Rindler, Wolfgang (1977). Essential Relativity: Special, General, and Cosmological (illustrated ed.). Springer Science & Business Media. p. §1,11 p. 7. ISBN 978-3-540-07970-5
- Kraus, Ute. "The Ball is Round". Space Time Travel: Relativity visualized. Institut für Physik Universität Hildesheim. Archived from the original on 16 April 2017. Retrieved 16 April 2017

Permissions

All chapters in this book are published with permission under the Creative Commons Attribution Share Alike License or equivalent. Every chapter published in this book has been scrutinized by our experts. Their significance has been extensively debated. The topics covered herein carry significant information for a comprehensive understanding. They may even be implemented as practical applications or may be referred to as a beginning point for further studies.

We would like to thank the editorial team for lending their expertise to make the book truly unique. They have played a crucial role in the development of this book. Without their invaluable contributions this book wouldn't have been possible. They have made vital efforts to compile up to date information on the varied aspects of this subject to make this book a valuable addition to the collection of many professionals and students.

This book was conceptualized with the vision of imparting up-to-date and integrated information in this field. To ensure the same, a matchless editorial board was set up. Every individual on the board went through rigorous rounds of assessment to prove their worth. After which they invested a large part of their time researching and compiling the most relevant data for our readers.

The editorial board has been involved in producing this book since its inception. They have spent rigorous hours researching and exploring the diverse topics which have resulted in the successful publishing of this book. They have passed on their knowledge of decades through this book. To expedite this challenging task, the publisher supported the team at every step. A small team of assistant editors was also appointed to further simplify the editing procedure and attain best results for the readers.

Apart from the editorial board, the designing team has also invested a significant amount of their time in understanding the subject and creating the most relevant covers. They scrutinized every image to scout for the most suitable representation of the subject and create an appropriate cover for the book.

The publishing team has been an ardent support to the editorial, designing and production team. Their endless efforts to recruit the best for this project, has resulted in the accomplishment of this book. They are a veteran in the field of academics and their pool of knowledge is as vast as their experience in printing. Their expertise and guidance has proved useful at every step. Their uncompromising quality standards have made this book an exceptional effort. Their encouragement from time to time has been an inspiration for everyone.

The publisher and the editorial board hope that this book will prove to be a valuable piece of knowledge for students, practitioners and scholars across the globe.

Index

A

Absolute Rotation, 120-121

Acceleration, 3-6, 14, 18, 28, 34, 36-41, 43-45, 47-48, 53, 59-60, 62, 65, 77, 83-90, 106, 122-131, 141-151, 178, 184, 196, 214-215, 223, 226, 240-241, 259-265

Analytical Mechanics, 159, 169, 171, 175, 183

Angular Acceleration, 3, 41, 44-45, 48, 53, 62, 83-89, 105-106, 125-126, 130, 149-150, 220

Angular Momentum, 3, 7, 11, 13, 16, 25, 31-32, 53, 58, 73, 79-80, 87-88, 131, 186, 190, 193, 195, 204, 225, 237, 253, 255

Angular Velocity, 3, 40-41, 43-45, 47-48, 53, 79, 83-89, 91, 93, 95-106, 108, 110, 114, 119, 125-126, 130-131, 136-137, 149-150, 195, 204, 220

Astrodynamics, 10, 18-19

B

Balance Laws, 30-33

Body Force, 25-26, 31, 52, 78

C

Canonical Transformation, 159, 176-177, 210

Celestial Mechanics, 1, 10, 12, 15-16, 18-19, 80

Centrifugal Force, 5, 81, 109, 112-113, 115-118, 120-122, 124, 131, 195-196, 246, 248, 252

Centripetal Force, 81, 115, 117-118, 122-124, 127, 129, 131, 137, 196, 252

Circular Motion, 1-2, 56, 81, 86, 96, 100, 124, 127-138, 151, 252

Clausius-duhem Inequality, 33

Conservation of Energy, 6-7, 21, 169

Constant Velocity, 106, 182, 195, 232, 242, 248-249, 253, 255

Constraints, 48-49, 60-61, 72, 159-162, 170-171, 182-184, 187, 200, 214-215, 218-219, 222-225, 265

Continuum Mechanics, 1, 10, 20-26, 30

Coriolis Effect, 113, 248

Coriolis Force, 5, 109, 112-114, 118, 120, 246, 248, 252

Curvilinear Motion, 89, 145, 148-150

D

D'alembert's Principle, 59-62, 170-171, 178, 183-185, 224

Deceleration, 4, 248

Displacement, 6, 27, 29, 34-37, 41, 60, 62, 83-86, 88, 90-93, 95-96, 101, 112, 132, 134, 140, 142-143, 151-157, 162, 183-184, 217-218, 222-224, 226

E

Elastic Collision, 67, 72

Electrodynamics, 58, 66, 237

Electromagnetism, 6, 8, 73, 179, 196, 241

Equation of Motion, 6, 56, 109, 120, 158, 164-165, 174, 178, 181, 195, 214, 223, 239

Euler Force, 109, 112, 114, 120, 248

Euler-lagrange Equation, 159, 188, 192, 195, 200, 202, 209, 226-227, 229-230

Eulerian Description, 20, 28-29

F

Fictitious Forces, 5, 109-110, 112-113, 117, 120-121, 196, 246-247, 251-252

Frames of Reference, 4, 35, 55, 66, 108, 139, 241, 243, 245-250, 252, 254

Friction, 6-7, 18, 54, 58, 123, 156, 160, 180, 224-226

G

Galilean Invariance, 55-56, 67, 70, 231, 237, 239, 241

Galilean Transformation, 5, 66, 70, 236, 239-240, 253-254

General Relativity, 8, 15, 58, 176, 179, 182, 198-199, 231, 237, 241, 245, 254, 263-264

Generalized Momentum, 72-74, 173, 189, 197, 204

Gravitational Field, 5, 25, 52, 241

H

Hamiltonian Mechanics, 8, 10, 73, 159, 169, 172, 177, 197-198, 204-206, 208, 211

Homogeneous Transform, 42

Hooke's Law, 7, 156

I

Inertia Forces, 5, 196

Inertial Frame, 4-5, 55, 70, 108, 112-114, 116-121, 194, 196, 235, 237-238, 240-243, 246, 248-256

Initial Velocity, 6, 140-143, 145, 155

K

Kinematics, 1, 10-11, 20, 26, 34, 41, 43, 45, 61, 83, 86, 125

Kinetic Energy, 6-9, 67-69, 73, 88, 132, 154, 156-157, 171-172, 175, 182, 185, 188, 190-191, 193, 199-200, 219, 239, 242

Kvn, 162-165, 167-168

L

Lagrangian Description, 20, 27-29

Lagrangian Mechanics, 8, 10, 73-74, 169, 171, 180-181, 197-200, 204, 211

Linear Dynamics, 53, 87-88

Liouville Mechanics, 165, 168

Lorentz Invariance, 70, 241

M

Magnetic Field, 5, 9-10, 26, 52, 74, 196, 253, 261-262

Magnitude, 4, 36, 45, 51-52, 56-57, 63-64, 69, 72, 79-80, 85, 87-89, 92, 95, 97, 99, 105-106, 110, 115, 122, 124, 127-138, 149-150, 156, 158, 161, 195, 242, 260, 264

Moment of Inertia, 3, 53, 62, 79, 86-88, 131, 158

N

Newton's Equations, 14-15, 215

Newton's First Law, 54-55, 80, 117, 124, 246, 250

Newton's Second Law, 5-6, 8, 45, 55-56, 59-60, 62, 65, 67, 69, 71, 75, 109-110, 112-113, 115-116, 124, 173, 181-182, 185, 192, 248-249, 264

Newton's Third Law, 6, 24, 57, 80, 117

Noether's Theorem, 74, 179-180, 189, 199-200

Non-inertial Reference Frame, 108, 112, 119, 255

O

Oscillatory Motion, 1-2, 140, 151-152

P

Partial Differential Equation, 22

Physical Laws, 4, 21, 66, 250, 252

Potential Energy, 7, 68, 73, 154, 156-157, 171, 185, 188, 190-191, 193, 198-200, 210

Q

Quantum Field Theory, 8, 58, 169, 179, 197, 237

Quantum Mechanics, 2, 8-9, 58, 76, 162, 164-165, 167-169, 173-174, 176, 197-199, 204-206, 209, 237

R

Reaction Force, 6, 58, 118, 242

Rectilinear Motion, 83, 140-144

Relative Motion, 5, 13, 195, 232, 241, 258

Relative Velocity, 5, 68, 237, 241, 249-250, 254

Rotating Reference Frame, 81, 108-110, 112-114, 117, 121, 195

Rotational Motion, 1-2, 81-82, 84, 86-90

Routhian Mechanics, 177, 198

S

Scalars, 36-37, 69

Simple Harmonic Motion, 140, 151-155, 157-158

Special Relativity, 5, 8-9, 58, 71, 198-199, 231, 234, 236-237, 240-241, 245-246, 250, 252, 254-255, 260-262, 264-265

Statics, 1, 10, 50-51, 53, 59

Statistical Mechanics, 8, 10, 21, 162-163

T

The Poisson Bracket, 173-174, 176, 206

Thermodynamics, 8, 30, 33, 59

Translational Motion, 1, 140-141

U

Udwadia-kalaba Equation, 221, 223-224

V

Variable Mass, 69

Vector Quantity, 36, 52, 54, 63-64, 69, 79, 88, 90, 105